Methods of Experimental Physics

VOLUME 13

SPECTROSCOPY

PART A

METHODS OF EXPERIMENTAL PHYSICS:

L. Marton, *Editor-in-Chief*
Claire Marton, *Assistant Editor*

1. Classical Methods
   Edited by Immanuel Estermann
2. Electronic Methods, Second Edition (in two parts)
   Edited by E. Bleuler and R. O. Haxby
3. Molecular Physics, Second Edition (in two parts)
   Edited by Dudley Williams
4. Atomic and Electron Physics—Part A: Atomic Sources and Detectors, Part B: Free Atoms
   Edited by Vernon W. Hughes and Howard L. Schultz
5. Nuclear Physics (in two parts)
   Edited by Luke C. L. Yuan and Chien-Shiung Wu
6. Solid State Physics (in two parts)
   Edited by K. Lark-Horovitz and Vivian A. Johnson
7. Atomic and Electron Physics—Atomic Interactions (in two parts)
   Edited by Benjamin Bederson and Wade L. Fite
8. Problems and Solutions for Students
   Edited by L. Marton and W. F. Hornyak
9. Plasma Physics (in two parts)
   Edited by Hans R. Griem and Ralph H. Lovberg
10. Physical Principles of Far-Infrared Radiation
    L. C. Robinson
11. Solid State Physics
    Edited by R. V. Coleman
12. Astrophysics—Part A: Optical and Infrared
    Edited by N. Carleton
    Part B: Radio Telescopes, Part C: Radio Observations
    Edited by M. L. Meeks
13. Spectroscopy (in two parts)
    Edited by Dudley Williams

Volume 13

Spectroscopy

PART A

Edited by

DUDLEY WILLIAMS

*Department of Physics
Kansas State University
Manhattan, Kansas*

1976

ACADEMIC PRESS · **New York   San Francisco   London**
A Subsidiary of Harcourt Brace Jovanovich, Publishers

Copyright © 1976, by Academic Press, Inc.
ALL RIGHTS RESERVED.
NO PART OF THIS PUBLICATION MAY BE REPRODUCED OR
TRANSMITTED IN ANY FORM OR BY ANY MEANS, ELECTRONIC
OR MECHANICAL, INCLUDING PHOTOCOPY, RECORDING, OR ANY
INFORMATION STORAGE AND RETRIEVAL SYSTEM, WITHOUT
PERMISSION IN WRITING FROM THE PUBLISHER.

ACADEMIC PRESS, INC.
111 Fifth Avenue, New York, New York 10003

United Kingdom Edition published by
ACADEMIC PRESS, INC. (LONDON) LTD.
24/28 Oval Road, London NW1

Library of Congress Cataloging in Publication Data

Main entry under title:

Spectroscopy.

(Methods of experimental physics ; v. 13)
Includes bibliographical references and index.
1.   Spectrum analysis.   I.   Williams, Dudley,
(date)   II.   Series.
QC451.S63   535'.84   76-6854
ISBN 0–12–475913–0 (pt. A)

PRINTED IN THE UNITED STATES OF AMERICA

CONTENTS

CONTRIBUTORS ix

FOREWORD xi

PREFACE xiii

CONTENTS OF VOLUME 13, PART B xv

CONTRIBUTORS TO VOLUME 13, PART B xvii

1. Introduction
by DUDLEY WILLIAMS

1.1. History of Spectroscopy 3
    1.1.1. Newton's Contributions 3
    1.1.2. Nineteenth-Century Developments 5
    1.1.3. Twentieth-Century Developments 7
    1.1.4. The Infrared Region 8
    1.1.5. The Submillimeter and
           Microwave Regions 10
    1.1.6. The Radio Frequency Region 12
    1.1.7. The Ultraviolet Region 13
    1.1.8. The X-Ray Region 14
    1.1.9. The Gamma-Ray Region 16
    1.1.10. The Role of Spectroscopy in
            Twentieth-Century Physics 17

1.2. General Methods of Spectroscopy 19
    1.2.1. Emission Spectra 20
    1.2.2. Absorption Spectra 22
    1.2.3. Sources 24
    1.2.4. Resolving Instruments 24
    1.2.5. Detectors 26
    1.2.6. Data Handling Techniques 28

2. Theory of Radiation and Radiative Transitions
   by BASIL CURNUTTE, JOHN SPANGLER, AND
   LARRY WEAVER

   2.1. Introduction 31
   2.2. Light 32
       2.2.1. Classical Picture 32
       2.2.2. Quantum Picture 64

   2.3. Interaction of Light and Matter 79
       2.3.1. Time-Dependent Perturbation Theory 79
       2.3.2. The Multipole Expansion 93
       2.3.3. Selection Rules 97

   2.4. Applications 100
       2.4.1. Atomic and Nuclear Decay Rates 100
       2.4.2. Molecular Transitions 106

   2.5. Conclusion 113

3. Nuclear and Atomic Spectroscopy
   3.1. Gamma-Ray Region 115
       by JAMES C. LEGG AND GREGORY G. SEAMAN

       3.1.1. Energy and Intensity 115
       3.1.2. Gamma-Ray Detectors 121
       3.1.3. Angular Correlations 134
       3.1.4. Transition Rate and Lifetime Measurements . 141

   3.2. X-Ray Region 148
       by ROBERT L. KAUFFMAN AND PATRICK RICHARD

       3.2.1. Introduction 148
       3.2.2. Detectors and Spectrometers 149
       3.2.3. X-Ray Spectra 166
       3.2.4. Selected Topics 191

3.3. Far Ultraviolet Region 204
by JAMES A. R. SAMSON

    3.3.1. Introduction 204
    3.3.2. Photon Sources 205
    3.3.3. Dispersive Devices 226
    3.3.4. Optical Windows and Filters 238
    3.3.5. Polarizers 239
    3.3.6. Detectors 241
    3.3.7. Wavelength Standards 246
    3.3.8. Experimental Applications 247

3.4. Optical Region 253
by P. F. A. KLINKENBERG

    3.4.1. Introduction 253
    3.4.2. Light Sources 259
    3.4.3. Spectroscopic Instruments 274
    3.4.4. Detection of Optical Radiation 314
    3.4.5. Evaluation of Spectra 325
    3.4.6. Analysis of Atomic Spectra 336

AUTHOR INDEX 347

SUBJECT INDEX FOR PART A 359

SUBJECT INDEX FOR PART B 363

CONTRIBUTORS

Numbers in parentheses indicate the pages on which the authors' contributions begin.

BASIL CURNUTTE, *Department of Physics, Kansas State University, Manhattan, Kansas* (31)

ROBERT L. KAUFFMAN, *Department of Physics, Kansas State University, Manhattan, Kansas* (148)

P. F. A. KLINKENBERG, *Zeeman-Laboratorium, University of Amsterdam, The Netherlands* (253)

JAMES C. LEGG, *Department of Physics, Kansas State University, Manhattan, Kansas* (115)

PATRICK RICHARD, *Department of Physics, Kansas State University, Manhattan, Kansas* (148)

JAMES A. R. SAMSON, *Behlen Laboratory of Physics, University of Nebraska, Lincoln, Nebraska* (204)

GREGORY G. SEAMAN, *Department of Physics, Kansas State University, Manhattan, Kansas* (115)

JOHN SPANGLER, *Department of Physics, Kansas State University, Manhattan, Kansas* (31)

LARRY WEAVER, *Department of Physics, Kansas State University, Manhattan, Kansas* (31)

DUDLEY WILLIAMS, *Department of Physics, Kansas State University, Manhattan, Kansas* (1)

FOREWORD

Several aspects of spectroscopy have been treated in some of our earlier volumes (see Volumes 3A and 3B, Molecular Physics, second edition; Volume 10, Far Infrared; Volume 12A, Astrophysics). The rapid expansion of physics made it desirable to issue a separate treatise devoted to spectroscopy only, emphasizing such aspects which may not have been treated adequately in the volumes dealing essentially with other facets of physics. The present volumes contain a much more thoroughgoing treatment of the spectroscopy of photons of all energies. It is our intention to follow this with a volume devoted to particle spectroscopy.

Professor Dudley Williams, who is already well known to readers of "Methods of Experimental Physics" as editor of our Molecular Physics volumes, was kind enough to accept the editorship of the Spectroscopy volumes. His knowledge of the field and his excellent judgment will, no doubt, be appreciated by the users of "Spectroscopy" methods. We wish to express our profound gratitude to him and to all contributors to these volumes for their untiring efforts.

L. Marton
C. Marton

PREFACE

Spectroscopy has been a method of prime importance in adding to our knowledge of the structure of matter and in providing a basis for quantum physics, relativistic physics, and quantum electrodynamics. However, spectroscopy has evolved into a group of specialties; practitioners of spectroscopic arts in one region of the electromagnetic spectrum feel little in common with practitioners studying other regions; in fact, some practitioners do not even realize that they are engaged in spectroscopy at all!

In the present volumes we attempt to cover the entire subject of spectroscopy from pair production in the gamma-ray region to dielectric loss in the low radio-frequency region. Defining spectroscopy as the study of the emission and absorption of electromagnetic radiation by matter, we present a general theory that is applicable throughout the entire range of the electromagnetic spectrum and show how the theory can be applied in gaining knowledge of the structure of matter from experimental measurements in all spectral regions.

The books are intended for graduate students interested in acquiring a general knowledge of spectroscopy, for spectroscopists interested in acquiring knowledge of spectroscopy outside the range of their own specialties, and for other physicists and chemists who may be curious as to "what those spectroscopists have been up to" and as to what spectroscopists find so interesting about their own work! The general methods of spectroscopy as practiced in various spectral regions are remarkably similar; the details of the techniques employed in various regions are remarkably different.

Volume A begins with a brief history of spectroscopy and a discussion of the general experimental methods of spectroscopy. This is followed by a general theory of radiative transitions that provides a basis for an understanding of and an interpretation of much that follows. The major portion of the volumes is devoted to chapters dealing with the spectroscopic methods as applied in various spectral regions and with typical results. Each chapter includes extensive references not only to the original literature but also to earlier books dealing with spectroscopy in various regions; the references to earlier books provide a guide to readers who may wish to go more deeply into various branches of spectroscopy. The final chapters of Volume B are devoted to new branches of spectroscopy involving beam foils and lasers.

The list of contributors covers a broad selection of competent active research workers. Some exhibit the fire and enthusiasm of youth; others are at the peak of the productive activity of their middle years; and still others are battled-scarred veterans of spectroscopy who hopefully draw effectively on long experience! All contributors join me in the hope that the present volumes will serve a useful purpose and will provide valuable insights into the general subject of spectroscopy.

DUDLEY WILLIAMS

CONTENTS OF VOLUME 13, PART B

4. Molecular Spectroscopy

    4.1. Infrared Region   by Dudley Williams
    4.2. Far-Infrared and Submillimeter-Wave Region   by D. Oepts
    4.3. Microwave Region   by Donald R. Johnson and Richard Pearson, Jr.
    4.4. Radio-Frequency Region   by J. B. Hasted

5. Recent Developments

    5.1. Beam-Foil Spectroscopy   by C. Lewis Cocke
    5.2. Tunable Laser Spectroscopy   by Marvin R. Querry

Author Index—Subject Indexes for Parts A and B

CONTRIBUTORS TO VOLUME 13, PART B

C. LEWIS COCKE, *Department of Physics, Kansas State University, Manhattan, Kansas*

J. B. HASTED, *Department of Physics, Birkbeck College, University of London, London, England*

DONALD R. JOHNSON, *National Bureau of Standards, Molecular Spectroscopy Section, Optical Physics Division, Washington, D.C.*

D. OEPTS, *Association Euratom-FOM, FOM-Instituut voor Plasmafysica, Rijnhuizen, Nieuwegein, The Netherlands*

RICHARD PEARSON, JR., *National Bureau of Standards, Molecular Spectroscopy Section, Optical Physics Division, Washington, D.C.*

MARVIN R. QUERRY, *Department of Physics, University of Missouri, Kansas City, Missouri*

DUDLEY WILLIAMS, *Department of Physics, Kansas State University, Manhattan, Kansas*

1. INTRODUCTION*

The success of spectroscopy has been so great that many of its terms have passed into common parlance; for example, the views expressed at a political meeting are sometimes described as representing a *broad spectrum* of opinion. In science, the term spectroscopy has come to mean the separation or classification of various items into groups; for example, the separation of the various isotopes of a chemical element is called *mass spectroscopy*. Similarly, the analysis of an acoustical wave train into sinusoidal components of different frequencies is called *acoustical spectroscopy*; a plot of the relative intensities of the components as a function of frequency is called the *acoustical spectrum* of the source. In nuclear physics, the study of resonances associated with bombarding particles of various energies has been termed *nuclear spectroscopy*. We even find the term spectroscopy being extended into high-energy physics, where the plots of energy-level diagrams for mesons and baryons have been termed a *new spectroscopy*!

In the present volumes, we shall restrict the term spectroscopy to the study of processes involving the emission and absorption of electromagnetic radiation. However, we shall attempt to cover the whole range of the electromagnetic spectrum, from the gamma-ray region to the low radio-frequency region. The general methods employed in the various spectral regions are remarkably similar; the details of the experimental techniques in the various regions are remarkably different!

Part 1 includes a brief chapter dealing with the history of spectroscopy from the Newtonian epoch to the present, and a second chapter dealing with the general methods of spectroscopy. Part 2 gives a general treatment of the theory of radiative transitions that is basic to an understanding of the phenomena treated in Parts 3 and 4, in which each chapter deals with special experimental methods employed in a particular spectral region; examples of the application of experimental methods to one or more problems of current interest and importance are presented. Part 5 covers the experimental methods being employed in two recently developed fields of spectroscopic investigation.

Since we shall be dealing with electromagnetic radiation, we can designate the various spectral regions in terms of frequency ν or vacuum wavelength λ,

---

* Part 1 is by Dudley Williams.

where these quantities are connected by the familiar relation $\nu\lambda = c$. In Fig. 1 we give a plot of the electromagnetic spectrum with various spectral regions labeled. We should emphasize that the boundaries between the regions designated in Fig. 1 do not represent Dedekind cuts; for our purposes, the regions overlap in the sense that spectra in the ranges of overlap can be

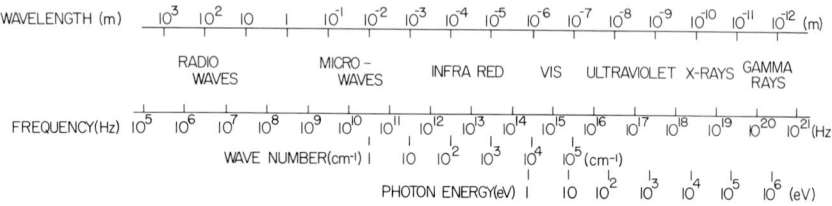

FIG. 1. The electromagnetic spectrum.

investigated by characteristic experimental methods used in either of the adjacent spectral regions; for example, spectra in the very near infrared can be mapped either by the photographic techniques used in the visible region or by means of the thermocouples frequently employed in the infrared. There is also overlap in the basic types of phenomena characteristic of the adjacent spectral regions, as labeled in Fig. 1; for example, the radiation produced in modern high-voltage x-ray machines can have frequencies that extend well into the gamma-ray region, as labeled in the figure.

We note that spectroscopists practicing their arts in various spectral regions have their own favorite special units for designation of frequency or wavelength. Those working in the radio frequency and microwave regions usually measure spectral frequencies in terms of standard radio frequencies broadcast by the National Bureau of Standards station WWV in the USA, the BBC in the UK, or other stations with similar missions; these spectroscopists prefer *frequency* units such as megahertz (MHz), gigahertz (GHz), or even terahertz (THz). In the infrared, visible, and near ultraviolet regions, spectroscopists measure *wavelengths* in terms of the spacing of the lines on gratings or in terms of distances in interferometers; these spectroscopists usually prefer designating *wavelengths* in terms of micrometers (μm) or nanometers (nm); although not recognized in modern SI units, the angstrom unit = 0.1 nm is still widely used in specifying wavelengths in the visible and near ultraviolet. Since there are no larger named SI multiples than tera = 10^{12}, there are no convenient frequency units for use in the infrared or in spectral regions of higher frequency; when frequencies are stated, they are usually specified indirectly in terms of wave number $\tilde{\nu}$ giving the number of waves per centimeter: $\tilde{\nu} = f/c = 1/\lambda_{\text{vac}}$ with c in centimeters per second and λ_{vac} in centimeters. In the hard x-ray and γ-ray regions, it is usually desirable to characterize radiation in terms of its quantum energy $E = hv$; the quantum energy is nearly always expressed in electron volts (eV).

The spectroscopist working in any spectral region wishes to express his designation of a given spectral feature in terms of several significant digits and some unit that he regards as *convenient*. Similarly, in making actual measurements, he prefers to base his measurements on some *convenient* secondary standard of wavelength, frequency, or quantum energy rather than going back for every measurement to the primary international standards themselves; thus, in nearly every spectral region, secondary standards have been well established. Of course the primary standards of length and frequency themselves are at present based on indestructible spectral standards: 1 m ≡ 1,650,763.73 times the wavelength of the orange light emitted by ^{86}Kr and 1 Hz = 1 cycle/sec, where 1 sec ≡ the time for 9,192,631,770 cycles of the Cs atom in the molecular-beam device known popularly as the "atomic clock."

1.1. History of Spectroscopy

In this chapter, we trace the development of spectroscopy in the visible region and then return to the history of spectroscopy outside the visible region. We close the chapter with a brief discussion of the influence of spectroscopy on the development of twentieth-century physics.

1.1.1. Newton's Contributions

The first dispersed spectrum to be observed was, of course, the rainbow. Unable to explain this beautiful phenomenon, primitive man was disposed to attribute to it a supernatural significance sometimes related to legends such as the one included in the Biblical account of the flood. Its relationship to the laws of refraction was not understood even after these laws were firmly established by Willebrord Snell of Leyden (1591–1626). Although Snell was able to establish the fact that the ratio of the sines of the angles of incidence and refraction was a constant now known as the refractive index of the medium, he did not perform experiments on light of different colors. In fact, up to the time of Newton, even the best scientists had extremely vague ideas regarding the nature of color.

The first and possibly the most important step in the development of spectroscopy was taken by Newton in 1665 when, at the age of twenty three, he purchased a glass *prism* with the stated purpose "to try therewith the phenomena of colors"! His simple but fundamental experiments were reported first in the *Transactions of the Royal Society* in 1672; this paper led to his famous controversy with Robert Hooke. The experiments were described more fully in the first edition of "Opticks" in 1704. He placed red and blue strips of paper side by side; when he viewed them through the prism, he found that their apparent displacements were different. This first experiment showed that the refractive indices for red and blue light were different.[1]

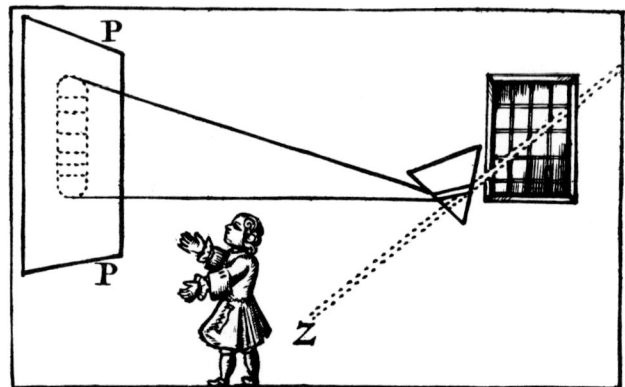

FIG. 2. "And so the true Cause of the Length of that Image was detected to be no other than that *Light* is not similar or Homogeneal, but consists of difform rays, some of which are more refrangible than others."—I. Newton. The cut in the figure is taken from Voltaire's "Elémens de la Philosophie de Neuton," published in Amsterdam in 1738.

Newton's second experiment involved a pencil of sunlight passing through a small hole in a shutter; after passing through the prism, the light reached a screen. Newton recognized that the resulting *spectrum* displayed on the screen was, in essentials, a series of colored images of the hole; the term spectrum was introduced by Newton. Subsequent experiments showed that light of a given color dispersed by his prism was further refracted but not further dispersed by a second prism. Such light, with all rays similarly *refrangible*, he termed as *homogeneal*, as distinguished from *heterogeneal* light with rays of differing *refrangibility*. He concluded that sunlight was "a heterogeneal mixture of difform rays, some of which are more refrangible than others." In the prism, the difform rays were "parted from one another."

Newton recognized that the separation of the rays and hence the purity of the spectrum could be improved by using a slit in the window shutter in combination with a lens to produce an image of the slit on the screen; when the light from the lens passed through the prism, a much purer spectrum was displayed on the screen. Although Newton was able to display a solar spectrum 25 cm long on the screen, he failed to observe the Fraunhofer

---

[1] I. Newton, "Opticks." London, 1703. Edition with commentary by E. T. Whitaker and I. B. Cohen is available from Dover Press, New York.

lines—presumably as a result of the poor quality of the glass in his prism and lens.

This work constituted Newton's contribution to spectroscopy. In his later treatment of the colors of various flames, he seems to have violated his credo, "*Hypotheses non fingo!*", and passed from experiment to speculation. His prestige was such that his unfortunate corpuscular theory of light served to stifle real progress in optics for nearly a century. His mistaken belief that dispersion was proportional to deviation for all types of glass delayed the development of achromatic lenses for many years.

1.1.2. Nineteenth-Century Developments

In the eighteenth century, the only noteworthy work in spectroscopy was Thomas Melvill's prism study of a sodium flame. In the "Physical and Literary Essays" (1752), Melvill gave the first description of a laboratory emission spectrum; this description was reprinted 162 years later.[2]

Early in the nineteenth century Thomas Young (1802) made the first wavelength determinations by applying his wave theory of light to the problem of interference colors in thin films; he showed with surprising accuracy that the range of wavelengths in the visible spectrum extends from 424 to 675 nm. In the same year, W. H. Wollaston[3] observed some of the dark lines that appear in the otherwise continuous spectrum of the sun but seems to have regarded them as natural boundaries between various pure colors. Wallaston also reported observations of flame spectra and the first investigations of spark spectra, but made no serious attempt to explain his observations, which were in fact hampered by crude apparatus and impure source samples.

The techniques of spectroscopy advanced rapidly as a result of the extraordinary work of Joseph Fraunhofer (1787–1826), a man with little formal education who was associated with a glass-making firm in Munich. By placing his flint-glass prism approximately 8 m from a slit in a window shutter, and by viewing the dispersed radiation with a theodolite telescope placed behind the prism, Fraunhofer was able to make highly precise angle measurements. He found that the solar spectrum was crossed by "an almost countless number of strong and weak dark lines"; experiments with different prisms demonstrated that those dark lines were actually characteristic of the solar spectrum. He mapped nearly 700 of those dark lines and assigned to the eight most prominent lines the letters A to H, by which they are still identified.[4] These lines provided standards that could be used for comparison

---

[2] T. Melvill, *J. Roy. Astron. Soc. Canada* **8**, 231 (1914).
[3] W. H. Wollaston, *Phil. Trans.* **92**, 365 (1802).
[4] J. Fraunhofer, *Ann. Phys.* **56**, 264 (1817).

of the dispersion of different types of glass, and provided the basis for exact spectroscopic measurements. In other work with prisms in combination with a telescope, Fraunhofer first applied spectroscopy to astronomy when he mapped the spectra of Sirius and other stars and the spectra of the planets.[5]

Fraunhofer invented the diffraction grating by extending the single-slit studies of Young to two and, later, to many slits. His first transmission grating was fabricated by winding fine wires upon two parallel fine screws; with relatively coarse gratings of this type he made remarkably accurate wavelength measurements of the D lines. Later, with a diamond point and a ruling engine of his own design and manufacture, Fraunhofer ruled the first glass transmission gratings.[6] By measuring the wavelengths of a number of the Fraunhofer lines, he was able to establish a basis which prism workers could employ in identifying spectral lines in terms of wavelength rather than in terms of prism angles and angles of deviation.

Considerable progress was made in the mid-nineteenth century in the study of flame and spark emission spectra and absorption spectra by J. F. W. Herschel, W. H. F. Talbot, C. Wheatstone, W. Crookes, L. Foucault, A. J. Ångström, and others. Foucault noted in 1848 that a sodium flame that *emitted* the D lines would also *absorb* the D lines from the brighter light emitted by an arc placed behind it. Enough work had been done to indicate that an observed emission spectrum has characteristics that depend on the chemical constitution of the source and on the method of source excitation; Foucault's work indicated a relationship between emission and absorption spectra. The stage was set for a broad generalization of these results.

This generalization, which showed the relationship between absorption and emission of light, was formulated in 1859 by G. R. Kirchhoff,[7–9] who was Professor of Physics at Heidelberg. Kirchhoff's law states that the ratio of the emissive power to the absorptivity for thermal radiation of the same wavelength is constant for all bodies at the same temperature. Thus, a perfectly transparent body cannot emit light; a body that emits a continuous spectrum must be opaque; a perfect absorber is also a perfect radiator! A gas that radiates a given set of spectral lines must absorb the same lines which it radiates at the same temperature.

Kirchhoff immediately proceeded to explain the Fraunhofer lines in the solar spectrum as due to absorption by vaporized elements in the relatively

---

[5] J. S. Ames, "Prismatic and Diffraction Spectra." Harper, New York, 1898. (A collection of early papers by Fraunhofer and Wollaston.)
[6] J. Fraunhofer, *Ann. Phys.* **74**, 337 (1823).
[7] G. R. Kirchhoff, *Ann. Phys.* **109**, 148, 275 (1860).
[8] G. R. Kirchhoff, *Phil. Mag.* **20**, 1 (1860).
[9] G. R. Kirchhoff, *Ann. Chim. Phys.* **58**, 254 (1860).

cool solar atmosphere. In attempting a chemical analysis of the sun's atmosphere, Kirchhoff worked in cooperation with R. Bunsen, Professor of Chemistry at Heidelberg. Working together, these two men are regarded as the founders of modern spectroscopy. They certainly established spectroscopy as a potent method of chemical analysis, and demonstrated that each chemical element has a unique spectrum.[10–12] Unfortunately, in their work they reported spectral lines not in terms of wavelength but in terms of the micrometer scale of the Heidelberg spectrometer—information useless to workers in other laboratories!

A development of great importance to spectroscopy was the introduction of the dry-gelatin photographic plate, which became available in the years after 1870. Whereas all earlier work had been done *visually* with instruments properly designated as *spectroscopes*, the availability of photographic plates made possible the development of the *spectrograph*, which provides permanent records that can be examined carefully at leisure.

Further work on precise wavelength measurement was done by A. J. Ångström, who employed several glass gratings; his results for the Fraunhofer lines were given to six significant figures and were expressed in a unit of 10^{-10} m, the unit that became known as the angstrom unit. Important improvements in the techniques of ruling plane gratings were made in the years just prior to 1887 by Henry A. Rowland of the Johns Hopkins University, who also invented the concave grating; he used these improved gratings to advantage in preparing new lists of solar wavelengths.[13] In 1893, A. A. Michelson, at Chicago, used his recently developed interferometer to measure wavelengths with an accuracy that greatly surpassed those of earlier investigators; in terms of the standard meter in Paris, he was able to measure three cadmium wavelengths to eight significant figures.[14] Fabry and Perot in Paris extended interferometric methods to the spectra of other elements, and led a movement for the adoption of the red cadmium line as a basic spectroscopic standard.

1.1.3. Twentieth-Century Developments

Recognition of the importance of spectroscopy in the study of atomic structure and in chemical analysis led to enormous activity in the first third of the present century. Most of the experimental techniques employed in the visible region during this period represented refinements of the prism,

---

[10] G. R. Kirchhoff and R. Bunsen, *Ann. Phys.* **110**, 160 (1860).
[11] G. R. Kirchhoff and R. Bunsen, *Phil. Mag.* **20**, 89 (1860).
[12] G. R. Kirchhoff and R. Bunsen, *Ann. Chim. Phys.* **62**, 452 (1861).
[13] H. A. Rowland, *Phil. Mag.* **23**, 287 (1887).
[14] A. A. Michelson, *Phil. Mag.* **24**, 463 (1887); **31**, 338 (1891); **34**, 280 (1892).

grating, and interferometric instruments developed during the nineteenth century. Recognition of the practical, industrial importance of spectroscopy as a tool of chemical analysis led to the commercial development of instruments of high quality. These instruments also proved useful in studies of molecular structure.

During the second third of the century, the nucleus replaced the atom as the subject of major interest to physicists. However, during this period, many advances in instrument design were made. We might mention the development of nonreflecting coatings for optical components, many new varieties of optical materials, new varieties of detectors, and many other devices made possible by advances in solid-state physics and related advances in chemistry. The most obvious change in spectroscopic work was the automation of many operations. The use of solid-state detectors, electronic amplifiers, and chart recorders replaced photographic methods in many laboratories; however, photographic techniques have been improved and are still being used to advantage in many important research and industrial laboratories. One very important development involved the manufacture of replica gratings of extremely high quality.

During the final third of the century, many nuclear physicists have transferred their attention to atomic spectroscopy. In turning to spectroscopic work, they have brought with them many of the techniques originally developed for use in nuclear physics. The use of photon-counting systems, multichannel analyzers, coincidence circuits, and on-line computers is becoming commonplace. The impact of modern digital computers on spectroscopic research has been enormous. The use of laser techniques in spectroscopy should also be noted.

The measurement of standard wavelengths has progressed to the point where the international meter was redefined[15] in 1960 in terms of the orange line of ^{86}Kr.

1.1.4. The Infrared Region†

In 1800, Sir William Herschel,[16] the British Astronomer Royal, discovered infrared radiation. While attempting to determine the distribution of "radiant heat" from the sun by means of sensitive thermometers with their blackened bulbs laid along a solar spectrum, Herschel found the greatest heating effects beyond the red end of the spectrum. He showed that the radiation

---

[15] "Units of Measurement," Nat. Bur. Std. Publ. 286 Washington, D.C., (1972).
[16] W. Herschel, *Phil. Trans.* **90**, 255, 284, 293, 437 (1800).

† See also Vol. 10, Chapter 1.2.

involved could be reflected and refracted but erroneously regarded this thermal radiation as something quite different from light. It is interesting to note that he made the first infrared absorption measurement when he showed that the absorption of thermal radiation by water is different from its absorption by certain alcoholic beverages! Sir William's son, J. F. W. Herschel, demonstrated in 1840 the spectral selectivity of absorption by showing that a black paper soaked in alcohol and placed in a dispersed solar spectrum dried more rapidly when exposed to some spectral regions than when exposed to others.

Rowland's measurements at the red end of the spectrum ended at 772 nm. However, with specially prepared photographic plates, Abney, in 1880, claimed to have extended photographic detection to 2000 nm; modern commercially available plates give good results to about 1250 nm. Further extension of investigation to longer wavelengths required other types of detectors.

The most sensitive of the early, so-called thermal detectors was the *bolometer*, an extremely sensitive resistance thermometer developed by S. P. Langley[17] in 1881; he used this device in following years to map the solar spectrum to 18 μm. In the years following 1892, H. Rubens and F. Paschen, together with their associates in Germany, undertook an extensive research program in all phases of infrared spectroscopy; they made effective use of thermocouples[18] and radiometers as detectors. Rubens is now remembered chiefly for his work on *reststrahlen*, and Paschen for his studies of the atomic hydrogen spectrum in the infrared. One of Rubens' students was E. L. Nichols, who later established an infrared laboratory at Cornell and was one of the founders of the *Physical Review*. Perhaps the most important of the early German investigations were the detailed studies of the emission spectrum of a blackbody operated at various temperatures[19]; interpretation of the observed spectra led M. Planck to the development of the quantum theory. Later, German workers such as Eva von Bahr, M. Czerny, and R. Mecke made important pioneer contributions.

On the basis of initial work at Cornell and subsequent studies at the National Bureau of Standards, W. W. Coblentz[20] made a monumental contribution to the development of experimental techniques for the study of the spectra of solids, liquids, and gases; his publication of numerous spectra provided a guide for many later workers. Other early centers of

[17] S. P. Langley, *Proc. Amer. Acad.* **16**, 342 (1881).

[18] H. Rubens, *Z. Instrkde.* **18**, 65 (1898).

[19] O. Lummer and E. Pringsheim, *Verh. Phys. Ges.* **1**, 23, 215 (1899), **2**, 163 (1900).

[20] W. W. Coblentz, "Investigations of Infrared Spectra," Parts I–VII. Carnegie Inst. of Washington, 1905, 1906, and 1908.

research in the USA were founded by H. M. Randall at Michigan and by R. W. Wood and A. H. Pfund at Johns Hopkins. Most of the early spectrographs employed Nernst glowers or globars as sources; original gratings or prisms fabricated from natural crystals of quartz, fluorite, and the alkali halides were used as dispersing elements; and vacuum thermocouples served as detectors. Elaborate systems for amplification of the dc signals from the detectors were devised.

Following World War II, progress in infrared spectroscopy has been rapid as a result of the development of automatically recording spectrographs or spectrophotometers by R. B. Barnes, V. Z. Williams, J. D. Hardy, N. Wright, J. Strong, and others. These instruments have made use of new synthetic crystals, improved detectors, excellent replica gratings, and electronic amplifiers. The development of high-quality spectrophotometers by commercial firms has led to numerous industrial applications of infrared spectroscopy. The availability of improved, more sensitive detectors involving "quantum" processes, as opposed to strictly thermal processes, has broadened the use of infrared methods in astronomy.

1.1.5. The Submillimeter and Microwave Regions

The conventional, thermally excited infrared sources such as the Nernst glower, globar, and gas mantle have emission spectra that closely approximate those to be expected for blackbodies. Since their spectral emissive powers thus decrease in the far infrared, the operation of conventional infrared spectrographs becomes increasingly difficult in the spectral range 300 μm $< \lambda <$ 1000 μm, the so-called *submillimeter region*. The quartz-enclosed mercury arc has considerably higher emissive power at still lower frequencies—primarily as a result of emission by the enclosed plasma. However, even with such a source, conventional infrared techniques usually fail.

There was much early work on the development of electrically excited sources to replace thermally excited sources for use in this region. In principle, the spark discharges of the kind used by Hertz can be used to produce extremely short-wave radiation by the excitation of small metallic spheres or short lengths of fine wires. Work of this type met with some initial success,[21-23] and, indeed, E. F. Nichols and J. W. Tear succeeded in bridging the gap between the thermal region and the radio region. However, it was not until the development of oscillators involving resonant cavities[24]

[21] A. Garbasso, *J. Phys.* **22**, 259 (1893).
[22] E. F. Nichols and J. W. Tear, *Phys. Rev.* **21**, 587 (1923).
[23] R. W. Wood, *Phil. Mag.* **25**, 440 (1913).
[24] W. W. Hanson and R. D. Richtmyer, *J. Appl. Phys.* **10**, 189 (1939).

1.1. HISTORY OF SPECTROSCOPY

that much progress was made in the submillimeter region. These cavity-resonator tubes, developed in the course of World War II, typically provide sources of radiation in the wavelength range 1–30 cm, the so-called *microwave region*.

In the immediate post-war years, the techniques of microwave spectroscopy involving klystron oscillators, waveguide and resonant-cavity absorption cells, and crystal-diode detectors were rapidly developed.[25, 26] These techniques have been widely used to enormous advantage in molecular spectroscopy† and in studies of electronic paramagnetic resonance.‡ The extension of the radiation from microwave generators to wavelengths in the millimeter and submillimeter ranges was first accomplished by use of crystal diodes as harmonic generators[27, 28]; the resulting radiation has been employed to advantage in investigations of molecular structure.[29] Other methods of harmonic generation have been developed by Froome.[30] The laboratory methods in the submillimeter region represent an interesting combination of the waveguide techniques developed for the microwave region and the classical infrared techniques involving mirrors.

While methods of harmonic generation of submillimeter waves were being developed, other developments were taking place in the far-infrared which led to the extension of *optical* methods to the submillimeter region. These methods involve what is now termed *Fourier-transform spectroscopy*, since the actual spectrum is the Fourier transform of an interferogram observed experimentally; the advantages of the general methods were pointed out by Fellgett.[31] One device that has been used effectively involves a Michelson interferometer, a commercial model of which was first developed in England with the cooperation of H.A. Gebbie and his associates at the National Physical Laboratory; this commercial instrument now operates satisfactorily in the spectral wavelength range between about 16 μm and 3 mm (660–3 cm^{-1}). Another more complicated instrument employing a

---

[25] W. Gordy, W. V. Smith, and R. F. Trambarulo, "Microwave Spectroscopy." Wiley, New York, 1953.

[26] C. H. Townes and A. L. Schawlow, "Microwave Spectroscopy." McGraw-Hill, New York, 1955.

[27] C. A. Burrus and W. Gordy, *Phys. Rev.* **93**, 897 (1954).

[28] W. C. King and W. Gordy, *Phys. Rev.* **93**, 407 (1954).

[29] W. Gordy and R. L. Cook, "Microwave Molecular Spectra." Wiley (Interscience), New York, 1970.

[30] K. D. Froome, *Proc. Int. Conf. Quantum Electron., 3rd, Paris, 1963*, pp. 1527–1539. Columbia Univ. Press, New York, 1964.

[31] P. Fellgett, *J. Phys. (Paris)* **28**, 165 (1967).

---

† See Lide in Vol. 3A, Chapter 2.1.

‡ See Memory and Parker in Vol. 3B, Part 4.

lamellar grating developed by Strong and Vanasse[32–34] offers certain advantages. However, the ultimate limitations of the two types of interferometers are essentially the same and occur as a result of diffraction effects that set in when the wavelength becomes comparable with the diameters of the radiation beams.

With the development of Fourier-transform methods and harmonic-generation techniques, the previous spectroscopic gap between the infrared region, where incoherent thermal sources are employed, and the microwave region, where electrically excited coherent sources are used, can be considered as completely bridged!

1.1.6. The Radio Frequency Region

The discovery of electrically produced electromagnetic waves by H. Hertz in 1887 and the study of their properties in terms of Clerk Maxwell's theory are parts of a familiar story. With the development of the thermionic vacuum tube, it became possible to use these waves in various systems of communication—and, incidentally, to employ them in a variety of ways in spectroscopy.

One of the earliest applications was in the determination of dielectric constants. It was found[35] in the case of polar molecules that the dielectric constant for very low frequencies was considerably higher than that for much higher frequencies; this was attributed to the inability of the polar molecules to follow the electric field at extremely high frequencies so far as orientation is concerned. At very low frequencies, measured polarization included effects due to both molecular orientation and molecular distortion, whereas at extremely high frequencies, the measured polarization is due to molecular distortion alone.[36] The dividing line between the two takes place at a frequency v defined by the equation $v\tau = 1$, where τ is called the *dielectric relaxation time*.

These effects are associated with an anomalous dispersion that occurs in the radio frequency region in the case of many liquids[37]; the energy dissipation as reflected in the imaginary part of the dielectric constant represents a direct conversion of electromagnetic energy to the thermal energy associated with random motion of the molecules in the fields of their neighbors. It is thus related directly to viscosity effects; in the case

---

[32] J. Strong and G. A. Vanasse, *J. Phys. Radium* **19**, 192 (1958).
[33] J. Strong and G. A. Vanasse, *J. Opt. Soc. Amer.* **50**, 113 (1960).
[34] P. L. Richards, *J. Opt. Soc. Amer.* **54**, 1474 (1964).
[35] P. Drude, *Z. Phys. Chem.* **23**, 267 (1897).
[36] P. Debye, *Berichte* **15**, 777 (1913).
[37] P. Debye, "Polar Molecules." Chemical Catalogue Co., New York, 1929.

of solids, other related phenomena are involved. This type of absorption is called *Debye absorption*; the absorption line shape is quite different from that of other absorption lines.

Following the discovery of nuclear magnetic resonance (NMR) by Bloch and Purcell, special radio frequency techniques have been developed for the study of the phenomena involved.† When a sample containing a given type of nucleus with nonzero spin is placed in a strong magnetic field, the degeneracy involving the spatial quantum number is removed. The frequency separation of the resulting energy levels is related to the Larmor precessional frequency of the nucleus in the strong external field; transitions between these levels can be observed in absorption. In the case of liquids with low viscosity, the resulting lines are exceedingly narrow and are split by various characteristic molecular effects; true molecular splittings as low as a few Hz have been observed. Nuclear magnetic resonance lines are, in general, much broader in the case of solids; the shape and observed splittings of such lines give valuable information regarding crystal structure. A related field of study in solids is called nuclear quadrupole resonance (NQR); absorption can be observed as a result of transitions between spatially quantized levels involving electrostatic interactions between the electric quadrupole moment of the nucleus and the resultant crystal field at its site.†

Radio frequency techniques are also employed to advantage in research involving molecular beams.‡

1.1.7. The Ultraviolet Region

In 1803, J. W. Ritter[38] of Jena, while studying the effects of the solar spectrum on the blackening of silver chloride, found that the blackening effects were greatest beyond the violet end of the solar spectrum; he had thus discovered ultraviolet radiation. As ultraviolet radiation is strongly absorbed by glass, the radiation Ritter observed was in the very near ultraviolet. The ultraviolet range was greatly extended in 1862 by G. G. Stokes,[39] who discovered the transparency of quartz in the near ultraviolet. By the use of a uranium phosphate fluorescent screen, he was able to observe the arc and spark spectra of several metals but made no wavelength measurements. The first wavelength measurements in the ultraviolet were made in

[38] J. W. Ritter, *Ann. Phys.* **12**, 409 (1803).
[39] G. G. Stokes, *Phil. Trans.* **152**, 599 (1862).

---

† See Memory and Parker in Vol. 3B, Part 4.
‡ See English and Zorn in Vol. 3B, Part 6.

1863 in France by Mascart in studies of the solar spectrum, which apparently ends abruptly at 295 nm as a result of the absorption of the atmospheric ozone layer. Accurate wavelength measurements were later made by Rowland[13] for wavelengths as short as 215 nm. Further extensions toward shorter wavelengths by the techniques then available were limited because of the absorption of the gelatin in the photographic emulsion for wavelengths shorter than 230 nm, and the absorption of atmospheric oxygen, which sets in at 185 nm.

The next extension of the ultraviolet was made in 1893 by Victor Schumann,[40] who made photographic plates nearly free of gelatin, employed fluorite optics, and operated his spectrograph in an evacuated enclosure. Schumann extended the ultraviolet spectrum to 120 nm, a limit set by the absorption of fluorite. The next great extension was achieved in 1906 by Theodore Lyman,[41] who eliminated the absorbing fluorite optics by the substitution of a concave grating; the only losses in the system were those involved in reflection at the grating surface. With this equipment, Lyman made wavelength measurements down to 50 nm. Subsequently, by making improvements in gratings, methods of using the gratings at oblique incidence, light sources, and photographic plates, R. A. Millikan, in collaboration with R. A. Sawyer and I. S. Bowen, was able to extend wavelength measurements to 3 nm, a limit which had been overlapped by the methods of x-ray spectroscopy.[42–44] More recent developments in methods of studying the extreme ultraviolet have been reviewed by Samson.[45]

1.1.8. The X-Ray Region

In the course of experiments with a gas discharge tube in 1895, W. K. Roentgen quite by accident discovered x rays by noting the fluorescence produced in a paper impregnated with barium platinocynide. Realizing the importance of his discovery, Roentgen proceeded with a qualitative investigation of many of the properties of this new phenomenon and described them in his initial paper.[46] Most of the early work done on x rays was done with gas discharge tubes essentially similar to the one employed by Roentgen. However, in 1913, Coolidge developed the highly

[40] V. Schumann, *Wien. Sitzbes.* **102**, 415, 625, 994 (1893).
[41] T. Lyman, *Astrophys. J.* **5**, 349 (1906).
[42] R.A. Millikan and R. A. Sawyer, *Phys. Rev.* **12**, 168 (1918).
[43] R. A. Millikan and I. S. Bowen, *Phys. Rev.* **23**, 1 (1924).
[44] R. A. Millikan, I. S. Bowen, and R. A. Sawyer, *Astrophys. J.* **53**, 150 (1921).
[45] J. A. R. Samson, "Techniques of Vacuum Ultraviolet Spectroscopy." Wiley, New York, 1967.
[46] W. K. Roentgen, *Sitzungsber. Phys. Med. Ges., Würzburg* (1895). Reprinted in *Ann. Phys. Chem.* **64**, 1 (1898).

evacuated x-ray tube, providing for the bombardment of a target by electrons of controlled energy; it has essentially the same form as the x-ray tubes used at present. In addition to the photographic and fluorescent methods of detection discovered by Roentgen, gas-filled ionization chambers were developed for detection and measurement purposes. Although the original form of these chambers has been superceded in recent times by solid-state devices, the essentials of the method remain unchanged.

Although the early efforts of Haga and Wind to measure x-ray wavelengths by observing the diffraction pattern of a single slit indicated that typical wavelengths were of the order of 0.1 nm, no precise measurements were made until 1912. Earlier, M. Laue had suggested that, for wavelengths of the order of 0.1 nm, a crystal could well serve as a three-dimensional diffraction grating. Acting on this suggestion, Friedrich and Knipping found that the diffraction pattern of a narrow pencil of x rays transmitted by a crystal was of the type predicted by Laue.[47] The pattern obtained by the use of the three-dimensional crystal grating consists of a set of bright spots surrounding a central image, and is now called a Laue pattern.

Shortly after the first diffraction patterns were obtained, W. H. Bragg[48] showed that there are certain planes in a crystal that scatter x radiation in such a way that at large angles of incidence they produce specular reflection in a certain directions that involve the glancing angle θ, the spacing d between the planes, and the wavelength of the incident radiation. W. H. Bragg and his son W. L. Bragg[49] used these ideas in the development of the crystal x-ray spectrometer, which in later, refined forms is still in use.

The development of the Bragg spectrometer made it possible to make precise x-ray wavelength measurements. In 1913, H. G. J. Moseley[50] began a systematic study of the x-ray emission lines of a number of elements and discovered the K, L, and M series characteristic of each element. The characteristic lines are superposed on a continuous background, which was studied in detail by W. Duane and H. L. Hunt[51] and interpreted in terms of the inverse photoelectric effect. Although the diffraction measurements of Bragg and others clearly demonstrated the wave characteristics of x rays, A. H. Compton's discovery[52] that the wavelengths of x rays are altered when they are scattered could be successfully interpreted only on the basis of the particle character of the x-ray *photons*, to which momentum h/λ as well as quantum energy $h\nu$ must be assigned. The particle nature

---

[47] M. Laue, W. Friedrich, and P. Knipping, *Ann. Phys.* **41**, 971 (1913).
[48] W. H. Bragg, *Proc. Cambridge Phil. Soc* **17**, 43 (1912).
[49] W. L. Bragg, "The Crystalline State." Bell, London, 1962.
[50] H. G. J. Moseley, *Phil. Mag.* **26**, 1024 (1913); **27**, 703 (1914).
[51] W. Duane and F. L. Hunt, *Phys. Rev.* **6**, 166 (1915).
[52] A. H. Compton, *Phys. Rev.* **21**, 207, 483 (1923).

becomes increasingly important in the hard x-ray region, where it is more convenient to characterize x rays in terms of quantum energy than in terms of wavelength or frequency.

1.1.9. The Gamma-Ray Region

Shortly after A. H. Becquerel's accidental discovery of natural radioactivity in 1896, ionizing positive α rays and negative β rays were distinguished by their markedly different penetrations of air and by their deflections in a magnetic field. A third much more penetrating component called γ radiation was identified by P. U. Villard in 1900. These γ rays cannot be deflected by a magnetic field and were quickly identified as electromagnetic radiation of extremely short wavelength. Although the wavelengths of "soft" γ rays can be measured by means of specially designed crystal spectrometers,[53] typical γ rays are usually characterized by their quantum energies. Since these energies are much larger than the energies required to ionize single atoms or molecules, individual γ-ray quanta, like x-ray quanta, are easily observable, and their energies can be measured in terms of the total ionization produced.

The initial interaction of an x-ray photon with matter usually involves the production of electrons by the photoelectric effect, which usually predominates at low quantum energies; by the Compton effect, which is predominant at intermediate energies; and by pair-production, for photon energies above the threshold of 1.02 MeV. Studies of these primary electrons by magnetic-deflection techniques have been employed in γ-ray spectroscopy.[54]

However, in more recent work, measurements of total photon energies have continued to be employed. In contrast to the earliest techniques involving the ionization of gases, modern techniques for measurement of photon energies have involved photometric detection of the light emitted by scintillators such as napthalene and sodium iodide,[55] or electrical measurement of the total number of electrons raised from the bound states to the conduction band of a doped semiconductor.

Since the entire quantum energy $h\nu$ of a γ ray never goes completely into light production in scintillators nor into charge separation in semiconductors, it is difficult even with modern techniques to obtain precise measurements of ν. However, it is well known that certain γ rays involve radiative transitions from excited nuclear states with extremely long lifetimes τ; the spread $\Delta\nu$ in the frequencies involved in such transitions, in accord

[53] J. W. M. Dumond, *Rev. Sci. Instrum.* **18**, 626 (1947).
[54] M. Deutsch, L. G. Elliott, and R. D. Evans, *Rev. Sci. Instrum.* **15**, 178 (1944).
[55] W. H. Jordan and P. R. Bell, *Nucleonics* **5**, 30 (1949).

with the uncertainty principle, is given by $\Delta v \approx 1/\tau$, and can be extremely small as compared with v. In fact, the "sharpness" of such a γ-ray resonance $v/\Delta v$ is the greatest found in nature! The discovery of the *Mössbauer effect*[56] has made it possible to use nuclear resonance radiation in studies of the Doppler effect at small speeds, gravitational shifts in frequency, and a variety of other phenomena.[57]

1.1.10. The Role of Spectroscopy in Twentieth-Century Physics

Now that we have given a short outline of the way in which the methods of spectroscopy and its several branches have developed, we shall consider briefly the question of the role played by spectroscopy in the development of twentieth-century physics. The two major developments of the present century have been the formulations of quantum mechanics and the theory of relativity. Spectroscopy has been involved in both developments, in supplying the original empirical knowledge on the one hand and in making crucial tests on the other. It has also been of great importance to the development of our understanding of atomic, molecular, and nuclear structure.

Spectroscopy was of prime importance to the early development of quantum theory. We recall that careful, quantitative measurements of the *blackbody spectrum* in the visible and infrared provided the information on which Planck based his original quantum theory of radiation. Spectroscopy also played a significant role in the investigation of the *photoelectric effect*, which Einstein interpreted by one of the first successful applications of quantum ideas. Empirical studies of atomic spectra led to the *Ritz combination principle*, which stated that the frequencies of the large numbers of observed spectral lines of a given element can be expressed as differences between a much smaller number of *spectroscopic term values* characteristic of the element. These term values were later interpreted by Bohr as representing the quantized stationary *energy levels* of the atom, radiative transitions between which resulted in observed emission and absorption lines. Spectroscopy was also applied in the Franck–Hertz experiment that verified the existence of atomic *energy levels*.

Prior to the development of the nuclear model of the atom, Pieter Zeeman's studies of the magnetic splitting of spectral lines yielded, in the case of the normal Zeeman effect, a value of e/m in such close agreement with J. J. Thomson's value that it established the electron as a constituent of the atom and as the constituent directly involved in the emission of atomic

[56] R. L. Mössbauer, *Z. Phys.* **151**, 124 (1958).
[57] G. K. Wertheim, "Mössbauer Effect: Principles and Applications." Academic Press, New York, 1964.

spectral lines.† Early studies of x-ray scattering indicated that the number of electrons in an atom is equal to the atomic number Z of the element in the periodic table. On the basis of the Rutherford nuclear model of the atom, the hydrogen atom thus served as the basic two-body problem to which new ideas could be applied. Spectroscopic studies of the hydrogen spectrum by Balmer, Paschen, and Lyman had resulted in the discovery of several series of lines in the visible, near infrared, and ultraviolet regions, the frequencies of which could be expressed in terms of simple integers and a single empirical frequency known as the *Rydberg constant*; by arbitrarily introducing the integers as *quantum numbers*, Bohr was able to set up a simple model of the hydrogen atom in terms of which the Rydberg constant could be *calculated* with amazing accuracy. Sommerfeld's extension of Bohr's ideas to interpret the spectra of more complicated atoms is a familiar story; the resulting "old quantum mechanics" became an increasingly complicated theoretical contraption that involved numerous arbitrary assumptions but worked fairly well in giving an account of various features of observed spectra. During this period, investigations of spectral line shapes prompted important developments in dispersion theory. With the development of *modern quantum mechanics* by Heisenberg and Schrödinger, a new era in physics began; the detailed application of quantum mechanics to the vast body of spectroscopic information that had been developed was made successfully by Condon and Shortley in their classic treatise "*The Theory of Atomic Spectra*" (Cambridge Univ. Press, London and New York, 1935).

Spectroscopic investigations as interpreted in terms of quantum mechanics have provided valuable, basic information regarding the structure of matter and the interactions of electromagnetic radiation with matter. Observations of so-called multiplet fine structure in atomic spectra were interpreted in terms of *electron spin* and *spin-orbit coupling*; related interpretations of the anomalous Zeeman effect provided information regarding the *magnetic moment of the electron*. The spectroscopic hyperfine structure provided evidence of *nuclear spin*; later applications of Zeeman methods in the radio frequency region have provided highly precise values of *magnetic moments of nuclei* by NMR techniques. Spectroscopic studies in the γ-ray region have revealed the existence of well-defined excited energy states in nuclei. Quantum mechanics has provided an understanding of the nature

---

† Zeeman's results represented a major triumph for the "electron theory" of matter formulated by H. A. Lorentz (1853–1928). Although Lorentz is still well known in connection with the *Lorentz transformation* and the *Lorentz force*, his important contributions to the theory of atomic structure tend to be forgotten. Richard Feynman has pointed out in Section 31–2 of his "Lectures on Physics" (Addison-Wesley, Reading, Massachusetts, 1963) that the Lorentz model still represents the best starting point for treatments of refractive indices and related dispersion phenomena.

of radiative transitions between energy levels in atoms, molecules, and nuclei.

Spectroscopic studies in the visible and ultraviolet regions have provided an understanding of the *nature of chemical bonds* in molecules. Studies in the infrared, submillimeter, and microwave regions have provided a wealth of information regarding the *sizes and shapes of molecules* and an understanding of *molecular vibrational and rotational motions*. Spectroscopy has been an extremely useful tool to solid-state physics in a wide variety of ways. Comparisons of x-ray spectra as observed by means of ruled gratings with those observed with crystal spectrometers have given highly precise values for the lattice spacings in crystals; these values have been used in obtaining improved values of Avogadro's number. Some of the symmetry considerations involved in the interpretation of x-ray diffraction studies led to group theory as an important method of treating a wide variety of physical problems.

Spectroscopic methods have also been applied in several crucial tests of the theory of relativity. The most famous of these involves the observation and measurement of the γ-ray photons involved in the *production and annihilation of electron–positron* pairs; these processes involve direct conversion of electromagnetic energy to matter and the reverse of this process. Another is the direct observation of the *quadratic Doppler effect* predicted by relativity theory and first observed by Ives in 1941. A third is Pound's laboratory measurement of the *gravitational red shift* predicted by the general theory of relativity.

Spectroscopic contributions to relativistic quantum electrodynamics include Lamb's measurement of the shift between the 2S and 2P states in the hydrogen atom. This beautiful experiment laid the basis for the initial formulation of the renormalization program of quantum electrodynamics and still remains one of the most delicate tests of more sophisticated formulations of the theory. Spectroscopic measurements of the anomalous magnetic moment of the electron have also provided an important test of quantum electrodynamics.

1.2. General Methods of Spectroscopy

In this chapter, we first give a general discussion of the meaning of such terms as *spectrum* and *spectral line*, and then proceed to a general discussion of *sources, resolving instruments*, and *detectors*. As in most scientific usage, the terminology employed in spectroscopy is, to some extent, an accident of history. For example, the term spectral line has its origin in the fact that in early spectroscopes and spectrographs the observer studied the colored images of the narrow entrance slit, which in spite of their finite widths superficially resembled short geometrical lines; in modern spectroscopy much more is involved in the term *spectral line*! Similarly, the term *band* was initially introduced to denote a region in which lines were so closely spaced that they could not be resolved; however, the term band has been retained even after improved techniques have made it possible to resolve the individual lines.

The list of general spectroscopic components to be discussed is also based on early history; when the spectrum of a flame was being dispersed by means of a prism instrument and recorded on a photographic plate, it was simple to distinguish between source, resolving instrument, and detector. However, in some fields of modern spectroscopy, resolution and detection processes are accomplished in a single device; in other branches of spectroscopy, a single device can serve as source and resolving instrument.

1.2.1. Emission Spectra

During the process of emission, energy is transferred from the source to the electromagnetic field. The radiant flux ϕ from the source is the time rate of energy transfer and thus has the dimensions of power, properly expressed in watts (W). The flux from the source is usually different in different directions; the proper description of an extended source therefore involves a statement of the radiant flux emitted per unit area of the source per unit solid angle in a specified direction. In quantitative measurements of radiative emission, proper attention should be given to these radiometric considerations, and careful attention should be devoted to proper radiometric units.[58] Most such measurements are referred to the flux from a blackbody cavity radiator, the emission of which can be stated in terms of the flux per unit area of the opening per steradian in a direction normal to the opening $(W/m^2 \cdot sr)$.

Spectroscopists are not usually interested in making absolute radiometric measurements; their primary interest usually concerns the distribution of radiant flux in various wavelength or frequency intervals in the electromagnetic spectrum. Their final presentation of an emission spectrum purports to give the plot of a quantity called *spectral intensity* $I(v) = d\phi/dv$ as a function of frequency v, as indicated schematically in Fig. 3; the total intensity $I = \int_0^\infty I(v)\, dv$ is thus *proportional* to the total flux accepted from the source as viewed in the direction of observation. As commonly employed by most spectroscopists, I and $I(v)$ are usually expressed in "arbitrary units" but can be expressed in SI units by referring the measurements to a comparison blackbody; extensive tables of blackbody data are readily avail-

---

[58] F. E. Nicodemus, *Appl. Opt.* **12**, 2960 (1973).

1.2. GENERAL METHODS OF SPECTROSCOPY

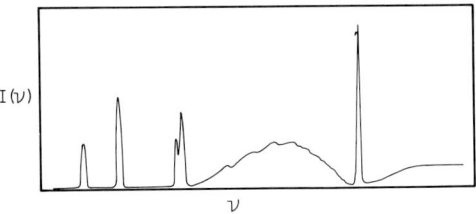

FIG. 3. An emission spectrum: spectral intensity $I(v)$ as a function of frequency v.

able.[59] Absolute spectroscopic measurements in spectral regions not covered by the blackbody tables must be established by other calibration techniques.

A single narrow peak in the plot of Fig. 3 is called a *spectral line* even though its width is finite. Similarly, a very broad peak or a group of closely spaced lines is called a *band*. A very broad region in which $I(v) \neq 0$, but in which the $I(v)$ versus v curve shows little or no structure, is called a *continuum*.

The spectra obtained experimentally with a given source actually provide only approximations of the emission curve plotted in the figure. In the first place, the detector response must be *strictly* proportional to the incident radiant flux, and the proportionality constant involved must be exactly the same for *all* frequencies if the ordinates of the experimental curve are to give close approximations of $I(v)$. Further, since a finite amount of radiant flux $\Delta\phi$ must reach the detector, Δv never really approaches zero; the resolving instrument must therefore pass flux in the frequency interval v to $v + \Delta v$ to the detectors. The detector can thus at best give a response that is proportional to $I(v)_{av} = \Delta\phi/\Delta v$ that represents an average value of $I(v)$ over the interval v to $v + \Delta v$; if changes in $I(v)$ are small in the frequency interval Δv called the "spectral slitwidth," $I(v)_{av}$ gives a fairly close approximation of $I(v)$. If Δv is small as compared with the width of the observed spectral features, and if it is the same for all frequencies, a plot of $I(v)_{av}$ versus v gives a close approximation of $I(v)$ versus v.

Although a spectroscopist working in a given spectral region usually states some value δv as a measure of the ability of his resolving instrument and detector to separate closely neighboring lines at frequencies v_1 and v_2, such a statement of $\delta v = v_1 - v_2$ does not represent a significant measure of the performance of his instrument; a given limiting resolution δv may represent excellent performance for frequencies in the ultraviolet but extremely poor performance in the far infrared. The proper characterization of a spectrograph is given by the resolving power $R - v/\delta v$ for a limiting

---

[59] M. Pivovonsky and M. R. Nagel, "Tables of Blackbody Radiation Functions." Macmillan, New York, 1961.

resolution δv for two lines in the vicinity of v. Although the value of R achieved in the actual operation of a spectrograph is sometimes limited by properties of the source and the detector, diffraction effects imposed by the wave nature of radiation provide the ultimate limit of R for a resolving instrument employing imaging processes; the value of the diffraction-limited resolving power of such an instrument is usually stated in terms of the familiar Rayleigh criterion.[60] The resolving powers of interferometers have analogous ultimate limitations.

The strength of an emission line can be expressed in terms of the integral $I = \int I(v)\, dv$, where the limits of the integral are set to include all frequencies at which the line in question has values of $I(v)$ measurably different from zero. Comparisons of the relative strengths of neighboring but completely resolved lines in a given spectrum is a relatively simple matter, and can provide ratios of radiative transition probabilities between the energy levels involved. Absolute determination of line strengths is a much more difficult process that involves reference to comparison blackbody radiation curves, calorimetric measurements, or other calibration procedures.

1.2.2. Absorption Spectra

Absorption studies typically employ a source, an absorbing sample, a resolving instrument, and a detector. In most conventional work, the source itself has a more or less continuous emission spectrum as represented by the curve labeled $I_0(v)$ in Fig. 4. When the absorbing sample is interposed

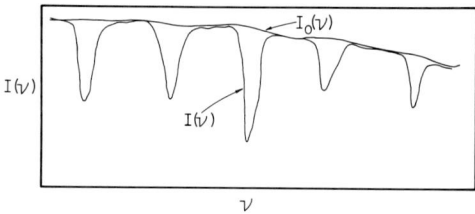

FIG. 4. An absorption spectrum: $I_0(v)$ represents the spectral intensity of the source; $I(v)$ represents the spectral intensity of the source as modified by the absorbing sample.

in a collimated beam of flux between the source and the resolving instrument, a modified spectrum symbolized by the $I(v)$ curve is observed. The relationship between $I(v)$ and $I_0(v)$ is given by Lambert's law: $I(v) = I_0(v) \exp[-\alpha(v)x]$, where $\alpha(v)$ is called the Lambert absorption coefficient and x is the path length traversed by the radiant flux in passing through the absorbing sample.

[60] R. H. Sawyer, "Experimental Spectroscopy." Prentice-Hall, Englewood Cliffs, New Jersey, 1946.

If the absorbing sample is enclosed in an absorption cell, $I_0(v)$ must represent the spectral emission of the source as observed through the empty absorption cell, or must be corrected in some other manner that takes account of the reflection and absorption processes associated with the absorption cell windows.

The absorption spectrum of the sample is usually given by a plot of the Lambert coefficient $\alpha(v)$ as a function of frequency v, as shown schematically in Fig. 5. The value of $\alpha(v)$ actually depends on the characteristics of the absorbing entities, such as atoms or molecules, and on their number density in the sample; an absorption coefficient $\sigma(v)$ per atom or molecule can be obtained from the relation $\sigma(v) = \alpha(v)/n$, where n is the number of absorbers per unit volume. Since $\sigma(v)$ has the dimensions of area, it is sometimes called the atomic or molecular absorption cross section.

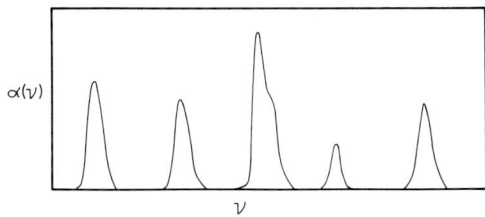

FIG. 5. An absorption spectrum: the Lambert absorption coefficient $\alpha(v)$ as a function of frequency v; the ordinates are also proportional to the absorption cross section $\sigma(v) = \alpha(v)/n$, where n is the number of absorbers per unit volume.

The strength S of an absorption line is given by the integral $S = \int \alpha(v)\,dv$, where the limits of the integral include all frequencies at which there is measurable absorption associated with the line; the corresponding integral $\sigma = \int \sigma(v)\,dv$ gives the total cross section for the transition involved. The line strength depends on the radiative transition probabilities between the energy levels involved and on the difference between the populations of the lower and upper energy level.

Like emission lines, absorption lines have finite widths even after corrections have been made for all spectral-slitwidth and other instrumental effects. The actual widths of spectral lines can be classified as:

(1) *natural broadening* with widths Δv determined by the uncertainty principle $\Delta v \cdot \tau \approx 1$, where τ is the lifetime of the excited level;

(2) *collision broadening* with widths Δv given by $\Delta v \cdot \tau_0 \approx 1$, where τ_0 is the mean time between collisions; and

(3) *Doppler broadening* with widths depending upon the motion of the atoms or molecules relative to the spectrograph.

The shapes of absorption lines subject to natural and collision broadening are adequately approximated by the Lorentz expression

$$\alpha(v) = (S/\pi)\{\gamma/[(v - v_0)^2 + \gamma^2]\} \quad \text{(Lorentz line shape)},$$

where $\gamma = 1/2\pi\tau$ or $1/2\pi\tau_0$ gives the half width of the line at half maximum and v_0 represents the central frequency of the line. The profile of a line subject only to Doppler broadening has the form

$$\alpha(v) = \text{const} \exp[-\beta c^2(v - v_0)^2/v^2] \quad \text{(Doppler line shape)}$$

with $\beta = \mu/2RT$, where μ is the molecular or atomic weight of the absorber and T is the sample temperature.

1.2.3. Sources

The sources of radiation employed in laboratory spectroscopy have a wide variety of forms; it is perhaps a truism that each form must provide the quantum energy hv involved in the emission of the radiation of interest. This can be done in a variety of ways.

Thermal sources can provide for emission in the infrared and visible regions provided $kT \geqslant hv$; thermal sources include such devices as Nernst glowers, hot metallic filaments, flames, and electric arcs. In order to excite emission spectra in the ultraviolet, spark discharges must be operated at voltages V sufficiently high to ensure that $eV \geqslant hv$; similar considerations apply to the voltages applied to x-ray tubes. In the γ-ray region, sufficient energy must be provided in some manner to put the nucleus of interest in the excited state involved; the energy may be supplied by charged particle bombardment, neutron excitation, γ-ray excitation, or other processes.

The sources of coherent radiation employed in the radio-frequency and microwave regions are usually operated at voltages far in excess of $eV = hv$; this statement is generally true for the operation of lasers, which provide coherent radiation at higher frequencies. It certainly applies to synchrotrons which can serve as sources of ultraviolet radiation!

1.2.4. Resolving Instruments

The earliest resolving instruments employed in spectroscopy were prism instruments; such instruments calibrated in terms of standard reference spectra are still in wide use in applications for which high resolution is not an important consideration. On the basis of the Rayleigh criterion, the resolving power of a prism instrument is directly proportional to the spectral dispersion dn/dv of the prism material and the width of the prism at its

1.2. GENERAL METHODS OF SPECTROSCOPY

base[60, 61]; with intense sources and optics of high quality the diffraction limit can be approached. A set of prisms constructed of fluorite (CaF_2), quartz, glass, and alkali halides can be used to cover the spectral region between the far ultraviolet and the far infrared regions.

Diffraction gratings in a variety of forms have been constructed for use in the even broader spectral range between the x-ray region and the submillimeter region. The Rayleigh criterion for the diffraction limit indicates that the resolving power of a grating is proportional to the total number of lines in the grating and to the diffraction order in which the grating is used.[60, 61] Under normal working conditions, resolving powers of one-half or one-third that given by the Rayleigh criterion can readily be achieved.

The resolving power of an interferometer[62] depends on the ratio of the maximum path length differences provided by the interferometer elements to the wavelength of the radiation being studied. The path length differences involved might be the differences in the actual geometrical path lengths traversed by the two beams in a Michelson interferometer, or the differences involved in multitraversals of a fixed path length in other devices.

In the microwave and radio frequency regions, the resolving power of a spectrograph is actually determined by the frequency control of the source itself. The resolving power is thus given in terms of the ratio $v/\delta v$, where the source frequency v can be maintained constant in comparison with a standard radio frequency (WWV) to within a frequency interval δv. Enormous resolving powers can be routinely attained in microwave studies and in the NMR studies conducted at radio frequencies. It is generally much easier to measure and control frequencies in terms of international time standards than to measure and control wavelengths in terms of the international standards of length!

The resolving powers of γ-ray spectrographs are usually limited by properties of detectors, which also serve as resolving instruments; the resolving powers usually attained are low as compared with those attained in other spectral regions. The one glaring exception to this generalization is the spectroscopy that can be accomplished by applications of the Mössbauer effect; since the γ-ray lines involved in the Mössbauer effect are the sharpest spectral lines observed, it is not surprising that they can be employed to great advantage as tools in certain spectroscopic investigations.

It should be noted that spectroscopic methods involving tunable sources are being applied to regions of the electromagnetic spectrum for which lasers have been developed. We also note that the characteristic spectral

---

[61] R. W. Ditchburn, "Light." Blackie, London, 1963.

[62] M. Rousseau and J. P. Mathieux, "Problems in Optics" (transl. by J. W. Baker). Pergamon, Oxford, 1973.

absorption and emission processes involved in Mössbauer studies, in which source and detector are "tuned," have an earlier infrared counterpart in the form of nondispersive gas analyzers[63] that can detect absorbing gases with concentration two parts per million without the use of any resolving instruments other than the characteristic absorption spectrum of the gas itself!

1.2.5. Detectors

In the earliest spectroscopic studies the human eye served as the detector. Although limited in its response to a rather narrow range of the electromagnetic spectrum, the eye provides an excellent method for observing line positions. However, it is rather ineffective in making estimates of spectral intensity.

The next detector to be employed was the photographic plate, which has much to recommend it. In the first place, it is a "multiplex device" that provides for the investigation of many spectral frequencies at once[†]; all spectral lines are observed at once, and sudden changes in source intensity do not introduce as many spurious effects as those involved when the lines are studied one at a time. In the second place, the photographic plate is a fairly satisfactory *integrating detector*; in a given region, within certain limits, photographic blackening is roughly proportional to the total radiant *energy* reaching the plate, and thus, by making long exposures, it is possible to observe weak emission lines that might well be missed by a nonintegrating detector. Scanning microphotometers have been developed to supply chart recordings of photographic blackening as a function of plate position; these charts can then be interpreted semiquantitatively in terms of time-integrated spectral intensity as a function of wavelength or frequency. The major shortcoming of the photographic plate is inherent in the rather complicated laws of photographic blackening in various spectral regions; the photographic process can at best supply limited information regarding absolute line intensities.

The thermal detectors initially developed for use in the infrared are nonintegrating detectors that provide accurate measures of spectral intensity over wide spectral ranges. Thus, the temperature rise caused by spectral flux entering these detectors will result in (1) a change in the electrical resistance of a bolometer, (2) an EMF in a thermocouple or thermopile, or

---

[63] W. G. Fastie and A. H. Pfund, *J. Opt. Soc. Amer.* **37**, 762 (1947).

[†] Multiplexing is a term introduced by Jacquinot; a spectral multiplex system is the analog of a telephonic multiplex system in which a single conductor can be used to transmit multiple messages.

(3) a change in gas pressure in a pneumatic detector like the Golay cell. Each of these thermal detectors is equipped with a receiver which should be *black* for the regions in which the detectors are employed; uniform spectral blackening is most closely approached by the receiver in the Golay cell. At a given frequency, the output signal provided by these detectors is very nearly proportional to the flux reaching the receiver.

More recently developed have been the solid-state devices, in one form of which incident flux serves to raise electrons from the valence band of a semiconductor to the conduction band. As employed in the visible and infrared regions, the incident flux serves to change the resistance of the device by an amount that depends on the incident flux; although the change in resistance is approximately proportional to $\Delta \phi(v) = I(v) \Delta v$ at a given frequency, the proportionality constant varies considerably with frequency. Such devices are useful only in regions for which the quantum energy of the radiation is greater than the energy gap E_G between the valence and conduction bands of the semiconductor. In order to maximize the relative change in resistance caused by incident flux, the detector is usually cooled in order to minimize the ambient population of the conduction band.

As normally used, the solid-state detectors employing quantum transitions of electrons from the valence band to the conduction band are nonintegrating detectors. However, in cases where weak light sources are involved, they can be employed with circuits which provide for counting responses of the detector to individual photons. Thus, photon-counting techniques can be employed to make the semiconducting detector an integrating device; the total photon count is closely proportional to the integrated flux associated with a given spectral line.

Semiconducting devices are employed in the x-ray and γ-ray regions in such a way as to perform the functions of resolving instruments as well as those of detectors. These devices are, in essentials, the solid-state analogs of the ionization chambers employed early in the century; just as a single γ-ray photon is capable of producing ion pairs in a gas, the passage of a single γ-ray photon through a semiconductor results in the formation of electron-hole pairs when electrons are raised through the energy gap between the valence and the conduction bands. The measurement of the total charge separation in the semiconductor provides an estimate of the quantum energy of the gamma ray in either case; in the detection circuit the output is given in the form of a voltage pulse that is directly proportional to the total charge collected; i.e., the output gives a pulse that is proportional to the number N of the electron-hole pairs created. The number of pulses can be counted over any desired time interval and classified by means of a multichannel analyzer; used in this manner, the semiconducting element thus acts as an integrating detector. A plot of total counts per unit energy interval as a function of quantum energy thus provides an emission spectrum of the source; by dividing the ordinates of the plot by the total observation time a spectral plot directly analogous to that in Fig. 3 can be obtained.

In its role as a resolving instrument, the semiconductor device has certain limitations that are somewhat analogous to the limitations encountered with instruments used in other spectral regions. The ultimate limitation is related to the conversion of γ-ray quantum energy to the total ionization energy; not *all* of the quantum energy of the primary γ ray goes into ionization processes. For example, if the γ ray first interacts with matter to produce electrons through Compton processes, these electrons along with secondary and tertiary electrons formed by collision can in turn produce further charge separations, but they can also fritter away portions of their total energies in processes *not* leading to charge separation. For a set of monoenergetic γ-ray photons, the *average* number of charge separations may be some number N, but the *individual* numbers of charge separations will have a distribution about this average value; the width of the distribution curve will be proportional to $\sqrt{N}$, and the fractional uncertainty $\Delta E/E$ in the quantum energy of the γ ray is thus proportional to $\sqrt{N}/N = 1/\sqrt{N}$. This fractional uncertainty can be minimized by making N as large as possible; this can be accomplished by employing a semiconductor with a small energy gap E_G since $N \sim h\nu/E_G$. Since the uncertainty in registration of quantum energy poses the ultimate limitation on the ability of the detector to distinguish between γ rays of nearly equal energy, the ultimate resolving power of the device $E/\delta E = \nu/\delta\nu$ is thus proportional to $\sqrt{h\nu/E_G}$.

The resolving power just stated is analogous in some respects to that imposed by the Rayleigh criterion in other types of resolving instruments. The practical limitation corresponding to the flux limit or power limit to the resolving power of a prism or grating instrument is involved with the counting rate attainable with a given source of γ rays. Since the solid-state detectors with the usual auxilliary counting circuits act as integrating detectors, this practical limitation can be avoided by the use of long counting times, as in the case of prolonged photographic exposure times in the visible region.

1.2.6. Data Handling Techniques

As with other branches of science, the advent of the digital computer has produced a revolution in spectroscopy. In addition to the mere handling of large quantities of data, computers are now being put to such diverse usages as the automatic microphotometry and analysis of spectra recorded on photographic plates and the deconvolution of observed spectral lines to remove the instrumental effects imposed by finite spectral slitwidths.† Michelson himself, in the early years of the present century, had foreseen the possibility of obtaining spectra from the interferograms obtained with his interferometer; however. Fourier-transform spectroscopy was impractical prior to the development of the digital computers. In the determination of the optical constants of solids and liquids from measurements of spectral reflectance or spectral absorption, Kramers and Kronig established general relationships that can now be readily employed through the use of modern computers.[64–66]

[64] M. Gottlieb, *J. Opt. Soc. Amer.* **50**, 343 (1960).
[65] C. W. Robertson, B. Curnutte, and D. Williams, *Mol. Phys.* **26**, 183 (1973).
[66] R. K. Ahrenkiel, *J. Opt. Soc. Amer.* **61**, 1651 (1971).

† See Blass and Nielsen in Vol. 3A, Chapter 2.2.

2. THEORY OF RADIATION AND RADIATIVE TRANSITIONS*

2.1. Introduction

This part is concerned with the interaction of electromagnetic radiation with matter. In particular, we will discuss the theory of those interactions associated with radiative transitions (photon absorption or emission) of individual nuclei, atoms, and molecules. We will not attempt to deal with the properties of matter in bulk or with the various kinds of magnetic resonance spectroscopy.

In characterizing the readers for this presentation, we imagine persons interested in the spectroscopy of one or more of the wavelength regions treated in subsequent chapters, and who have reached the level of a year of graduate study in physics or physical chemistry in an American university. Such persons will likely have completed a year-long introductory course in quantum mechanics, and we assume that our readers can profitably use quantum mechanics texts at the level of Merzbacher[1] as references. Individual backgrounds in classical electrodynamics are likely to be more varied, but our experience indicates it is not reasonable to assume more than one-half year of study at the level of, say, the first half of Corson and Lorrain.[2] Therefore, we have made our discussion of the classical picture of light more extensive and detailed than would have been necessary had we assumed a background of advanced study in electrodynamics.

The traditional approach in introductory quantum mechanics texts has emphasized the energy eigenvalues and eigenfunctions of various systems of interest. In our discussion we will emphasize the interaction between electromagnetic fields and matter. This sort of emphasis does exist in a few texts;

---

[1] E. Merzbacher, "Quantum Mechanics," 2nd ed. Wiley, New York, 1970. Among the many other useful texts at about the same level are A. Messiah, "Quantum Mechanics" (two volumes). North-Holland Publ., Amsterdam, 1961; L. I. Schiff, "Quantum Mechanics," 3rd ed. McGraw-Hill, New York, 1968.

[2] D. R. Corson and P. Lorrain, "Electromagnetic Fields and Waves," 2nd ed. Freeman, San Francisco, California, 1970. Among the many other useful texts at about the same level are J. R. Reitz and F. J. Milford, "Foundations of Electromagnetic Theory." Addison-Wesley, Reading, Massachusetts, 1960; E. M. Pugh and E. W. Pugh, "Principles of Electricity and Magnetism." Addison-Wesley, Reading, Massachusetts, 1960.

---

* Part 2 is by Basil Curnutte, John Spangler, and Larry Weaver.

we mention in particular that by Hameka.[3] The standard treatise on the quantum theory of electromagnetic radiation and its interactions with matter has long been that by Heitler.[4] While this is a valuable reference with which a serious student should become familiar, it is not particularly easy reading and does not have a major spectroscopic emphasis.

The limitation of available space requires that many possible topics and examples not be included in our discussion. In particular, we will not be able to discuss the important and beautiful applications of group theory in labeling states and predicting selection rules. A few of the many valuable treatises on group theory are listed below.[5]

We adopt the SI units.[6] The reader is warned that other systems of units are popular in discussions of quantum mechanics and electrodynamics. This leads to differences in the appearance of equations that can be confusing even though they are not fundamental.

The purpose of our discussion is expository and tutorial. Little in our presentation is original except for the particular organization and language. We acknowledge a debt owed to many texts and articles, to our own teachers, and to discussions with colleagues that have helped shape our ideas in this area.

2.2. Light

2.2.1. Classical Picture

2.2.1.1. *Electromagnetic Fields, Potentials, and Gauge Transformations.* The basic equations relating electric and magnetic fields in vacuum to the source densities are Maxwell's equations[†]:

$$\mathbf{V} \cdot \mathscr{E} = \rho/\varepsilon_0, \qquad (2.2.1)$$

---

[3] H. F. Hameka, "Advanced Quantum Chemistry." Addison-Wesley, Reading, Massachusetts, 1965.

[4] W. Heitler, "The Quantum Theory of Radiation," 3rd ed. Oxford Univ. Press, London and New York, 1954.

[5] M. Hamermesh, "Group Theory and Its Applications to Physical Problems." Addison-Wesley, Reading, Massachusetts, 1962; L. Jansen and M. Boon, "Theory of Finite Groups. Applications in Physics." North-Holland Publ., Amsterdam, 1967; M. Tinkham, "Group Theory and Quantum Mechanics." McGraw-Hill, New York, 1964; R. McWeeny, "Symmetry—An Introduction to Group Theory and Its Applications." Pergamon, Oxford, 1963; E. P. Wigner, "Group Theory and Its Application to the Quantum Mechanics of Atomic Spectra." Academic Press, New York, 1959.

[6] "Units of Measurement." Nat. Bur. of Std. Publ. 286, Washington, D.C., 1972.

[†]There are many books that develop the experimental and theoretical foundations of classical electrodynamics. A popular text at the intermediate level is that by Corson and Lorrain.[2] At a

2.2. LIGHT

$$\nabla \cdot \mathcal{B} = 0, \tag{2.2.2}$$

$$\nabla \times \mathcal{E} + \partial \mathcal{B}/\partial t = 0, \tag{2.2.3}$$

$$\nabla \times \mathcal{B} - \mu_0 \varepsilon_0 \, \partial \mathcal{E}/\partial t = \mu_0 \mathbf{j}. \tag{2.2.4}$$

Here, $\mathcal{E}$ is the electric field strength, $\mathcal{B}$ the magnetic field strength, ρ the volume charge density, and $\mathbf{j}$ the current density. We consider equations for the fields in vacuum since we will be concerned primarily with individual molecules, atoms, and nuclei, and not with matter in bulk.

The current density $\mathbf{j}$ associated with a charge density ρ moving with a local velocity $\mathbf{v}$ is

$$\mathbf{j} = \rho \mathbf{v}. \tag{2.2.5}$$

The charge and current densities are also related by the equation of continuity

$$\nabla \cdot \mathbf{j} + \partial \rho/\partial t = 0, \tag{2.2.6}$$

which is a consequence of charge conservation. Maxwell's equations are supplemented by the Lorentz force equation, which says that the force per unit volume $\mathbf{f}$ on a charge distribution described by a charge density ρ is given by

$$\mathbf{f} = \rho(\mathcal{E} + \mathbf{v} \times \mathcal{B}). \tag{2.2.7}$$

For the case of a point charge, where ρ is nonzero only within some infinitely small volume at the position $\mathbf{r}'$ and the total charge is q, we have $\rho = q \, \delta(\mathbf{r} - \mathbf{r}').$[†] The force equation becomes

$$\mathbf{F} = q(\mathcal{E} + \mathbf{v} \times \mathcal{B}), \tag{2.2.8}$$

where $\mathbf{F}$ is the total electromagnetic force on the charge q.

---

[7] R. H. Good, Jr., and T. J. Nelson, "Classical Theory of Electric and Magnetic Fields." Academic Press, New York, 1971.

[8] J. D. Jackson, "Classical Electrodynamics." Wiley, New York, 1962.

[9] E. Butkov, "Mathematical Physics," Chapter 6. Addison-Wesley, Reading, Massachusetts, 1968; P. Dennery and A. Krzywicki, "Mathematics for Physicists," Section 13. Harper, New York, 1967; I. Stakgold, "Boundary Value Problems of Mathematical Physics," Vol. I, Sections 1.2 and 1.3. Macmillian, New York, 1967.

---

more advanced level we recommend the texts by Good and Nelson[7] and by Jackson.[8] The reader is reminded that Gaussian units are often used in discussions of electrodynamics rather than the SI units, which we adopt here. This leads to superficial differences in many equations. A brief, useful discussion of units and dimensions in electrodynamics is given in the appendix of Jackson. The SI system for electrodynamics corresponds to what Jackson calls the rationalized mks system

[†] We introduce here the Dirac delta function, which is widely used in physics but deserves to be better understood by many students. Some careful discussions of the delta function can be found in a number of sources.[9]

The linearity of Maxwell's equations implies that the fields arising from several sets of sources can be obtained by superposing the fields arising from each set of sources treated independently. The linearity also implies that we can Fourier analyze the oscillations in time of a source distribution and consider each frequency component separately. The total fields are obtained by superposing the contributions for the various frequencies.

The two homogeneous Maxwell equations, Eqs. (2.2.2) and (2.2.3), imply the existence of auxiliary functions, called potentials, which are valuable aids in the development of the theory. By a standard theorem of the calculus of vector fields,[10] Eq. (2.2.2) implies the existence of a vector field $\mathscr{A}$, called a vector potential, such that

$$\mathscr{B} = \nabla \times \mathscr{A}. \tag{2.2.9}$$

Equation (2.2.3) then becomes

$$0 = \nabla \times \mathscr{E} + \partial(\nabla \times \mathscr{A})/\partial t = \nabla \times (\mathscr{E} + \partial \mathscr{A}/\partial t).$$

This in turn implies the existence of a scalar field Φ, called a scalar potential, such that $\mathscr{E} + \partial \mathscr{A}/\partial t = -\nabla\Phi$,[11] or

$$\mathscr{E} = -\nabla\Phi - \partial \mathscr{A}/\partial t. \tag{2.2.10}$$

The scalar and vector potentials that give the physically correct electric and magnetic field strengths for a given situation via Eqs. (2.2.9) and (2.2.10) are not unique. However, the various sets of potentials that can be defined all lead to the same physically observable consequences. If $\mathscr{A}_1$ and Φ_1 are potentials satisfying $\mathscr{B} = \nabla \times \mathscr{A}_1$ and $\mathscr{E} = -\nabla\Phi_1 - \partial \mathscr{A}_1/\partial t$, then it is easy to see that the same field strengths are represented by the potentials

$$\Phi_2 = \Phi_1 - \partial\Lambda/\partial t \tag{2.2.11}$$

and

$$\mathscr{A}_2 = \mathscr{A}_1 + \nabla\Lambda, \tag{2.2.12}$$

where Λ is any function having well-defined second-order space and time derivatives. Conversely, if we assume that $(\mathscr{A}_1, \Phi_1)$ and $(\mathscr{A}_2, \Phi_2)$ are both sets of potentials that give the physically correct fields, then it is easy to show that these sets of potentials must be related by equations of the form of (2.2.11) and (2.2.12). Sets of potentials satisfying Eqs. (2.2.11) and (2.2.12) are said to be related by a gauge transformation[†] with respect to the function Λ.

---

[10] H. Jeffreys and B. S. Jeffreys, "Methods of Mathematical Physics," 3rd ed., p. 224. Cambridge Univ. Press, London and New York, 1956.

[11] L. Brand, "Advanced Calculus," p. 336. Wiley, New York, 1955.

[†] The fact that physically observable predictions of the theory are unchanged if the potentials are subjected to a gauge transformation is called gauge invariance. The term seems to have

2.2. LIGHT

By substituting Eqs. (2.2.9) and (2.2.10) into Eqs. (2.2.1) and (2.2.4), we obtain differential equations that must be satisfied by the potentials

$$\nabla^2\Phi + \partial(\nabla \cdot \mathscr{A})/\partial t = -\rho/\varepsilon_0 \quad (2.2.13)$$

and

$$\nabla^2\mathscr{A} - \mu_0\varepsilon_0\,\partial^2\mathscr{A}/\partial t^2 - \nabla(\nabla \cdot \mathscr{A} + \mu_0\varepsilon_0\,\partial\Phi/\partial t) = -\mu_0\mathbf{j}. \quad (2.2.14)$$

We can use the freedom of choice of Λ in the gauge transformation equations to simplify these equations. The imposition of additional conditions on the potentials (beyond Eqs. (2.2.9) and (2.2.10)) that can always be satisfied by making a gauge transformation is called choosing a particular gauge.

2.2.1.2. *The Lorentz Gauge.* Equations (2.2.13) and (2.2.14) can be uncoupled if we require that

$$\nabla \cdot \mathscr{A} + \mu_0\varepsilon_0\,\partial\Phi/\partial t = 0. \quad (2.2.15)$$

This is called the Lorentz condition, and potentials satisfying this condition are said to belong to the Lorentz gauge. If the Lorentz condition is satisfied, Eqs. (2.2.13) and (2.2.14) become

$$\nabla^2\Phi - \mu_0\varepsilon_0\,\partial^2\Phi/\partial t^2 = -\rho/\varepsilon_0 \quad (2.2.16)$$

and

$$\nabla^2\mathscr{A} - \mu_0\varepsilon_0\,\partial^2\mathscr{A}/\partial t^2 = -\mu_0\mathbf{j}. \quad (2.2.17)$$

Techniques for the solution of inhomogeneous wave equations of this form are well-known.[13]

We must verify that Eq. (2.2.15) can always be satisfied. Suppose that $\mathscr{A}_1$ and Φ_1 are any potentials that give the correct fields. Make a gauge transformation using Eqs. (2.2.11) and (2.2.12), and require that the new potentials satisfy the Lorentz condition. It is easy to see that this is true if

$$\nabla^2\Lambda - \mu_0\varepsilon_0\,\partial^2\Lambda/\partial t^2 = -\nabla \cdot \mathscr{A}_1 - \mu_0\varepsilon_0\,\partial\Phi_1/\partial t. \quad (2.2.18)$$

Since $\mathscr{A}_1$ and Φ_1 are assumed known, this inhomogenous wave equation can be solved for Λ, so a gauge transformation producing Lorentz-gauge potentials exists. If $\mathscr{A}_1$ and Φ_1 belong to the Lorentz gauge at the onset, Eq. (2.2.18) becomes $\nabla^2\Lambda - \mu_0\varepsilon_0\,\partial^2\Lambda/\partial t^2 = 0$. A gauge transformation with respect to a function Λ satisfying this homogeneous wave equation then

---

[12] H. Weyl, "Raum-Zeit-Materie," Sechste unveränderte Auflage, p. 123. Springer-Verlag, Berlin and New York, 1970. (The first German edition appeared in 1918.) See also H. Weyl, "The Theory of Groups and Quantum Mechanics" [translated from the 2nd (rev.) German ed. by H. P. Robertson], p. 100. Dover, New York, 1950.
[13] Good and Nelson,[7] Section 24.

---

been originated in a somewhat different physical context by Weyl,[12] who introduced the German term *eichinvarianz*, which can be alternatively translated as calibration invariance.

yields new potentials also belonging to the Lorentz gauge. Within this gauge the potentials are not unique until boundary conditions are specified.

2.2.1.3. The Coulomb Gauge. The Lorentz gauge is advantageous when it is important to explicitly display the relativistic invariance of the theory, since it is covariant under a Lorentz transformation.[14] However, for our purposes, simplicity is gained by using an alternative known as the Coulomb gauge. The requirement defining the Coulomb gauge, replacing Eq. (2.2.15), is

$$\nabla \cdot \mathscr{A} = 0. \qquad (2.2.19)$$

With this condition, Eqs. (2.2.13) and (2.2.14) become

$$\nabla^2 \Phi = -\rho/\varepsilon_0 \qquad (2.2.20)$$

and

$$\nabla^2 \mathscr{A} - \mu_0 \varepsilon_0 \, \partial^2 \mathscr{A}/\partial t^2 = -\mu_0 \mathbf{j} + \mu_0 \varepsilon_0 \, \nabla \, \partial \Phi/\partial t. \qquad (2.2.21)$$

Equation (2.2.20) is Poisson's equation, which is familiar from electrostatics, except that the time appears here as a parameter. It is well-known that a solution of this equation can be expressed in the form[15]

$$\Phi(\mathbf{r}, t) = (4\pi\varepsilon_0)^{-1} \int d\tau' \, [\rho(\mathbf{r}', t)/|\mathbf{r} - \mathbf{r}'|]. \qquad (2.2.22)$$

By substituting Eq. (2.2.22) into Eq. (2.2.21) and using the equation of continuity, Eq. (2.2.6), we obtain

$$\nabla^2 \mathscr{A} - \mu_0 \varepsilon_0 \, \partial^2 \mathscr{A}/\partial t^2 = -\mu_0 \mathbf{j} - (\mu_0/4\pi) \, \nabla \int d\tau' \, [\nabla' \cdot \mathbf{j}(\mathbf{r}', t)/|\mathbf{r} - \mathbf{r}'|]. \qquad (2.2.23)$$

Straightforward mathematical manipulation[16] shows that

$$\mathbf{j} = \mathbf{j}_\mathrm{T} + \mathbf{j}_\mathrm{L}, \qquad (2.2.24)$$

where

$$\mathbf{j}_\mathrm{T} = (4\pi)^{-1} \, \nabla \times \int d\tau' \, [\nabla' \times \mathbf{j}(\mathbf{r}', t)/|\mathbf{r} - \mathbf{r}'|] \qquad (2.2.25)$$

and

$$\mathbf{j}_\mathrm{L} = (-4\pi)^{-1} \, \nabla \int d\tau' \, [\nabla' \cdot \mathbf{j}(\mathbf{r}', t)/|\mathbf{r} - \mathbf{r}'|]. \qquad (2.2.26)$$

Note that $\nabla \cdot \mathbf{j}_\mathrm{T} = 0$ and $\nabla \times \mathbf{j}_\mathrm{L} = 0$, and Eq. (2.2.6) becomes $\nabla \cdot \mathbf{j}_\mathrm{L} = -\partial\rho/\partial t$. The quantities $\mathbf{j}_\mathrm{T}$ and $\mathbf{j}_\mathrm{L}$ are called, respectively, the transverse and longitudinal parts of $\mathbf{j}$. These names are used, in general, to denote that part of a vector field having vanishing divergence (transverse part) and that part of a vector field having vanishing curl (longitudinal part). By using

---

[14] Jackson,[8] p. 378.
[15] Jeffreys and Jeffreys,[10] p. 210.
[16] Good and Nelson,[7] p. 302; note also the technique of their Eq. (2.59).

Eqs. (2.2.24)–(2.2.26), we can write Eq. (2.2.23) in the form

$$\nabla^2 \mathcal{A} - \mu_0 \varepsilon_0 \, \partial^2 \mathcal{A}/\partial t^2 = -\mu_0 \mathbf{j}_T. \quad (2.2.27)$$

It is easy to show that Eq. (2.2.19), which defines the Coulomb gauge, can always be satisfied. Suppose that $\mathcal{A}_1$ and Φ_1 are any potentials that give the correct fields. Make a gauge transformation using Eqs. (2.2.11) and (2.2.12), and require that the new potentials satisfy Eq. (2.2.19). This leads directly to the requirement

$$\nabla^2 \Lambda = -\nabla \cdot \mathcal{A}_1. \quad (2.2.28)$$

Since a solution of this Poisson equation will exist for any physically reasonable function $\mathcal{A}_1$, a gauge transformation producing Coulomb-gauge potentials exists. If $\mathcal{A}_1$ already belongs to the Coulomb gauge, Eq. (2.2.28) becomes Laplace's equation $\nabla^2 \Lambda = 0$. If we start in the Coulomb gauge, a gauge transformation with respect to a function satisfying Laplace's equation gives new potentials also belonging to the Coulomb gauge. Imposition of this gauge requirement does not define unique potentials until boundary conditions are specified.

It is often convenient to separate the electric and magnetic field strengths into transverse and longitudinal parts. Since $\nabla \cdot \mathcal{B} = 0$, the magnetic field strength is purely transverse. If we write the electric field strength as $\mathcal{E} = \mathcal{E}_T + \mathcal{E}_L = -\nabla\Phi - \partial\mathcal{A}/\partial t$ and impose the Coulomb-gauge condition, Eq. (2.2.19), we have that

$$\nabla \times \mathcal{E} = \nabla \times \mathcal{E}_T = \nabla \times (-\partial\mathcal{A}/\partial t) \quad (2.2.29)$$

and

$$\nabla \cdot \mathcal{E} = \nabla \cdot \mathcal{E}_L = \nabla \cdot (-\nabla\Phi). \quad (2.2.30)$$

From these results we make the Coulomb-gauge identifications

$$\mathcal{E}_T = -\partial\mathcal{A}/\partial t \quad (2.2.31)$$

and

$$\mathcal{E}_L = -\nabla\Phi. \quad (2.2.32)$$

We can, in general, eliminate the scalar potential via a gauge transformation, using $\Lambda(\mathbf{r}, t) = \int^t \Phi_1(\mathbf{r}, t') \, dt'$. If we want to make such a transformation within the Coulomb gauge, we must satisfy Laplace's equation $0 = \nabla^2 \Lambda = -\int^t [\rho(\mathbf{r}, t')/\varepsilon_0] \, dt'$, where we have used Eq. (2.2.20). In particular, this will be true if the charge density vanishes throughout the region of interest. The field strengths are then entirely transverse in this region. If $\mathbf{j}_T$ also vanishes throughout the region of interest, we are guaranteed that the electric and magnetic field strengths in this region can be computed from a Coulomb-gauge vector potential through the relations $\mathcal{B} = \nabla \times \mathcal{A}$ and $\mathcal{E} = -\partial\mathcal{A}/\partial t$, and from Eq. (2.2.27) the vector potential will satisfy the homogeneous wave equation

$$\nabla^2 \mathcal{A} - \mu_0 \varepsilon_0 \, \partial^2 \mathcal{A}/\partial t^2 = 0. \tag{2.2.33}$$

For the particular situation just considered, the Lorentz-gauge condition, Eq. (2.2.15), reduces to just the Coulomb-gauge condition. Thus, for the physically important situation of an electromagnetic wave in free space, we can represent the electric and magnetic fields using only a vector potential, and both the Coulomb-gauge and the Lorentz-gauge conditions are simultaneously satisfied.

2.2.1.4. Mechanical Properties of Electromagnetic Fields.

2.2.1.4.1. ENERGY. It is possible to develop relations involving the electromagnetic field strengths and their sources that can be interpreted as generalizations of the conservation relations of classical mechanics for energy, linear momentum, and angular momentum. We start from Eqs. (2.2.1)–(2.2.4) and construct

$$\mathcal{B} \cdot (\nabla \times \mathcal{E} + \partial \mathcal{B}/\partial t) - \mathcal{E} \cdot (\nabla \times \mathcal{B} - \mu_0 \varepsilon_0 \, \partial \mathcal{E}/\partial t) = 0 - \mu_0 \mathbf{j} \cdot \mathcal{E}. \tag{2.2.34}$$

By using Eqs. (2.2.5) and (2.2.7) on the right-hand side and a standard vector identity for the divergence of a vector product[17] on the left-hand side, we see that Eq. (2.2.34) is equivalent to

$$\nabla \cdot (\mathcal{E} \times \mathcal{B}) + \tfrac{1}{2} \partial(\mu_0 \varepsilon_0 \mathcal{E}^2 + \mathcal{B}^2)/\partial t = -\mu_0 \mathbf{v} \cdot \mathbf{f}. \tag{2.2.35}$$

Rearranging, integrating over a volume ⊛ bounded by a stationary surface ⊶,[18] and using the divergence theorem,[19] we obtain

$$-\int_{\circlearrowright} d\mathbf{a} \cdot (\mathcal{E} \times \mathcal{B}/\mu_0) = d/dt \int_{\circledast} d\tau \, [\tfrac{1}{2}\varepsilon_0 \mathcal{E}^2 + \tfrac{1}{2}(\mathcal{B}^2/\mu_0)] + \int_{\circledast} d\tau \, \mathbf{v} \cdot \mathbf{f}. \tag{2.2.36}$$

We identify $\mathbf{v} \cdot \mathbf{f}$ as the power per unit volume leaving the electromagnetic field and entering the mechanical system of moving charges. The volume integral of this expression is then just the time rate of change of the total mechanical energy within the volume ⊛ due to the electromagnetic forces, dE_{mech}/dt. If we isolate the system by expanding the volume ⊛ until the

---

[17] A useful compilation of formulas from vector analysis can be found, e.g., in M. R. Spiegel, "Mathematical Handbook of Formulas and Tables" (Schaum's Outline Series). McGraw-Hill, New York, 1968.
[18] We introduce a convenient notation due to Good and Nelson,[7] p. 12.
[19] Jeffreys and Jeffreys,[10] p. 193.

surface integral vanishes,† we obtain

$$0 = d/dt \int_{\circledast} d\tau \left[\tfrac{1}{2}\varepsilon_0 \mathscr{E}^2 + \tfrac{1}{2}(\mathscr{B}^2/\mu_0)\right] + dE_{\text{mech}}/dt. \quad (2.2.37)$$

We interpret this equation as an energy-conservation relation of the form

$$d(E_{\text{field}} + E_{\text{mech}})/dt = 0, \quad (2.2.38)$$

and we identify

$$U = \tfrac{1}{2}\varepsilon_0 \mathscr{E}^2 + \tfrac{1}{2}(\mathscr{B}^2/\mu_0) \quad (2.2.39)$$

as the volume density of energy associated with the electromagnetic field. It then follows that the surface integral in Eq. (2.2.36) represents the rate of flow of energy carried by the electromagnetic field across the surface ⊙→ out of the volume ⊛. The vector

$$\mathbf{S} = \mathscr{E} \times \mathscr{B}/\mu_0 \quad (2.2.40)$$

is called the Poynting vector. It has dimensions of power per unit area, and can be interpreted as giving the local energy flux of the electromagnetic field.‡

2.2.1.4.2. LINEAR MOMENTUM. In order to obtain a relation associated with the conservation of linear momentum, we start from Eqs. (2.2.1)–(2.2.4) and construct

$$-\varepsilon_0 \mathscr{E} \times (\nabla \times \mathscr{E} + \partial \mathscr{B}/\partial t) - (\mathscr{B}/\mu_0) \times (\nabla \times \mathscr{B} - \mu_0 \varepsilon_0\, \partial \mathscr{E}/\partial t)$$
$$+ \varepsilon_0 \mathscr{E}(\nabla \cdot \mathscr{E}) + (\mathscr{B}/\mu_0)(\nabla \cdot \mathscr{B}) = 0 - \mathscr{B} \times \mathbf{j} + \mathscr{E}\rho + 0 = \mathbf{f}. \quad (2.2.41)$$

The final equality follows from Eqs. (2.2.5) and (2.2.7). We next introduce the Maxwell stress tensor,[22] which has components defined by

$$T_{ij} = \varepsilon_0 \mathscr{E}_i \mathscr{E}_j - \tfrac{1}{2}\varepsilon_0 \mathscr{E}^2\, \delta_{i,j} + (\mathscr{B}_i \mathscr{B}_j/\mu_0) - \tfrac{1}{2}(\mathscr{B}^2/\mu_0)\, \delta_{i,j} = T_{ji}. \quad (2.2.42)$$

---

[20] J. A. Stratton, "Electromagnetic Theory," p. 134. McGraw-Hill, New York, 1941; R. P. Feynman, R. B. Leighton, and M. Sands, "The Feynman Lectures on Physics," Vol. II, Section 27–4. Addison-Wesley, Reading, Massachusetts, 1964.

[21] M. Mason and W. Weaver, "The Electromagnetic Field," p. 226. Univ. of Chicago Press, Chicago, Illinois, 1929.

[22] Stratton,[20] Chapter II.

---

† For static fields this is no problem. For radiation fields, we can suppose that the sources started oscillating at some definite time in the past, and that we take the surface ⊙→ at such a large distance from the sources that the electromagnetic radiation has not yet reached it in any direction.

‡ There has been some controversy concerning the extrapolation from the integrals involved in Eq. (2.2.36) to the point relations represented by Eqs. (2.2.39) and (2.2.40). We have given the commonly accepted interpretation[20]; discussion of some alternative points of view can be found in Mason and Weaver.[21]

Here, $\delta_{i,j}$ is the Kronecker delta symbol.[23] Straightforward manipulation of the left-hand side of Eq. (2.2.41) and rearrangement yield[†]

$$\partial T_{ij}/\partial x_j = f_i + \varepsilon_0 \, \partial(\mathscr{E} \times \mathscr{B})_i/\partial t. \tag{2.2.43}$$

Again, we integrate over a volume ⊛ bounded by a stationary surface ⟲ and use the divergence theorem. This yields

$$\int_{\circlearrowright} da_j \, T_{ij} = \int_{\circledast} d\tau \, f_i + \varepsilon_0 \, d/dt \int_{\circledast} d\tau \, (\mathscr{E} \times \mathscr{B})_i. \tag{2.2.44}$$

We identify the first term on the right-hand side of Eq. (2.2.44) with the components of the time rate of change of the total mechanical linear momentum due to electromagnetic forces, $d\mathbf{P}_{\text{mech}}/dt$. Again, we isolate the system by expanding the volume ⊛ until the surface integral vanishes, which gives

$$0 = d\mathbf{P}_{\text{mech}}/dt + \varepsilon_0 \, d/dt \int_{\circledast} d\tau \, (\mathscr{E} \times \mathscr{B}). \tag{2.2.45}$$

We interpret this as a linear-momentum conservation relation of the form

$$d(\mathbf{P}_{\text{field}} + \mathbf{P}_{\text{mech}})/dt = 0, \tag{2.2.46}$$

and we identify

$$\mathbf{g} = \varepsilon_0(\mathscr{E} \times \mathscr{B}) \tag{2.2.47}$$

as the volume density of linear momentum associated with the electromagnetic field. It then follows that the surface integral in Eq. (2.2.44) represents the rate of flow of linear momentum carried by the electromagnetic field across the surface ⟲ into the volume ⊛.

In a static situation, where $d\mathbf{P}_{\text{field}}/dt = 0$, it is often convenient to compute the total electromagnetic force on an object by calculating the surface integral of the stress tensor. A simple example of such a computation is given by Good and Nelson.[25] It is also worth noting explicitly that the linear-momentum density $\mathbf{g}$ and the Poynting vector $\mathbf{S}$ differ by only a scalar factor of $\varepsilon_0 \mu_0 = 1/c^2$, where c is the speed of light in vacuum.

2.2.1.4.3. ANGULAR MOMENTUM. In order to develop a relation associated with the conservation of angular momentum, we fix an origin and introduce the position vector $\mathbf{r}$ relative to that origin. The existence of a linear-momentum density associated with the electromagnetic field implies there also

---

[23] Good and Nelson,[7] p. 72.

[24] J. W. Dettman, "Mathematical Methods in Physics and Engineering," 2nd ed., p. 2. McGraw-Hill, New York, 1969.

[25] Good and Nelson,[7] p. 263.

[†] We use the convention that lower case Latin indices take on the values 1, 2, and 3; x_1, x_2, and x_3 represent the three Cartesian coordinates x, y, and z, respectively. Repeated Latin indices imply summation from one to three (Einstein summation convention).[24]

exists an angular-momentum density associated with the electromagnetic field of amount

$$\mathscr{L} = \mathbf{r} \times \mathbf{g}, \tag{2.2.48}$$

and a total angular momentum

$$\mathbf{L}_{\text{field}} = \int_{\circledast} d\tau \, (\mathbf{r} \times \mathbf{g}). \tag{2.2.49}$$

We can construct a conservation relation by starting from Eq. (2.2.43). Expressing the vector product in terms of the Levi-Civita symbol,[26] we have

$$(\mathbf{r} \times \mathbf{f})_k + \partial(\mathbf{r} \times \mathbf{g})_k/\partial t = \varepsilon_{kli} x_l \, \partial T_{ij}/\partial x_j. \tag{2.2.50}$$

Using the defining properties of ε_{kli} and the symmetry of the stress tensor, it is a simple exercise to show

$$\varepsilon_{kli} x_l \, \partial T_{ij}/\partial x_j = \partial(\varepsilon_{kli} x_l T_{ij})/\partial x_j. \tag{2.2.51}$$

We define a new tensor

$$N_{kj} = \varepsilon_{kli} x_l T_{ij}. \tag{2.2.52}$$

We next integrate Eq. (2.2.50) over a volume $\circledast$ bounded by a stationary surface $\circlearrowright$, substitute from Eqs. (2.2.51) and (2.2.52), and use the divergence theorem. This yields

$$\int_{\circlearrowright} da_j \, N_{kj} = \int_{\circledast} d\tau \, (\mathbf{r} \times \mathbf{f})_k + d/dt \int_{\circledast} d\tau \, (\mathbf{r} \times \mathbf{g})_k$$

$$= d(\mathbf{L}_{\text{mech}} + \mathbf{L}_{\text{field}})_k/dt. \tag{2.2.53}$$

If, as before, we isolate the system by expanding the volume $\circledast$ until the surface integral vanishes, we have a conservation relation for total angular momentum of the form

$$d(\mathbf{L}_{\text{mech}} + \mathbf{L}_{\text{field}})/dt = 0. \tag{2.2.54}$$

The surface integral in Eq. (2.2.53) represents the rate of flow of angular momentum carried by the electromagnetic field across the surface $\circlearrowright$ into the volume $\circledast$.

2.2.1.5. *Plane Waves.* We noted that in a region devoid of sources we can represent the electric and magnetic field strengths using only a vector potential: $\mathscr{B} = \nabla \times \mathscr{A}$ and $\mathscr{E} = -\partial \mathscr{A}/\partial t$, where $\mathscr{A}$ will satisfy Eqs. (2.2.19) and (2.2.33). The simplest solutions of Eq. (2.2.33) are $\hat{\mathbf{e}} \sin(\mathbf{k} \cdot \mathbf{r} - \omega t)$ and $\hat{\mathbf{e}} \cos(\mathbf{k} \cdot \mathbf{r} - \omega t)$, where

$$k^2 = \mu_0 \varepsilon_0 \omega^2 = \omega^2/c^2, \tag{2.2.55}$$

and $\hat{\mathbf{e}}$ is an arbitrary constant unit vector. We will find it convenient to

---

[26] Good and Nelson,[7] p. 73; Jeffreys and Jeffreys,[10] p. 69.

make a linear combination of these solutions and write[†]

$$\mathscr{A} = \text{Re}\{A\hat{\mathbf{e}} \exp[i(\mathbf{k}\cdot\mathbf{r} - \omega t + \phi)]\}. \quad (2.2.56)$$

Here, A and ϕ are real constants determined by the boundary conditions. The quantity $\mathbf{k}\cdot\mathbf{r} - \omega t + \phi$ is called the phase of the wave represented by Eq. (2.2.56). Imposing the condition that $\mathscr{A}$ satisfy Eq. (2.2.19), we find

$$0 = \nabla\cdot\mathscr{A} = A\hat{\mathbf{e}}\cdot\mathbf{k}\exp[i(\mathbf{k}\cdot\mathbf{r} - \omega t + \phi)], \quad (2.2.57)$$

which implies that $\hat{\mathbf{e}} \perp \hat{\mathbf{k}}$. We use $\hat{\mathbf{k}}$ to represent a unit vector in the direction of $\mathbf{k}$; thus, $\mathbf{k} = k\hat{\mathbf{k}}$.

The function $\mathscr{A}(\mathbf{r}, t)$ has the same values everywhere on a surface of constant phase $\mathbf{k}\cdot\mathbf{r} - \omega t + \phi = C$. For any given time t, this surface is a plane normal to $\mathbf{k}$ and a distance $D = (C + \omega t - \phi)/k$ from the origin, as shown in Fig. 1. For all position vectors $\mathbf{r}$ locating points in this plane,

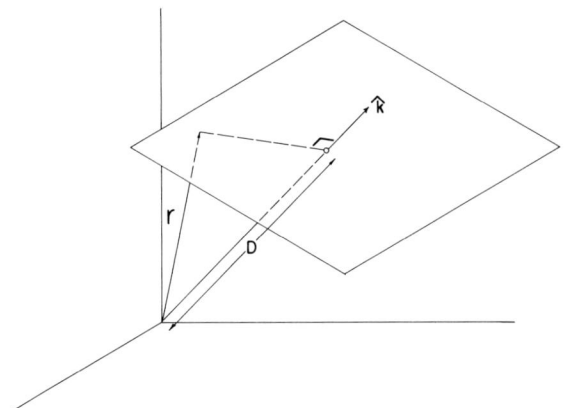

FIG. 1. Surface of constant phase for a plane wave.

$D = \hat{\mathbf{k}}\cdot\mathbf{r}$. In general, a surface of contiguous points of the same phase is called a wavefront, and wave functions having planar wavefronts are called plane waves. A particular wavefront, labeled by a particular value of C,

---

[†] Re{ } means take the real part of the expression inside the brackets. So long as all our mathematical operations are linear with respect to complex numbers, we can work with complex representatives of physical quantities and take the real part only at the end of the calculation. In the sequel, whenever a complex expression is used to represent a physical quantity without further comment, it is to be understood that the actual physical quantity is the real part of the complex expression.

moves in the direction $\hat{\mathbf{k}}$ with speed (phase velocity)

$$v_p = dD/dt = \omega/k = c, \qquad (2.2.58)$$

where we have used Eq. (2.2.55).

Not only the vector potential of Eq. (2.2.56), but also the electric and magnetic field strengths obtained from that potential are plane waves. We have

$$\mathscr{E} = -\partial \mathscr{A}/\partial t = i\omega A \hat{\mathbf{e}} \exp[i(\mathbf{k}\cdot\mathbf{r} - \omega t + \phi)] \qquad (2.2.59)$$

and

$$\mathscr{B} = \nabla \times \mathscr{A} = (\hat{\mathbf{k}} \times \hat{\mathbf{e}}) ikA \exp[i(\mathbf{k}\cdot\mathbf{r} - \omega t + \phi)]. \qquad (2.2.60)$$

We introduce the complex amplitudes $\mathscr{E} = i\omega A e^{i\phi} = E e^{i\delta}$ and $\mathscr{B} = ikA e^{i\phi} = B e^{i\delta}$, where E, B, and δ are real, so that Eqs. (2.2.59) and (2.2.60) become

$$\mathscr{E} = E\hat{\mathbf{e}} \exp[i(\mathbf{k}\cdot\mathbf{r} - \omega t + \delta)] \qquad (2.2.61)$$

and

$$\mathscr{B} = B(\hat{\mathbf{k}} \times \hat{\mathbf{e}}) \exp[i(\mathbf{k}\cdot\mathbf{r} - \omega t + \delta)]. \qquad (2.2.62)$$

The electric and magnetic field strengths are transverse ($\mathscr{E} \perp \hat{\mathbf{k}}$ and $\mathscr{B} \perp \hat{\mathbf{k}}$), perpendicular to one another ($\mathscr{E} \perp \mathscr{B}$), in phase, and have magnitudes related by

$$E/B = \omega/k = c. \qquad (2.2.63)$$

If we want to compute the Poynting vector or the linear-momentum density vector for this plane wave, we must be cautious with the complex representation because multiplication is not linear with respect to complex numbers. If the instantaneous values of $\mathbf{S} = \mathscr{E} \times \mathscr{B}/\mu_0$ or $\mathbf{g} = \varepsilon_0 \mathscr{E} \times \mathscr{B}$ are desired, we must take the real parts of the expressions given in Eqs. (2.2.61) and (2.2.62) before computing the vector product. Frequently, we are not concerned with instantaneous values, but only with time averages over some whole number of oscillations. In this case, the following theorem is valuable[27]:

Theorem. If $\mathbf{F}$ and $\mathbf{G}$ are two complex vector fields both having the time dependence $e^{-i\omega t}$, and the symbol $\otimes$ represents either the operation of scalar product or of vector product, and the symbol $\langle \rangle$ means to take the time average of the included quantity over a whole number of complete oscillations, then

$$\langle (\text{Re } \mathbf{F}) \otimes (\text{Re } \mathbf{G}) \rangle = \tfrac{1}{2} \text{Re}\{\mathbf{F} \otimes \mathbf{G}^*\} = \tfrac{1}{2} \text{Re}\{\mathbf{F}^* \otimes \mathbf{G}\}. \qquad (2.2.64)$$

[27] A proof can be found, e.g., in R. K. Wangsness, "Introduction to Theoretical Physics, Classical Mechanics and Electrodynamics," p. 290. Wiley, New York, 1963.

Applying this result to the plane wave of Eqs. (2.2.61) and (2.2.62), we find that

$$\langle \mathbf{S} \rangle = \tfrac{1}{2}\operatorname{Re}\{\mathscr{E} \times \mathscr{B}^*/\mu_0\} = \tfrac{1}{2}(E^2 \hat{\mathbf{k}}/c\mu_0) = \tfrac{1}{2}c\varepsilon_0 E^2 \hat{\mathbf{k}} \quad (2.2.65)$$

and

$$\langle \mathbf{g} \rangle = \tfrac{1}{2}\operatorname{Re}\{\varepsilon_0 \mathscr{E} \times \mathscr{B}^*\} = \tfrac{1}{2}(\varepsilon_0 E^2 \hat{\mathbf{k}}/c). \quad (2.2.66)$$

Both of these vectors point in the direction of propagation of the plane wave.

2.2.1.6. *Polarization.* Because of the linearity of Maxwell's equations, we can construct more complicated waves by superposition of plane waves. First, consider waves having a definite propagation direction $\hat{\mathbf{k}}$. Choose a pair of basis unit vectors $\hat{\mathbf{e}}_1$ and $\hat{\mathbf{e}}_2$ such that $\hat{\mathbf{e}}_1$, $\hat{\mathbf{e}}_2$, and $\hat{\mathbf{k}}$ form an orthogonal, right-handed triad: $\hat{\mathbf{e}}_1 \times \hat{\mathbf{e}}_2 = \hat{\mathbf{k}}$, etc. Suppose there are simultaneously present several waves having the same $\mathbf{k}$ and ω, and hence all having field strengths proportional to $\exp[i(\mathbf{k} \cdot \mathbf{r} - \omega t)]$. Denote the vector potential for the jth wave by $\mathscr{A}(j)$. Resolve each of these vectors into components along $\hat{\mathbf{e}}_1$ and $\hat{\mathbf{e}}_2$ so that, from Eq. (2.2.56),

$$\mathscr{A}(j) = \{\hat{\mathbf{e}}_1 A_1(j)e^{i\phi(j)} + \hat{\mathbf{e}}_2 A_2(j)e^{i\phi(j)}\} \exp[i(\mathbf{k} \cdot \mathbf{r} - \omega t)]. \quad (2.2.67)$$

For the total vector potential we then have

$$\mathscr{A}_{\text{tot}} = \sum_j \mathscr{A}(j) = \left\{\left(\sum_j A_1(j)e^{i\phi(j)}\right)\hat{\mathbf{e}}_1 + \left(\sum_j A_2(j)e^{i\phi(j)}\right)\hat{\mathbf{e}}_2\right\} \exp[i(\mathbf{k} \cdot \mathbf{r} - \omega t)]$$

$$= \{A_1 e^{i\gamma_1}\hat{\mathbf{e}}_1 + A_2 e^{i\gamma_2}\hat{\mathbf{e}}_2\} \exp[i(\mathbf{k} \cdot \mathbf{r} - \omega t)]. \quad (2.2.68)$$

Here, $A_1 e^{i\gamma_1} = \sum_j A_1(j)e^{i\phi(j)}$ and $A_2 e^{i\gamma_2} = \sum_j A_2(j)e^{i\phi(j)}$. The field strengths associated with this vector potential are

$$\mathscr{E} = -\partial \mathscr{A}_{\text{tot}}/\partial t = i\omega \mathscr{A}_{\text{tot}}$$

$$= E_1 \hat{\mathbf{e}}_1 \exp[i(\mathbf{k} \cdot \mathbf{r} - \omega t + \delta_1)] + E_2 \hat{\mathbf{e}}_2 \exp[i(\mathbf{k} \cdot \mathbf{r} - \omega t + \delta_2)], \quad (2.2.69)$$

where $E_1 e^{i\delta_1} = i\omega A_1 e^{i\gamma_1}$ and $E_2 e^{i\delta_2} = i\omega A_2 e^{i\gamma_2}$, and

$$\mathscr{B} = \nabla \times \mathscr{A}_{\text{tot}} = \{ikA_1 e^{i\gamma_1}(\hat{\mathbf{k}} \times \hat{\mathbf{e}}_1) + ikA_2 e^{i\gamma_2}(\hat{\mathbf{k}} \times \hat{\mathbf{e}}_2)\} \exp[i(\mathbf{k} \cdot \mathbf{r} - \omega t)]$$

$$= (1/c)\hat{\mathbf{k}} \times \mathscr{E}. \quad (2.2.70)$$

The actual physical components of the transverse electric field strengths are:

$$\mathscr{E}_1(\text{act}) = \operatorname{Re} \mathscr{E}_1 = E_1 \cos(\mathbf{k} \cdot \mathbf{r} - \omega t + \delta_1) \quad (2.2.71\text{a})$$

and

$$\mathscr{E}_2(\text{act}) = \operatorname{Re} \mathscr{E}_2 = E_2 \cos(\mathbf{k} \cdot \mathbf{r} - \omega t + \delta_2). \quad (2.2.71\text{b})$$

It is a straightforward exercise to show from Eq. (2.2.71) that, for a fixed value of $\mathbf{r}$, the tip of the electric-field-strength vector traces out an ellipse in the plane normal to $\hat{\mathbf{k}}$ as a function of time. This situation is called an elliptically polarized wave; it is, of course, always possible to focus attention on the two components displayed in Eq. (2.2.71) and not consider the ellipse explicitly. The equation of the polarization ellipse is[†]

$$(\mathscr{E}_1/E_1)^2 - (\mathscr{E}_2/E_2)^2 - (2\mathscr{E}_1\mathscr{E}_2/E_1 E_2)\cos(\delta_2 - \delta_1) = \sin^2(\delta_2 - \delta_1). \quad (2.2.72)$$

A similar result holds for the magnetic field strength, except the ellipse is rotated by 90°. Once the electric field strength is known for the wave, the magnetic field strength can be obtained from Eq. (2.2.70). It is, therefore, permissible (and conventional) to focus attention on the electric-field-strength vector when discussing polarization.

It is important to specify the sense in which the tip of the electric-field-strength vector traces out the ellipse of Eq. (2.2.72). There are several different conventions used for this specification; the most important are "handedness" and "helicity," which are defined in Table I.

TABLE I. Definitions of Handedness and Helicity

Sense of rotation of tip of electric-field-strength vector as seen by an observer *receiving* the radiation	Condition	Handedness	Helicity
Counterclockwise	$\sin(\delta_2 - \delta_1) > 0$	Left-handed	Positive
Clockwise	$\sin(\delta_2 - \delta_1) < 0$	Right-handed	Negative

It is possible to express the size and orientation of the polarization ellipse in terms of the parameters E_1, E_2, δ_1, and δ_2, though the algebraic manipulations involved are not simple. Let a be the semimajor axis, b the semiminor axis, and ζ the azimuth of the major axis with respect to the $\hat{\mathbf{e}}_1$-direction ($0 \leq \zeta < \pi$). Also, define a parameter η by $|\eta| = b/a$, and sgn η = {positive if positive helicity, negative if negative helicity}. One can then deduce the

---

[28] M. Born and E. Wolf, "Principles of Optics," 3rd ed. (rev.). Pergamon, Oxford, 1965.
[29] D. Clarke and J. F. Grainger, "Polarized Light and Optical Measurement." Pergamon, Oxford, 1971.
[30] L. Leithold, "The Calculus with Analytic Geometry," 2nd ed. Chapter 12. Harper, New York, 1972.

[†] For discussion of the polarization properties of electromagnetic radiation, see Born and Wolf[28] or Clarke and Grainger.[29] Equation (2.2.72) is derived by Born and Wolf on p. 25. Discussion of the geometrical properties of an ellipse, and the technique for showing that Eq. (2.2.72) really represents an ellipse, can be found in numerous texts on analytic geometry; for example, that by Leithold.[30]

relations[31]

$$a^2 + b^2 = E_1^2 + E_2^2, \tag{2.2.73}$$

and

$$2\eta/(1 + \eta^2) = 2E_1 E_2 \sin(\delta_2 - \delta_1)/(E_1^2 + E_2^2), \tag{2.2.74}$$

$$\tan 2\zeta = 2E_1 E_2 \cos(\delta_2 - \delta_1)/(E_1^2 - E_2^2). \tag{2.2.75}$$

For given E_1, E_2, δ_1, and δ_2, a and b can be determined from Eqs. (2.2.73) and (2.2.74). Equation (2.2.75) can be used to determine the orientation of the polarization ellipse if we can specify the quadrant of 2ζ. To do this, note that if $E_1^2 - E_2^2$ and $\cos(\delta_2 - \delta_1)$ have the same sign, $\tan 2\zeta$ is positive, and so 2ζ is either first or third quadrant. Also, $E_1^2 > E_2^2$ implies that the major axis makes an angle of less than 45° with the $\hat{\mathbf{e}}_1$-direction, and so 2ζ is either first or fourth quadrant. This gives us enough information to construct Table II, which allows determination of the quadrant of 2ζ for any particular case.

TABLE II. Determination of the Quadrant of 2ζ

	Quadrant of 2ζ			
	1	2	3	4
sgn$(E_1^2 - E_2^2)$	+	−	−	+
sgn$(\cos(\delta_2 - \delta_1))$	+	+	−	−

There are two important special cases. If $(\delta_2 - \delta_1) = m\pi$, $m = 0, \pm 1, \pm 2, \ldots$, we have a linearly polarized wave. The polarization ellipse reduces to a straight line. By a rotation of axes putting $\hat{\mathbf{e}}_1$ along the direction of this line, we return to the situation of a single plane wave. The general case represented by Eqs. (2.2.68)–(2.2.70) can always be considered as a superposition of two linearly polarized plane waves rather than as a single elliptically polarized wave.

The other special case occurs when $E_1 = E_2$ and $(\delta_2 - \delta_1) = n\pi/2$, $n = \pm 1, \pm 3, \pm 5, \ldots$. Then, the ellipse reduces to a circle. This is called a circularly polarized wave. It is possible to express the general case of Eqs. (2.2.68)–(2.2.70) as a superposition of circularly polarized waves by making a change of basis. We write Eq. (2.2.69) in the form

$$\mathscr{E} = \{E_1 \hat{\mathbf{e}}_1 + E_2 e^{i(\delta_2 - \delta_1)} \hat{\mathbf{e}}_2\} e^{i\delta_1} \exp[i(\mathbf{k} \cdot \mathbf{r} - \omega t)]. \tag{2.2.76}$$

For a circularly polarized wave of positive helicity this becomes

$$\mathscr{E}(+) = E_1 \{\hat{\mathbf{e}}_1 + i\hat{\mathbf{e}}_2\} e^{i\delta_1} \exp[i(\mathbf{k} \cdot \mathbf{r} - \omega t)], \tag{2.2.77}$$

[31] Born and Wolf,[28] p. 27.

and for a circularly polarized wave of negative helicity it becomes

$$\mathscr{E}(-) = E_1\{\hat{\mathbf{e}}_1 - i\hat{\mathbf{e}}_2\}e^{i\delta_1} \exp[i(\mathbf{k}\cdot\mathbf{r} - \omega t)]. \qquad (2.2.78)$$

This suggests that we should introduce the complex basis vectors, called the helicity basis, defined by

$$\hat{\mathbf{e}}_{(+)} = (1/\sqrt{2})(\hat{\mathbf{e}}_1 + i\hat{\mathbf{e}}_2) \qquad (2.2.79\text{a})$$

and

$$\hat{\mathbf{e}}_{(-)} = (1/\sqrt{2})(\hat{\mathbf{e}}_1 - i\hat{\mathbf{e}}_2). \qquad (2.2.79\text{b})$$

We then have

$$\hat{\mathbf{e}}_1 = (1/\sqrt{2})(\hat{\mathbf{e}}_{(+)} + \hat{\mathbf{e}}_{(-)}) \qquad (2.2.80\text{a})$$

and

$$\hat{\mathbf{e}}_2 = (-i/\sqrt{2})(\hat{\mathbf{e}}_{(+)} - \hat{\mathbf{e}}_{(-)}), \qquad (2.2.80\text{b})$$

and Eq. (2.2.69) can be written in the form

$$\mathscr{E} = E_{(+)}\hat{\mathbf{e}}_{(+)} \exp[i(\mathbf{k}\cdot\mathbf{r} - \omega t)] + E_{(-)}\hat{\mathbf{e}}_{(-)} \exp[i(\mathbf{k}\cdot\mathbf{r} - \omega t)], \qquad (2.2.81)$$

where $E_{(+)} = (1/\sqrt{2})(E_1 e^{i\delta_1} - iE_2 e^{i\delta_2})$ and $E_{(-)} = (1/\sqrt{2})(E_1 e^{i\delta_1} + iE_2 e^{i\delta_2})$. Equation (2.2.81) represents the desired superposition of circularly polarized waves.

For most frequencies, the polarization ellipse itself is not directly observable. It is useful to have parameters describing the polarization that are related to actual measurements. Such a description is provided by the Stokes parameters,[†] which are defined in terms of ideal experiments. Determination of the Stokes parameters for a wave requires a perfect polarizer, a quarter-wave plate,[33] and a device to measure intensity. We consider first the strictly monochromatic wave of Eqs. (2.2.68)–(2.2.70). Set the polarizer so that it transmits electric field oscillations in a direction making an angle α with the $\hat{\mathbf{e}}$-direction, and measure the total transmitted intensity. The electric field strength of the transmitted wave will have the form $\mathscr{E}(\alpha)\hat{\mathbf{e}}_\alpha$, where

$$\mathscr{E}(\alpha) = \mathscr{E}_1 \cos\alpha + \mathscr{E}_2 \sin\alpha = E_1 \exp[i(\mathbf{k}\cdot\mathbf{r} - \omega t + \delta_1)]\cos\alpha$$
$$+ E_2 \exp[i(\mathbf{k}\cdot\mathbf{r} - \omega t + \delta_2)]\sin\alpha; \qquad (2.2.82)$$

there will be a similar expression for the transmitted magnetic field. From

---

[32] Born and Wolf,[28] p. 554.
[33] M. V. Klein, "Optics," p. 500. Wiley, New York, 1970; Born and Wolf,[28] p. 691.

[†] References to the original paper by Stokes and to a number of other papers concerning use of the Stokes parameters in optics are given by Born and Wolf.[32] They also give a number of references to the use of Stokes's parameters in quantum mechanical treatments of polarization of particles.

Eq. (2.2.65), we will have for the measured intensity

$$\langle S(\alpha)\rangle = \tfrac{1}{2}c\varepsilon_0 \operatorname{Re}\{\mathscr{E}(\alpha)\hat{\mathbf{e}}_\alpha \times (\hat{\mathbf{k}} \times \mathscr{E}(\alpha)\hat{\mathbf{e}}_\alpha)^*\}$$
$$= \tfrac{1}{2}c\varepsilon_0 \hat{\mathbf{k}}\{\tfrac{1}{2}(E_1^2 + E_2^2) + \tfrac{1}{2}(E_1^2 - E_2^2)\cos 2\alpha$$
$$+ E_1 E_2 \cos(\delta_2 - \delta_1)\sin 2\alpha\}$$
$$= \tfrac{1}{2}c\varepsilon_0 \hat{\mathbf{k}}\{\tfrac{1}{2}I + \tfrac{1}{2}M\cos 2\alpha + \tfrac{1}{2}C\sin 2\alpha\}, \qquad (2.2.83)$$

where we have introduced the first three Stokes parameters, which are defined by

$$I = E_1^2 + E_2^2, \qquad (2.2.84)$$
$$M = E_1^2 - E_2^2, \qquad (2.2.85)$$

and

$$C = 2E_1 E_2 \cos(\delta_2 - \delta_1). \qquad (2.2.86)$$

Intensity measurements for three different values of α allow determination of I, M, and C, and hence of E_1, E_2, and $\cos(\delta_2 - \delta_1)$, but not of $\operatorname{sgn}(\delta_2 - \delta_1)$ (helicity) since the cosine is an even function.

In order to complete determination of the polarization state of the wave, we must make one more measurement. We pass the wave through the quarter-wave plate, which is oriented such that the phase difference for the emerging components, $\delta_2' - \delta_1'$, is $\pi/2$ radians less than the phase difference for the entering components; i.e.,

$$\delta_2' - \delta_1' = \delta_2 - \delta_1 - \pi/2. \qquad (2.2.87)$$

The wave is then passed through the polarizer, which is set to transmit electric field oscillations making an angle α' with the $\hat{\mathbf{e}}_1$-direction, and we then measure the total transmitted intensity. The result is analogous to Eq. (2.2.83):

$$\langle S(\alpha')\rangle = \tfrac{1}{2}c\varepsilon_0 \hat{\mathbf{k}}\{\tfrac{1}{2}I + \tfrac{1}{2}M\cos 2\alpha' + E_1 E_2 \cos(\delta_2' - \delta_1')\sin 2\alpha'\}$$
$$= \tfrac{1}{2}c\varepsilon_0 \hat{\mathbf{k}}\{\tfrac{1}{2}I + \tfrac{1}{2}M\cos 2\alpha' + \tfrac{1}{2}\mathscr{S}\sin 2\alpha'\}, \qquad (2.2.88)$$

where we have introduced the fourth Stokes parameter defined by

$$\mathscr{S} = 2E_1 E_2 \cos(\delta_2' - \delta_1') = 2E_1 E_2 \sin(\delta_2 - \delta_1). \qquad (2.2.89)$$

The four intensity measurements just described allow complete determination of the parameters of the polarization ellipse. Equations (2.2.73)–(2.2.75) can be written as

$$a^2 + b^2 = I, \qquad (2.2.90)$$
$$2\eta/(1 + \eta^2) = \mathscr{S}/I, \qquad (2.2.91)$$

and
$$\tan 2\zeta = C/M. \tag{2.2.92}$$
We also note from Eqs. (2.2.84)–(2.2.86) and (2.2.89) that
$$I^2 = M^2 + C^2 + \mathscr{S}^2. \tag{2.2.93}$$

The results deduced so far depend on the wave being of the simple (perfectly polarized) type represented by Eqs. (2.2.68)–(2.2.70). For any practical realization of a monochromatic wave, the amplitudes E_1 and E_2 and phases δ_1 and δ_2 will at best be functions of time that vary slowly compared with the period $T = 2\pi/\omega$. In this case, the measurements just described give averages over the time dependences of the various functions $E_1(t)$, $E_2(t)$, $\delta_1(t)$, and $\delta_2(t)$.[34] More generally, Stokes's parameters can be used to describe radiation that is a superposition of perfectly polarized waves, but where the relative amplitudes and phases of at least some subset of the component waves undergo random fluctuations in times short compared with the response time of the detector. Such waves, which we consider at some length in the following section, are called incoherent. We will discover in the next section that for a superposition of incoherent waves, the total intensity is just the sum of the intensities of the component waves. Suppose we have a superposition of two incoherent waves that, if considered separately, would have Stokes parameters $(I_1, M_1, C_1, \mathscr{S}_1)$ and $(I_2, M_2, C_2, \mathscr{S}_2)$, respectively. Since for the superposed waves the intensities add, it follows directly from the definitions in terms of intensity measurements that the Stokes parameters determined by experiments on the total wave will be

$$\begin{aligned} I_{\text{tot}} &= I_1 + I_2, & M_{\text{tot}} &= M_1 + M_2, \\ C_{\text{tot}} &= C_1 + C_2, & \mathscr{S}_{\text{tot}} &= \mathscr{S}_1 + \mathscr{S}_2. \end{aligned} \tag{2.2.94}$$

In a general situation, the radiation under consideration will be a superposition of a set of waves coherent to the extent that the phase difference $\delta_2 - \delta_1$ is stable over time intervals greater than the response time of the detector, and an additional superposition of a set of waves for which $\delta_2 - \delta_1$ undergoes random fluctuations in times less than the response time of the detector. These two contributions to the radiation are called, respectively, the polarized and the unpolarized components; the total wave is said to be partially polarized. If the polarized component were present alone, it would have Stokes's parameters $(I_p, M_p, C_p, \mathscr{S}_p)$. If the unpolarized component were present alone, the measured Stokes parameters would represent time averages of Eqs. (2.2.84)–(2.2.86) and (2.2.89). Since $\delta_2 - \delta_1$ undergoes random fluctuations in times less than the measurement time, the measured

[34] E. Hecht, *Amer. J. Phys.* **38**, 1156 (1970).

Stokes parameters C_u and $\mathscr{S}_u$ would both be zero. Further, Eq. (2.2.75) shows that the orientation of the polarization ellipse would undergo random fluctuations on the same time scale as $\delta_2 - \delta_1$. This means $\langle E_{1u}^2 \rangle = \langle E_{2u}^2 \rangle$, whence $M_u = 0$. Only the first Stokes parameter $I_u = \langle E_{1u}^2 \rangle + \langle E_{2u}^2 \rangle$ will not vanish for the unpolarized component. Then, from Eq. (2.2.94), the measured Stokes parameters for the total wave will be $(I_p + I_u, M_p, C_p, \mathscr{S}_p)$. Equation (2.2.93) is applicable to the polarized component, and so

$$I_p^2 = M_p^2 + C_p^2 + \mathscr{S}_p^2. \tag{2.2.95}$$

This relation allows separate determination of I_p and I_u, and so one set of four measurements allows complete characterization of both the polarized and the unpolarized components of the total wave.

2.2.1.7. *Coherence.* The electromagnetic radiation emitted by an actual source will involve neither a single frequency nor a single propagation direction. We can appreciate this by considering the discrete nature of the microscopic emission process. The electromagnetic radiation emitted by a typical source will consist of a superposition of wave packets having finite lengths in space and time emitted by various microscopic entities at different times. The Fourier representation[35] of any single wave packet will require superposition of numerous propagation directions and frequencies distributed about central values $\hat{\mathbf{k}}_c$ and ω_c. The details of this distribution will depend on things like the natural widths of the energy levels of the emitting systems, Doppler shifts, and collisions and other perturbations from the environment.[†]

Consider a detector located at some fixed point P. Suppose that this detector can analyze incoming electromagnetic radiation with regard to propagation direction, polarization, and frequency. Incoming radiation associated with a particular frequency and propagation direction will be represented by a vector potential of the form of Eq. (2.2.68). At some particular time t, each of the complex amplitudes $A_1 e^{i\gamma_1}$ and $A_2 e^{i\gamma_2}$ can conveniently be represented by a point p on an Argand diagram. At some later time t', where $c(t' - t)$ is greater than the mean length in space of a wave packet, the radiation at P will involve a completely different set of wave packets. If the various wave packets are generated by independent emission events so that there is no definite relation between the complex amplitudes for different packets, we expect to see random fluctuations in the location of

---

[35] For a review of Fourier integrals see, e.g., Butkov,[9] Chapter 7.
[36] G. Peach, *Contemp. Phys.* **16**, 17 (1975).
[37] R. G. Breene, Jr., "The Shift and Shape of Spectral Lines." Pergamon, Oxford, 1961.

[†] A recent elementary review of the theory of the shape of spectral lines is that by Peach[36]; a useful treatise on the subject is that by Breene.[37]

the point p on the Argand diagram over time intervals longer than $(t' - t)$. In such a case, we say that we have chaotic or incoherent radiation. In contrast, if there does exist some definite relation between the complex amplitudes of successive wave packets for at least some subset of the totality of packets composing the radiation, we say the radiation has a certain degree of coherence.

Let us consider a more quantitative discussion of these ideas. Let $\mathscr{E}_1(t)$ be a particular component of the electric field strength detected at P due to some arbitrarily chosen single wave packet, which we will call the first or standard wave packet. Choose some well-defined point on the wave packet, say the point where the electric field strength has its maximum amplitude, as a fiducial point with respect to which we can describe the location of the packet as a function of time. Choose the time origin so that the fiducial point of the standard wave packet reaches P at time $t = 0$. Assume that all wave packets have finite length so that $\mathscr{E}(t) = 0$ for $|t| \geq t_0$ for some suitably chosen t_0. Fourier analyze the real function $\mathscr{E}_1(t)$:

$$\mathscr{E}_1(t) = (2\pi)^{-1/2} \int_{-\infty}^{\infty} \phi_1(\omega) e^{-i\omega t} \, d\omega, \qquad (2.2.96)$$

$$\phi_1(\omega) = (2\pi)^{-1/2} \int_{-\infty}^{\infty} \mathscr{E}_1(t) e^{i\omega t} \, dt. \qquad (2.2.97)$$

The condition that $\mathscr{E}_1(t)$ is real leads to the requirement that its Fourier transform satisfy[38]

$$\phi_1(-\omega) = \phi_1(\omega)^*. \qquad (2.2.98)$$

Any physical detector will not instantaneously follow the oscillations of the field components; it will average over some time interval $2T$ during which, say, N wave packets reach P. Let t_n be the time at which the fiducial point of the nth wave packet reaches P. (Thus, for the standard wave packet, $t_1 = 0$.) Let $\mathscr{E}_n(t')$ be the description of the nth wave packet electric-field-strength component with respect to time coordinates having origin $t' = 0$ when the fiducial point of the nth wave packet reaches P. Then, the description of the nth wave packet in terms of time coordinates having origin when the fiducial point of the standard packet reaches P is

$$\mathscr{E}_n'(t) = \mathscr{E}_n(t - t_n). \qquad (2.2.99)$$

The relation of these two descriptions is illustrated in Fig. 2. The total electric-field-strength component at P at a time t during the measurement time interval is

$$\mathscr{E}(t) = \sum_{n=1}^{N} \mathscr{E}_n(t - t_n). \qquad (2.2.100)$$

---

[38] Butkov,[9] p. 266.

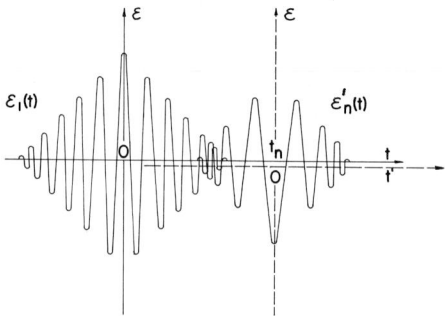

FIG. 2. Electric-field-strength component at a fixed point P for the standard wave packet and the nth wave packet. The diagram is drawn assuming that the nth wave packet reached P a time t_n after the standard wave packet reached P.

Suppose that the detector measures the intensity associated with the particular field component averaged over the time interval $2T$:

$$I_{\text{obs}} = c\varepsilon_0/2T \int_{-T}^{+T} \mathscr{E}(t)^2 \, dt. \tag{2.2.101}$$

We cannot readily proceed with the integration limits appearing in Eq. (2.2.101). However, it is not unreasonable to assume that $2Tc$ is much greater than the length of any of the N wave packets. Then, essentially all of the contribution to I_{obs} from the N wave packets comes during the time interval $-T \leqslant t \leqslant T$; extending the integration limits to $\pm \infty$ gives negligible additional contribution from these wave packets. Thus,

$$I_{\text{obs}} = c\varepsilon_0/2T \int_{-\infty}^{\infty} \mathscr{E}(t)^2 \, dt. \tag{2.2.102}$$

We now Fourier analyze the total electric-field-strength component as given by Eq. (2.2.100):

$$\mathscr{E}(t) = (2\pi)^{-1/2} \int_{-\infty}^{\infty} \Phi(\omega) e^{-i\omega t} \, d\omega, \tag{2.2.103}$$

and, from Eqs. (2.2.100) and (2.2.96), we have

$$\mathscr{E}(t) = \sum_{n=1}^{N} (2\pi)^{-1/2} \int_{-\infty}^{\infty} \phi_n(\omega) e^{-i\omega(t-t_n)} \, d\omega$$

$$= (2\pi)^{-1/2} \int_{-\infty}^{\infty} \left\{ \sum_{n=1}^{N} \phi_n(\omega) e^{i\omega t_n} \right\} e^{-i\omega t} \, d\omega. \tag{2.2.104}$$

Thus,

$$\Phi(\omega) = \sum_{n=1}^{N} \phi_n(\omega) e^{i\omega t_n}. \tag{2.2.105}$$

By Parseval's theorem[39] and Eq. (2.2.105) we have

$$\int_{-\infty}^{\infty} \mathscr{E}(t)^2 \, dt = \int_{-\infty}^{\infty} |\Phi(\omega)|^2 \, d\omega = \int_{-\infty}^{\infty} \left\{ \sum_{m=1}^{N} \sum_{n=1}^{N} \phi_m(\omega)^* \phi_n(\omega) e^{i\omega(t_n - t_m)} \right\} d\omega. \quad (2.2.106)$$

Let $\phi_p = \phi_{p0} e^{i\delta_p}$, where ϕ_{p0} and δ_p are real. Then, the double sum appearing in the integrand in Eq. (2.2.106) can be written

$$\sum_{m=1}^{N} \sum_{n=1}^{N} \phi_m^* \phi_n e^{i\omega(t_n - t_m)}$$

$$= \sum_{n=1}^{N} \phi_{n0}^2 + \sum_{n \neq m} \phi_{m0} \phi_{n0} \exp[i\omega(t_n - t_m) + i(\delta_n - \delta_m)]$$

$$= \sum_{n=1}^{N} \phi_{n0}^2 + 2 \sum_{n < m} \phi_{m0} \phi_{n0} \cos[\omega(t_n - t_m) + (\delta_n - \delta_m)]. \quad (2.2.107)$$

If the wave packets arise from independent emission processes, there will be a random distribution of time differences $(t_n - t_m)$ and phase differences $(\delta_n - \delta_m)$. There will then be equal likelihood for the cosine in Eq. (2.2.107) to be positive or negative, and as the number of wave packets involved becomes large, the last term in Eq. (2.2.107) becomes zero. Thus, for chaotic light, we have from Eqs. (2.2.102), (2.2.106), and (2.2.107) that

$$I_{\text{obs}} \cong c\varepsilon_0/2T \sum_{n=1}^{N} \int_{-\infty}^{\infty} |\phi_n(\omega)|^2 \, d\omega = c\varepsilon_0/2T \sum_{n=1}^{N} \int_{-\infty}^{\infty} \mathscr{E}_n(t)^2 \, dt, \quad (2.2.108)$$

where in the last step we have used Parseval's theorem. This equation represents a superposition of the intensities associated with the various wave packets.

If the radiation has some degree of coherence, there will be a definite correlation between any given wave packet and some subset of the other wave packets. As an example, let us consider a Michelson interferometer[40] having a perfect beam splitter and identical mirrors so that each incident wave packet has exactly half its energy sent down each arm of the interferometer, and each part of the split wave packet undergoes the same modifications before reaching the detector. Assume that one arm of the interferometer is a length d longer than the other arm so that the path difference for the two parts of the split wave packet is $2d$. Assume that the radiation originally incident on the interferometer is chaotic. Then, correlation at the detector

---

[39] Butkov,[9] p. 267.
[40] Born and Wolf,[28] p. 300.

exists only between the pairs of wave packets split from a single incident wave packet. In the last term of Eq. (2.2.107), only these pairs will give a finite contribution; the contributions from other pairs of wave packets will tend to cancel as discussed for completely chaotic light. If wave packets n and m are a pair split from a single incident wave packet, because of the assumed symmetry of the arms of the interferometer we will have in Eq. (2.2.107) that $\phi_{n0} = \phi_{m0}$, $\delta_n = \delta_m$, and $t_m = t_n \pm 2d/c$. For N wave packets reaching the detector there will be $N/2$ such pairs, and so Eq. (2.2.107) becomes

$$\sum_{m=1}^{N}\sum_{n=1}^{N} \phi_m^* \phi_n e^{i\omega(t_n - t_m)} = 2\sum_{n=1}^{N/2} \phi_{n0}^2 + 2\sum_{n=1}^{N/2} \phi_{n0}^2 \cos(2\omega d/c)$$

$$= \Theta(\omega)\{1 + \cos(2\omega d/c)\}, \qquad (2.2.109)$$

where $\Theta(\omega) = 2\sum_{n=1}^{N/2} \phi_{n0}^2(\omega)$. Then, from Eqs. (2.2.106) and (2.2.101), we have

$$I_{\text{obs}} = c\varepsilon_0/2T \int_{-\infty}^{\infty} \Theta(\omega)\{1 + \cos(2\omega d/c)\}\, d\omega$$

$$= c\varepsilon_0/T \int_{0}^{\infty} \Theta(\omega)\{1 + \cos(2\omega d/c)\}\, d\omega. \qquad (2.2.110)$$

Here, we have used Eq. (2.2.98), which implies that $\phi_{n0}(-\omega) = \phi_{n0}(\omega)$ and $\delta_n(-\omega) = -\delta_n(\omega)$. We can write Eq. (2.2.110) as

$$I_{\text{obs}} = \int_{0}^{\infty} I(\omega)\{1 + \cos(2\omega d/c)\}\, d\omega, \qquad (2.2.111)$$

where $I(\omega)\, d\omega$ represents the intensity of the incident radiation associated with the angular frequency interval ω to $\omega + d\omega$.[†]

If we want to have the greatest possible intensity at some frequency for a collection of wave packets, which requires the greatest degree of coherence, we must have the phase differences $(\delta_n - \delta_m)$ related to the time differences $(t_n - t_m)$ in such a way that all the cosines in Eq. (2.2.107) equal 1. Each emission process would have to be governed by the wave packets already existing in the vicinity in such a way that, relative to the standard wave packet, $\omega t_n + (\delta_n - \delta_1) = P_n 2\pi$, where P_n is an integer. Then, the argument

[41] Klein,[33] Chapter 6.
[42] Born and Wolf,[28] Chapter X.
[43] L. Mandel and E. Wolf, Rev. Mod. Phys. **37**, 231 (1965).

[†] A detailed general development of the theory of coherence is an intricate task. A useful treatment at the general level of our discussion, including a number of applications, is given by Klein.[41] Our Eq. (2.2.111) should be compared with Klein's development that culminates on p. 230. An introduction to the general theoretical treatment of coherence can be found in Born and Wolf[42] and in Mandel and Wolf.[43]

of the cosine in Eq. (2.2.107) would be $\omega(t_n - t_m) + (\delta_n - \delta_1) - (\delta_m - \delta_1) = (P_n - P_m)2\pi$, and the cosine itself would be 1. Consideration of how the emission process can be governed by the radiation already present in the vicinity would lead us to the theory of lasers, which is outside the purview of the present discussion.[†]

2.2.1.8. Fourier Representation of Electromagnetic Radiation. We will now consider Fourier representation of electromagnetic radiation from a somewhat different point of view. As a special case, we will consider the situation discussed at the end of Section 2.2.1.3, where the source densities vanish throughout the region of interest. There then exists a vector potential $\mathscr{A}(\mathbf{r}, t)$ satisfying Eqs. (2.2.19) and (2.2.23) that gives the field strengths in the region of interest via the equations $\mathscr{B} = \nabla \times \mathscr{A}$ and $\mathscr{E} = -\partial \mathscr{A}/\partial t$.

We Fourier analyze the space dependence of the vector potential, treating time as a parameter:

$$\mathscr{A}(\mathbf{r}, t) = (2\pi)^{-3/2} \int d^3k \, \mathbf{a}(\mathbf{k}, t) e^{i\mathbf{k} \cdot \mathbf{r}}. \quad (2.2.112)$$

This equation is an expression for the real vector potential, even though complex quantities appear on the right-hand side. Imposing the condition $\mathscr{A}(\mathbf{r}, t) = \mathscr{A}(\mathbf{r}, t)^*$ yields[38]

$$\mathbf{a}(\mathbf{k}, t) = \mathbf{a}(-\mathbf{k}, t)^*. \quad (2.2.113)$$

Equation (2.2.112) is generally valid. For the Coulomb gauge, Eq. (2.2.19) leads to the requirement $\mathbf{a}(\mathbf{k}, t) \cdot \hat{\mathbf{k}} = 0$. If we are in a source-free region, Eq. (2.2.33) requires that

$$0 = \nabla^2 \mathscr{A} - (1/c^2) \, \partial^2 \mathscr{A}/\partial t^2 = (2\pi)^{-3/2} \int d^3k \, \{-k^2 \mathbf{a} - (1/c^2) \, \partial^2 \mathbf{a}/\partial t^2\} e^{i\mathbf{k} \cdot \mathbf{r}}, \quad (2.2.114)$$

from which we conclude that

$$\partial^2 \mathbf{a}/\partial t^2 + k^2 c^2 \mathbf{a} = 0. \quad (2.2.115)$$

A general solution of Eq. (2.2.115) will have the form[‡]

$$\mathbf{a}(\mathbf{k}, t) = \mathbf{a}_P(\mathbf{k}) e^{i\omega t} + \mathbf{a}_M(\mathbf{k}) e^{-i\omega t}, \quad (2.2.116)$$

with

$$\omega = kc. \quad (2.2.117)$$

---

[44] A. E. Siegman, *Appl. Opt.* **10**, No. 12, A38 (1971).
[45] F. T. Arecchi and E. O. Schutz-Dubois (ed.), "Laser Handbook." North-Holland Publ., Amsterdam, 1972.

[†] A reader wishing to pursue the theory of lasers is advised to consult the annotated bibliography prepared by Siegman.[44] A wealth of useful information can also be found in the two-volume "Handbook" edited by Arecchi and Schulz-Dubois.[45]
[‡] The subscripts P and M used here refer to the plus or minus sign in the exponent, respectively.

In terms of this solution, Eq. (2.2.112) becomes

$$\mathscr{A}(\mathbf{r}, t) = (2\pi)^{-3/2} \left\{ \int d^3k \, \mathbf{a}_P(\mathbf{k}) \exp[i(\mathbf{k} \cdot \mathbf{r} + \omega t)] \right.$$
$$\left. + \int d^3k \, \mathbf{a}_M(\mathbf{k}) \exp[i(\mathbf{k} \cdot \mathbf{r} - \omega t)] \right\}. \quad (2.2.118)$$

When applied to Eq. (2.2.116), Eq. (2.2.113) requires that

$$\mathbf{a}_P(-\mathbf{k})^* = \mathbf{a}_M(\mathbf{k}) \quad \text{and} \quad \mathbf{a}_M(-\mathbf{k})^* = \mathbf{a}_P(\mathbf{k}). \quad (2.2.119)$$

Using these relations in Eq. (2.2.118) yields

$$\mathscr{A}(\mathbf{r}, t) = (2\pi)^{-3/2} \int d^3k \, \{\mathbf{a}_M(\mathbf{k}) \exp[i(\mathbf{k} \cdot \mathbf{r} - \omega t)]\}^*$$
$$+ (2\pi)^{-3/2} \int d^3k \, \mathbf{a}_M(\mathbf{k}) \exp[i(\mathbf{k} \cdot \mathbf{r} - \omega t)]$$
$$= 2(2\pi)^{-3/2} \, \text{Re} \left\{ \int d^3k \, \mathbf{a}_M(\mathbf{k}) \exp[i(\mathbf{k} \cdot \mathbf{r} - \omega t)] \right\}. \quad (2.2.120)$$

The condition $\nabla \cdot \mathscr{A} = 0$ is easily seen to lead to the requirement that $\hat{\mathbf{k}} \cdot \mathbf{a}_M(\mathbf{k}) = 0$. Thus, in the source-free (free-field) case, the vector potential, and hence the electric and magnetic field strengths, can always be represented as the real part of a superposition of complex, transverse plane waves.

Taking the Fourier inversion of Eq. (2.2.112), and using Eq. (2.2.116), we find that

$$\mathbf{a}(\mathbf{k}, t) = (2\pi)^{-3/2} \int d\tau \, \mathscr{A}(\mathbf{r}, t) e^{-i\mathbf{k} \cdot \mathbf{r}} = \mathbf{a}_P(\mathbf{k}) e^{i\omega t} + \mathbf{a}_M(\mathbf{k}) e^{-i\omega t}, \quad (2.2.121)$$

from which we have

$$\mathbf{a}_P(\mathbf{k}) + \mathbf{a}_M(\mathbf{k}) = (2\pi)^{-3/2} \int d\tau \, \mathscr{A}(\mathbf{r}, 0) e^{-i\mathbf{k} \cdot \mathbf{r}}. \quad (2.2.122)$$

Equation (2.2.121) can be differentiated with respect to time to give

$$i\omega(\mathbf{a}_P(\mathbf{k}) - \mathbf{a}_M(\mathbf{k})) = (2\pi)^{-3/2} \int d\tau \, \partial \mathscr{A}/\partial t|_{t=0} \, e^{-i\mathbf{k} \cdot \mathbf{r}}. \quad (2.2.123)$$

Equations (2.2.122) and (2.2.123) can easily be solved to express $\mathbf{a}_M(\mathbf{k})$ in terms of the initial values $\mathscr{A}(\mathbf{r}, 0)$ and $\partial \mathscr{A}/\partial t|_{t=0}$.

For many purposes, it is convenient to express the vector potential as a discrete sum rather than as a continuous superposition like Eq. (2.2.112). One way to accomplish this is to subject the basis functions $e^{i\mathbf{k} \cdot \mathbf{r}}$ to periodic boundary conditions within a cubical box of side L. In the limit that L is very large compared to the dimensions of the region of interest, this should have a negligible effect on our results. We locate the cube in the first octant with one corner at the origin and sides parallel to the axes. The conditions we want to impose then are $e^{ik_xL} = 1$, $e^{ik_yL} = 1$, and $e^{ik_zL} = 1$, which will be true if the components of **k** have the form

$$k_j = 2\pi n_j/L; \quad j = 1, 2, 3, \quad n_j = 0, \pm 1, \pm 2, \ldots. \quad (2.2.124)$$

Instead of a continuous range of propagation vectors, we then have the discrete set

$$\mathbf{k} = (2\pi n_1/L)\hat{\mathbf{e}}_1 + (2\pi n_2/L)\hat{\mathbf{e}}_2 + (2\pi n_3/L)\hat{\mathbf{e}}_3. \quad (2.2.125)$$

The various discrete values of **k** are each labeled by an ordered trio of integers (n_1, n_2, n_3).

The discrete basis functions are orthogonal with respect to integration over the cube used to define the periodic boundary conditions in the sense that

$$\int_{\mathcal{O}} d\tau \, \{e^{i\mathbf{k}\cdot\mathbf{r}}\}^* e^{i\mathbf{k}'\cdot\mathbf{r}} = L^3 \, \delta_{n_1,n_1'} \, \delta_{n_2,n_2'} \, \delta_{n_3,n_3'} = L^3 \, \delta_{\mathbf{k},\mathbf{k}'}. \quad (2.2.126)$$

Thus, the functions $L^{-3/2}e^{i\mathbf{k}\cdot\mathbf{r}}$, with the values of **k** as given by Eq. (2.2.125), form a discrete, orthonormal basis set.

With respect to this discrete basis, the analog of Eq. (2.2.112) for the expansion of the real vector potential is

$$\mathscr{A}(\mathbf{r}, t) = L^{-3/2} \sum_{\mathbf{k}} \mathbf{a}_{\mathbf{k}}(t) e^{i\mathbf{k}\cdot\mathbf{r}}. \quad (2.2.127)$$

The sum is over the discrete propagation vectors given by Eq. (2.2.125).

By use of Eq. (2.2.126) we obtain

$$\mathbf{a}_{\mathbf{k}}(t) = L^{-3/2} \int_{\mathcal{O}} d\tau \, e^{-i\mathbf{k}\cdot\mathbf{r}} \mathscr{A}(\mathbf{r}, t). \quad (2.2.128)$$

Comparing this result with Eq. (2.2.121), we see that the relation between the expansion coefficients for the discrete basis and the expansion coefficients for the continuous basis for corresponding **k** values is, in the limit of large L,

$$\mathbf{a}_{\mathbf{k}}(t) = (2\pi/L)^{3/2}\mathbf{a}(\mathbf{k}, t). \quad (2.2.129)$$

For the Coulomb gauge, Eq. (2.2.19) requires that $\hat{\mathbf{k}} \cdot \mathbf{a}_{\mathbf{k}}(t) = 0$.

It will be convenient to introduce polarization basis vectors $\hat{\mathbf{e}}_\lambda(\mathbf{k})$, $\lambda = 1, 2$, such that $\hat{\mathbf{e}}_1(\mathbf{k})$, $\hat{\mathbf{e}}_2(\mathbf{k})$, and $\hat{\mathbf{k}}$ form an orthogonal, right-handed triad, as was done in Section 2.2.1.6. When we express the vector expansion coefficients $\mathbf{a}_{\mathbf{k}}(t)$ in terms of these basis vectors, Eq. (2.2.127) for the expansion of the real vector potential becomes

$$\mathscr{A}(\mathbf{r}, t) = L^{-3/2} \sum_{\mathbf{k},\lambda} a_{\mathbf{k},\lambda}(t)\hat{\mathbf{e}}_\lambda(\mathbf{k})e^{i\mathbf{k}\cdot\mathbf{r}} = L^{-3/2} \sum_{\mathbf{k},\lambda} a_{\mathbf{k},\lambda}(t)^*\hat{\mathbf{e}}_\lambda(\mathbf{k})e^{-i\mathbf{k}\cdot\mathbf{r}}.$$
$$(2.2.130)$$

The electric and magnetic field strengths in the free-field case are given by

$$\mathscr{E}(\mathbf{r}, t) = -\partial \mathscr{A}/\partial t = -L^{-3/2} \sum_{\mathbf{k},\lambda} \dot{a}_{\mathbf{k},\lambda} \hat{\mathbf{e}}_\lambda(\mathbf{k}) e^{i\mathbf{k}\cdot\mathbf{r}}$$

$$= -L^{-3/2} \sum_{\mathbf{k},\lambda} \dot{a}^*_{\mathbf{k},\lambda} \hat{\mathbf{e}}_\lambda(\mathbf{k}) e^{-i\mathbf{k}\cdot\mathbf{r}} \quad (2.2.131)$$

and

$$\mathscr{B}(\mathbf{r}, t) = \nabla \times \mathscr{A} = L^{-3/2} \sum_{\mathbf{k},\lambda} ik[\hat{\mathbf{k}} \times \hat{\mathbf{e}}_\lambda(\mathbf{k})] a_{\mathbf{k},\lambda} e^{i\mathbf{k}\cdot\mathbf{r}}$$

$$= L^{-3/2} \sum_{\mathbf{k},\lambda} -ik[\hat{\mathbf{k}} \times \hat{\mathbf{e}}_\lambda(\mathbf{k})] a^*_{\mathbf{k},\lambda} e^{-i\mathbf{k}\cdot\mathbf{r}}. \quad (2.2.132)$$

The electromagnetic-field energy inside the cube of side L is, from Eq. (2.2.39),

$$E_{\text{field}} = \int_{\mathscr{O}} d\tau \, (\tfrac{1}{2}\varepsilon_0 \mathscr{E}^2 + \tfrac{1}{2}\mathscr{B}^2/\mu_0). \quad (2.2.133)$$

Using Eqs. (2.2.126) and (2.2.131), we have

$$\tfrac{1}{2}\varepsilon_0 \int_{\mathscr{O}} d\tau \, \mathscr{E}^2 = \tfrac{1}{2}(\varepsilon_0/L^3) \int_{\mathscr{O}} d\tau \sum_{\mathbf{k},\lambda} \sum_{\mathbf{k}',\lambda'} \dot{a}_{\mathbf{k},\lambda} \dot{a}^*_{\mathbf{k}',\lambda'} \hat{\mathbf{e}}_\lambda(\mathbf{k}) \cdot \hat{\mathbf{e}}_{\lambda'}(\mathbf{k}') e^{i\mathbf{k}\cdot\mathbf{r}} e^{-i\mathbf{k}'\cdot\mathbf{r}}$$

$$= \tfrac{1}{2}\varepsilon_0 \sum_{\mathbf{k},\lambda} |\dot{a}_{\mathbf{k},\lambda}|^2, \quad (2.2.134)$$

and, from Eq. (2.2.132),

$$1/2\mu_0 \int_{\mathscr{O}} d\tau \, \mathscr{B}^2$$

$$= 1/2\mu_0 L^3 \int_{\mathscr{O}} d\tau \sum_{\mathbf{k},\lambda} \sum_{\mathbf{k}',\lambda'} a_{\mathbf{k},\lambda} a^*_{\mathbf{k}',\lambda'} kk' [\hat{\mathbf{k}} \times \hat{\mathbf{e}}_\lambda(\mathbf{k})] \cdot [\hat{\mathbf{k}}' \times \hat{\mathbf{e}}_{\lambda'}(\mathbf{k}')] e^{i\mathbf{k}\cdot\mathbf{r}} e^{-i\mathbf{k}'\cdot\mathbf{r}}$$

$$= 1/2\mu_0 \sum_{\mathbf{k},\lambda} k^2 |a_{\mathbf{k},\lambda}|^2. \quad (2.2.135)$$

Thus, Eq. (2.2.133) becomes

$$E_{\text{field}} = \tfrac{1}{2}\varepsilon_0 \sum_{\mathbf{k},\lambda} \{|\dot{a}_{\mathbf{k},\lambda}|^2 + k^2 c^2 |a_{\mathbf{k},\lambda}|^2\}, \quad (2.2.136)$$

where we have used $\varepsilon_0 \mu_0 = 1/c^2$.

Using Eq. (2.2.126), we can construct an argument that parallels Eqs. (2.2.112)–(2.2.120), but with summation over the discrete propagation vectors replacing integration over **k**-space. The result, which is the analog of Eq. (2.2.120), is that, for a source-free region,

$$\mathscr{A}(\mathbf{r}, t) = L^{-3/2} \sum_{\mathbf{k}} \{a_{\mathbf{k},M} \exp[i(\mathbf{k}\cdot\mathbf{r} - \omega t)] + \text{c.c.}\}. \quad (2.2.137)$$

It will prove convenient in the sequel to express the vector potential in terms of the helicity basis defined by Eqs. (2.2.79) and (2.2.80). If, for example, we express the vector expansion coefficients of Eq. (2.2.137) in terms of this basis, $\mathbf{a}_{\mathbf{k},M} = a_{\mathbf{k},(+)}\hat{\mathbf{e}}_{(+)} + a_{\mathbf{k},(-)}\hat{\mathbf{e}}_{(-)}$, we have

$$\mathscr{A}(\mathbf{r}, t) = L^{-3/2} \sum_{\mathbf{k}} \{(a_{\mathbf{k},(+)}\hat{\mathbf{e}}_{(+)} + a_{\mathbf{k},(-)}\hat{\mathbf{e}}_{(-)}) \exp[i(\mathbf{k}\cdot\mathbf{r} - \omega t)] + \text{c.c.}\}. \quad (2.2.138)$$

The helicity basis has the advantage of removing the arbitrariness in the orientation of the 1 and 2 axes for each propagation direction.

It is also sometimes convenient to express the vector potential and the field energy in terms of real expansion coefficients. To do this, we define the real quantities

$$\alpha^{(1)}_{\mathbf{k},\lambda}(t) = \tfrac{1}{2}(a_{\mathbf{k},\lambda}(t) + a_{\mathbf{k},\lambda}(t)^*) = \operatorname{Re}\{a_{\mathbf{k},\lambda}(t)\} \quad (2.2.139a)$$

and

$$\alpha^{(2)}_{\mathbf{k},\lambda}(t) = (1/2i)(a_{\mathbf{k},\lambda}(t) - a_{\mathbf{k},\lambda}(t)^*) = \operatorname{Im}\{a_{\mathbf{k},\lambda}(t)\}. \quad (2.2.139b)$$

We then have

$$a_{\mathbf{k},\lambda} = \alpha^{(1)}_{\mathbf{k},\lambda} + i\alpha^{(2)}_{\mathbf{k},\lambda}, \quad (2.2.140)$$

and Eq. (2.2.130) for the vector potential can be written

$$\mathscr{A}(\mathbf{r}, t) = \tfrac{1}{2}L^{-3/2} \sum_{\mathbf{k},\lambda} \{(\alpha^{(1)}_{\mathbf{k},\lambda} + i\alpha^{(2)}_{\mathbf{k},\lambda})\hat{\mathbf{e}}_{\lambda}(\mathbf{k})e^{i\mathbf{k}\cdot\mathbf{r}} + \text{c.c.}\}$$

$$= L^{-3/2} \sum_{\mathbf{k},\lambda} \{\alpha^{(1)}_{\mathbf{k},\lambda} \cos(\mathbf{k}\cdot\mathbf{r}) - \alpha^{(2)}_{\mathbf{k},\lambda} \sin(\mathbf{k}\cdot\mathbf{r})\}\hat{\mathbf{e}}_{\lambda}(\mathbf{k}). \quad (2.2.141)$$

Equation (2.2.136) for the field energy in the free-field case becomes

$$E_{\text{field}} = \tfrac{1}{2}\varepsilon_0 \sum_{\mathbf{k},\lambda,\gamma} \{(\dot{\alpha}^{(\gamma)}_{\mathbf{k},\lambda})^2 + k^2c^2(\alpha^{(\gamma)}_{\mathbf{k},\lambda})^2\}, \quad (2.2.142)$$

where $\gamma = 1, 2$. In the case of a free field, the real expansion coefficients satisfy Eq. (2.2.115):

$$\partial^2 \alpha^{(\gamma)}_{\mathbf{k},\lambda}/\partial t^2 + k^2c^2 \alpha^{(\gamma)}_{\mathbf{k},\lambda} = 0. \quad (2.2.143)$$

The general real solution of this equation is

$$\alpha^{(\gamma)}_{\mathbf{k},\lambda} = C^{(\gamma)}_{\mathbf{k},\lambda} \cos(\omega t + \phi^{(\gamma)}_{\mathbf{k},\lambda}), \quad (2.2.144)$$

where $\omega = kc$, and $C^{(\gamma)}_{\mathbf{k},\lambda}$ and $\phi^{(\gamma)}_{\mathbf{k},\lambda}$ are real constants.

Thus,

$$\dot{\alpha}^{(\gamma)}_{\mathbf{k},\lambda} = -\omega C^{(\gamma)}_{\mathbf{k},\lambda} \sin(\omega t + \phi^{(\gamma)}_{\mathbf{k},\lambda}), \quad (2.2.145)$$

and Eq. (2.2.142) becomes

$$E_{\text{field}} = \tfrac{1}{2}\varepsilon_0 \sum_{\mathbf{k},\lambda,\gamma} \omega^2 (C^{(\gamma)}_{\mathbf{k},\lambda})^2. \quad (2.2.146)$$

In a more general situation where there are sources, the electric field strength will have both transverse and longitudinal parts, which, for the Coulomb gauge, are given by Eqs. (2.2.31) and (2.2.32). Thus,

$$\mathscr{E} = \mathscr{E}_T + \mathscr{E}_L = -\partial \mathscr{A}/\partial t - \nabla\Phi. \quad (2.2.147)$$

If we expand the scalar potential in terms of the discrete basis, we obtain
$$\Phi(\mathbf{r}, t) = L^{-3/2} \sum_{\mathbf{k}} a_{\mathbf{k}}'(t) e^{i\mathbf{k}\cdot\mathbf{r}}. \tag{2.2.148}$$

Using Eqs. (2.2.130) and (2.2.148) in Eq. (2.2.147) yields
$$\mathscr{E} = -L^{-3/2} \sum_{\mathbf{k}} \{\dot{a}_{\mathbf{k},1}\hat{\mathbf{e}}_1(\mathbf{k}) + \dot{a}_{\mathbf{k},2}\hat{\mathbf{e}}_2(\mathbf{k}) + ika_{\mathbf{k}}'\hat{\mathbf{k}}\} e^{i\mathbf{k}\cdot\mathbf{r}}. \tag{2.2.149}$$

Because of the mutual orthogonality of $\hat{\mathbf{e}}_1$, $\hat{\mathbf{e}}_2$, and $\hat{\mathbf{k}}$, we find that
$$\int_{\mathcal{O}} d\tau\, \mathscr{E}^2 = \sum_{\mathbf{k}} \{|\dot{a}_{\mathbf{k},1}|^2 + |\dot{a}_{\mathbf{k},2}|^2 + k^2|a_{\mathbf{k}}'|^2\}. \tag{2.2.150}$$

Equation (2.2.135) for the magnetic contribution to the field energy is unchanged. Thus, for the field energy of Eq. (2.2.133), we obtain Eq. (2.2.136) plus an additional term arising from the longitudinal part of the electric field strength.

2.2.1.9. Hamiltonian Formulation of Classical Electromagnetism.

2.2.1.9.1. FREE FIELDS. The path to quantization for a system having a classical analog is generally via the classical Hamiltonian function H. In the classical formulation, the canonical equations of Hamilton are

$$\partial H/\partial q_j = -\dot{p}_j \quad \text{and} \quad \partial H/\partial p_j = \dot{q}_j, \tag{2.2.151}$$

where the index j labels the degrees of freedom, and q_j and p_j are, respectively, the generalized coordinate and conjugate momentum associated with the jth degree of freedom.[46] It is possible to use the Hamiltonian approach in introducing quantization of the electromagnetic field. There is not a unique way to put the equations of classical electrodynamics into Hamiltonian form; we will consider one workable scheme.

We first consider the case of a free field, where the source densities vanish throughout the region of interest. The vector potential is then given by Eq. (2.2.130), or, alternatively, by Eq. (2.2.141). The electromagnetic-field energy is given by Eq. (2.2.136), or, alternatively, by Eq. (2.2.142). If the energy expression is to guide us to a suitable Hamiltonian function, it must be represented in terms of generalized coordinates and momenta. We define these quantities in terms of the real, time-dependent expansion coefficients:

$$Q_{\mathbf{k},\lambda}^{(\gamma)} = \varepsilon_0^{1/2} \alpha_{\mathbf{k},\lambda}^{(\gamma)} \tag{2.2.152}$$

and

$$P_{\mathbf{k},\lambda}^{(\gamma)} = \dot{Q}_{\mathbf{k},\lambda}^{(\gamma)}. \tag{2.2.153}$$

We assert that a suitable Hamiltonian function is obtained by expressing

---

[46] A reader wishing to review the Hamiltonian formulation of classical mechanics is advised to peruse H. Goldstein, "Classical Mechanics." Addison-Wesley, Reading, Massachusetts, 1950.

Eq. (2.2.142) in terms of these quantities. Thus,

$$H = \sum_{\mathbf{k},\lambda,\gamma} \tfrac{1}{2}\{(P_{\mathbf{k},\lambda}^{(\gamma)})^2 + k^2 c^2 (Q_{\mathbf{k},\lambda}^{(\gamma)})^2\}. \tag{2.2.154}$$

We can readily check that we recover the correct electrodynamic relations if we start from Eqs. (2.2.154) and (2.2.151). We have

$$\partial H / \partial Q_{\mathbf{k},\lambda}^{(\gamma)} = k^2 c^2 Q_{\mathbf{k},\lambda}^{(\gamma)} = -\dot{P}_{\mathbf{k},\lambda}^{(\gamma)} \tag{2.2.155}$$

and

$$\partial H / \partial P_{\mathbf{k},\lambda}^{(\gamma)} = P_{\mathbf{k},\lambda}^{(\gamma)} = \dot{Q}_{\mathbf{k},\lambda}^{(\gamma)}. \tag{2.2.156}$$

Equation (2.2.156) just reproduces the definition of Eq. (2.2.153). Differentiating Eq. (2.2.156) with respect to the time and using Eq. (2.2.155) gives

$$\ddot{Q}_{\mathbf{k},\lambda}^{(\gamma)} = \dot{P}_{\mathbf{k},\lambda}^{(\gamma)} = -k^2 c^2 Q_{\mathbf{k},\lambda}^{(\gamma)}, \tag{2.2.157}$$

from which we recover Eq. (2.2.143) and hence all the electrodynamic relations.

2.2.1.9.2. HAMILTONIAN FOR FIELDS AND SOURCES. Establishing a suitable classical Hamiltonian function is more complicated in the case where there are sources present in the region of interest.[47] In treating molecules, atoms, and nuclei, we will be concerned with situations where the sources are a finite collection of point charges. Therefore, we consider a charge density

$$\rho = \sum_s q_s \, \delta(\mathbf{r} - \mathbf{r}_s). \tag{2.2.158}$$

Here, the index s labels the various point charges, which have charges q_s and locations $\mathbf{r}_s$. For the current density, we have, from Eq. (2.2.5), that

$$\mathbf{j} = \sum_s q_s \mathbf{v}_s \, \delta(\mathbf{r} - \mathbf{r}_s). \tag{2.2.159}$$

We will again work in the Coulomb gauge, where $\nabla \cdot \mathscr{A} = 0$ [Eq. (2.2.19)]. The basic differential equations for the scalar and vector potentials are then

$$\nabla^2 \Phi = -\rho/\varepsilon_0 \tag{2.2.20}$$

and

$$-\nabla^2 \mathscr{A} + \mu_0 \varepsilon_0 \, \partial^2 \mathscr{A}/\partial t^2 + \mu_0 \varepsilon_0 \, \nabla(\partial \Phi/\partial t) = \mu_0 \mathbf{j}. \tag{2.2.21}$$

For a charge density of the form of Eq. (2.2.158), the solution of Eq. (2.2.20) is well-known from electrostatics:

$$\Phi(\mathbf{r},t) = (4\pi\varepsilon_0)^{-1} \sum_s q_s / |\mathbf{r} - \mathbf{r}_s(t)|. \tag{2.2.160}$$

---

[47] Heitler,[4] Chapter I, Section 6; J. C. Slater, "Quantum Theory of Atomic Structure," Vol. I, Appendix 12. McGraw-Hill, New York, 1960; W. T. Grandy, Jr., "Introduction to Electrodynamics and Radiation," Chapter VIII. Academic Press, New York, 1970; R. Loudon, "The Quantum Theory of Light." Oxford Univ. Press, London and New York, 1973.

2. THEORY OF RADIATION AND RADIATIVE TRANSITIONS

We expect that the field coordinates and conjugate momenta will again be related to the expansion coefficients of the potentials with respect to the discrete basis $\{L^{-3/2}e^{i\mathbf{k}\cdot\mathbf{r}}\}$, where the allowed propagation vectors are given by Eq. (2.2.125). Thus, we introduce the expansions of Eqs. (2.2.130) and (2.2.148). When we substitute these expansions into Eq. (2.2.21) we obtain

$$L^{-3/2}\left\{\sum_{\mathbf{k}',\lambda'}\hat{\mathbf{e}}_{\lambda'}(\mathbf{k}')((k')^2 a_{\mathbf{k}',\lambda'} + \mu_0\varepsilon_0 \ddot{a}_{\mathbf{k}',\lambda'})e^{i\mathbf{k}'\cdot\mathbf{r}} + \mu_0\varepsilon_0\sum_{\mathbf{k}'}\dot{a}'_{\mathbf{k}'}i\mathbf{k}'e^{i\mathbf{k}'\cdot\mathbf{r}}\right\}$$
$$= \mu_0\sum_s q_s\mathbf{v}_s\,\delta(\mathbf{r}-\mathbf{r}_s), \qquad (2.2.161)$$

where we have used Eq. (2.2.159). If we take the scalar product of this equation by $\hat{\mathbf{e}}_{\lambda}(\mathbf{k})L^{-3/2}e^{-i\mathbf{k}\cdot\mathbf{r}}$, integrate over the cube used to define the periodic boundary conditions, and use Eq. (2.2.126) and the fact that $\hat{\mathbf{e}}_{\lambda}(\mathbf{k})\cdot\mathbf{k} = 0$, we obtain

$$k^2 a_{\mathbf{k},\lambda} + \mu_0\varepsilon_0 \ddot{a}_{\mathbf{k},\lambda} + 0 = L^{-3/2}\mu_0\sum_s q_s e^{-i\mathbf{k}\cdot\mathbf{r}_s}\hat{\mathbf{e}}_{\lambda}(\mathbf{k})\cdot\mathbf{v}_s. \qquad (2.2.162)$$

Similarly, if we take the scalar product of Eq. (2.2.161) by $\hat{\mathbf{k}}L^{-3/2}e^{-i\mathbf{k}\cdot\mathbf{r}}$, and integrate over the cube used to define the periodic boundary conditions, we obtain

$$\varepsilon_0\dot{a}'_{\mathbf{k}}ik = L^{-3/2}\sum_s q_s e^{-i\mathbf{k}\cdot\mathbf{r}_s}\hat{\mathbf{k}}\cdot\mathbf{v}_s. \qquad (2.2.163)$$

We can also substitute Eq. (2.2.148) into Eq. (2.2.20), multiply by $L^{-3/2}e^{-i\mathbf{k}\cdot\mathbf{r}}$, and integrate. This yields

$$k^2 a'_{\mathbf{k}}(t) = (1/L^{3/2}\varepsilon_0)\sum_s q_s e^{-i\mathbf{k}\cdot\mathbf{r}_s(t)}. \qquad (2.2.164)$$

If we differentiate Eq. (2.2.164) with respect to time, we obtain

$$\dot{a}'_{\mathbf{k}} = (-i/L^{3/2}k\varepsilon_0)\sum_s q_s(\hat{\mathbf{k}}\cdot\mathbf{v}_s)e^{-i\mathbf{k}\cdot\mathbf{r}_s}, \qquad (2.2.165)$$

which is equivalent to Eq. (2.2.163). Thus, Eq. (2.2.163) is a consequence of Eq. (2.2.164), and the former need not be maintained as a separate equation. Moreover, Eq. (2.2.164) establishes $a_{\mathbf{k}}'(t)$ as a unique function of the particle position vectors $\mathbf{r}_s(t)$ at the same time. We will identify the coordinates that locate the particles as canonical variables. Since the $a_{\mathbf{k}}'(t)$ are completely determined by these coordinates, we will not have independent canonical variables associated with the expansion coefficients of the scalar potential.

We can easily deduce the differential equations satisfied by the canonical field coordinates of Eq. (2.2.152). Taking the sum of Eq. (2.2.162) and its complex conjugate, noting that $\mu_0\varepsilon_0 = 1/c^2$, and using Eq. (2.2.139), we obtain

$$k^2 c^2 Q^{(1)}_{\mathbf{k},\lambda} + \ddot{Q}^{(1)}_{\mathbf{k},\lambda} = (\varepsilon_0 L^3)^{-1/2}\sum_s q_s\hat{\mathbf{e}}_{\lambda}(\mathbf{k})\cdot\mathbf{v}_s\cos(\mathbf{k}\cdot\mathbf{r}_s). \qquad (2.2.166)$$

2.2. LIGHT

Similarly, taking the difference of Eq. (2.2.162) and its complex conjugate and using Eq. (2.2.139), we get

$$k^2 c^2 \ddot{Q}^{(2)}_{\mathbf{k},\lambda} + Q^{(2)}_{\mathbf{k},\lambda} = -(\varepsilon_0 L^3)^{-1/2} \sum_s q_s \hat{\mathbf{e}}_\lambda(\mathbf{k}) \cdot \mathbf{v}_s \sin(\mathbf{k} \cdot \mathbf{r}_s). \quad (2.2.167)$$

It is known from classical mechanics that the correct nonrelativistic equation of motion for a charged particle in a specified electromagnetic field, Eq. (2.2.8), is produced from the Hamiltonian[48]

$$H_s = [(\mathbf{p}_s - q_s \mathscr{A}(\mathbf{r}_s))^2/2m_s] + q_s \Phi(\mathbf{r}_s), \quad (2.2.168)$$

where $\mathscr{A}(\mathbf{r}_s)$ and $\Phi(\mathbf{r}_s)$ are the vector and scalar potentials at the location of the particle, and the generalized momentum conjugate to the position coordinates is

$$\mathbf{p}_s = m\mathbf{v}_s + q_s \mathscr{A}(\mathbf{r}_s). \quad (2.2.169)$$

We expect that the Hamiltonian for the combined system of particles and fields might be some combination of terms like Eq. (2.2.168) for the particles, and a term like Eq. (2.2.154) for the fields, with the energy contribution from the longitudinal part of the electric-field strength being represented by the term involving the scalar potential at the various particle locations. Thus, we construct a nonrelativistic Hamiltonian for the total system by combining a term like Eq. (2.2.154) with a sum of terms like Eq. (2.2.168) for each particle, where, in the latter terms, Eqs. (2.2.141), (2.2.152), and (2.2.160) are used to express the potentials. In using Eq. (2.2.160), we count each pair of particles only once, and we avoid the embarrassment of the infinite self-energy associated with a point charge by excluding the interaction of each particle with itself.[†] We thus have for our Hamiltonian

$$\begin{aligned} H = &\sum_{\mathbf{k},\lambda,\gamma} \tfrac{1}{2}\{(P^{(\gamma)}_{\mathbf{k},\lambda})^2 + k^2 c^2 (Q^{(\gamma)}_{\mathbf{k},\lambda})^2\} \\ &+ \sum_s (1/2m_s)[\mathbf{p}_s - q_s(\varepsilon_0 L^3)^{-1/2} \sum_{\mathbf{k},\lambda} \{Q^{(1)}_{\mathbf{k},\lambda} \cos(\mathbf{k} \cdot \mathbf{r}_s) - Q^{(2)}_{\mathbf{k},\lambda} \sin(\mathbf{k} \cdot \mathbf{r}_s)\}\hat{\mathbf{e}}_\lambda(\mathbf{k})]^2 \\ &+ \tfrac{1}{2} \sum_{s,j;s \neq j} [q_s q_j / 4\pi \varepsilon_0 |\mathbf{r}_s - \mathbf{r}_j|]. \end{aligned} \quad (2.2.170)$$

It is a straightforward exercise to show that application of the canonical equations to this Hamiltonian leads to the correct equations of motion of the field coordinates, Eqs. (2.2.166) and (2.2.167), and for the particles, Eq. (2.2.8).

---

[48] Slater,[47] Appendix 4; Goldstein,[46] p. 222.

[†] This does not totally eliminate problems with infinities associated with classical point charged particles, but we will not encounter difficulties for the applications we want to make. Further discussion can be found in the works by Slater and by Heitler.[47]

2.2.2. Quantum Picture

2.2.2.1. Electric and Magnetic Fields in Quantum Theory. A lesson we learn from the quantum theory of matter is that observable quantities such as momentum or energy are to be represented by Hermitian operators. These operators satisfy several conditions, the most important being equations of motion and commutation relations. When we come to quantize the electromagnetic field,[49] therefore, we first look for the observable quantities that characterize this field. We then seek the conditions that operators representing these quantities must satisfy. Finally, we characterize the vectors on which these operators act, the vectors that specify the state of the field.

Electric and magnetic field strengths are the fundamental observables of classical electromagnetism; they retain this status in quantum theory.[50] To see this, we consider the possibility of measuring a component of the electric field strength $\mathscr{E}$.

Imagine that in some region of space there is an electric field whose strength we wish to measure. To make this measurement, we bring a test body (whose mass and charge are m and q, respectively) into the region and measure its momentum $\mathbf{p}$, first at time t_1 and then at time t_2. The average field strength is then given by

$$\langle \mathscr{E} \rangle = [\mathbf{p}(t_2) - \mathbf{p}(t_1)]/(t_2 - t_1)q.$$

This $\langle \mathscr{E} \rangle$ is an average over the volume of the test body and the time interval $\delta t = t_2 - t_1$. However, the measurement of the change in momentum is inevitably imprecise by some amount $\Delta \mathbf{p}$, which can be made smaller only by accepting a larger uncertainty in the position of the test body, and so of the position at which the strength of the field is measured. Thus, there is an uncertainty in the field strength

$$\Delta \langle \mathscr{E} \rangle = \Delta \mathbf{p}/q \, \delta t. \tag{2.2.171}$$

We see that this uncertainty can be made arbitrarily small by using a large charge on the test body. The electric field strength is, therefore, measurable,

---

[49] P. A. M. Dirac, *Proc. Roy. Soc.* **A114**, 243 (1927). This reference is reprinted, together with other papers fundamental to the development of modern quantum field theory, in J. Schwinger (ed.), "Selected Papers on Quantum Electrodynamics." Dover, New York, 1958. See also E. Fermi, *Rev. Mod. Phys.* **4**, 87 (1932). Recent textbooks that offer an introduction to the subject of this section include Merzbacher[1] and K. Gottfried, "Quantum Mechanics," Benjamin, New York, 1966. The books by Heitler,[4] by F. Mandl, "Introduction to Quantum Field Theory," Wiley (Interscience), New York, 1960, and by J. D. Bjorken and S. D. Drell, "Relativistic Quantum Fields," McGraw-Hill, New York, 1965, are representative of the many books containing thorough discussions of field theory at a more advanced level.

[50] N. Bohr and L. Rosenfeld, *Kgl. Dan. Vidensk. Selsk. Mat. Fys. Medd.* **12**, No. 8 (1933); *Phys. Rev.* **78**, 749 (1950). See also W. Heitler[4] and G. Källén, "Quantum Electrodynamics." Springer-Verlag, Berlin and New York, 1972.

in principle, and is represented in quantum theory by a Hermitian operator. Strictly speaking, we have found that an *average* of $\mathscr{E}$ over the volume of the test body and the duration of the measurement is observable. We *assume* that nothing goes wrong when we extrapolate to very small volumes and short time, so that the field strength *at a point*, $\mathscr{E}(\mathbf{r}, t)$, is also represented by a Hermitian operator.[51] The magnetic field strength, $\mathscr{B}(\mathbf{r}, t)$, is also measurable with arbitrary precision; the measurement can, in principle, be made with a large current. We have now reached the starting point for the quantum theory of the electromagnetic field: the field strengths $\mathscr{E}(\mathbf{r}, t)$ and $\mathscr{B}(\mathbf{r}, t)$ are observable and represented by Hermitian operators. Next, we seek the conditions these operators must satisfy.

2.2.2.2. Quantum Conditions. The classical quantities $\mathscr{E}$ and $\mathscr{B}$ vary in time and space in accordance with Maxwell's equations. It is therefore convenient, though not essential, to let the quantum operators $\mathscr{E}$ and $\mathscr{B}$ also be time dependent and satisfy Maxwell's equations. This means that we do most of our work in the Heisenberg picture[52] in which the *vectors* that specify the state of the field are constant in time while the *operators* vary. We have concentrated on the operators that represent observables and have not yet considered how the state of the field can be specified. We will soon learn that the state of the field can be specified, for example, by the number of photons in each mode—momentum $\hbar\mathbf{k}$ and polarization λ—of the field. This illustrates a distinction between classical and quantum theory. In classical electromagnetism, $\mathscr{E}$ and $\mathscr{B}$ are observable *and* specify the state of the field, while in quantum theory, these roles may be separated. For the moment, we continue to concentrate on $\mathscr{E}$ and $\mathscr{B}$.

Maxwell's equations for these operators are, in free space,

$$\mathbf{\nabla} \cdot \mathscr{E} = 0, \quad \mathbf{\nabla} \cdot \mathscr{B} = 0, \quad c^2 \mathbf{\nabla} \times \mathscr{B} = \dot{\mathscr{E}}, \quad \mathbf{\nabla} \times \mathscr{E} = -\dot{\mathscr{B}}. \quad (2.2.172)$$

To compute the time derivatives, we use the Heisenberg equation of motion for any operator Q and Hamiltonian H:

$$\dot{Q} = [Q, H]/i\hbar. \quad (2.2.173)$$

The Hamiltonian for the field is taken over directly from the classical theory, Eq. (2.2.133), as

$$H = \tfrac{1}{2}\varepsilon_0 \int d\tau \, [\mathscr{E}^2 + c^2 \mathscr{B}^2]. \quad (2.2.174)$$

---

[51] R. F. Streater and A. S. Wightman, "PCT, Spin and Statistics, and All That." Benjamin, New York, 1964. Chapter 3 of this reference makes precise this notion by using operators smeared in time and space.

[52] The Heisenberg picture is discussed in most texts on quantum mechanics, e.g., those given in ref. 1.

Using this H we find, for example, that

$$\dot{\mathscr{E}}_1(\mathbf{r}, t) = \varepsilon_0/2i\hbar \int d\tau' \{\mathscr{E}(\mathbf{r}', t) \cdot [\mathscr{E}_1(\mathbf{r}, t), \mathscr{E}(\mathbf{r}', t)]$$
$$+ [\mathscr{E}_1(\mathbf{r}, t), \mathscr{E}(\mathbf{r}', t)] \cdot \mathscr{E}(\mathbf{r}', t) + c^2 \mathscr{B}(\mathbf{r}', t) \cdot [\mathscr{E}_1(\mathbf{r}, t), \mathscr{B}(\mathbf{r}', t)]$$
$$+ c^2 [\mathscr{E}_1(\mathbf{r}, t), \mathscr{B}(\mathbf{r}', t)] \cdot \mathscr{B}(\mathbf{r}', t)\}. \qquad (2.2.175)$$

From the appropriate one of Maxwell's equations, this is also

$$\dot{\mathscr{E}}_1(\mathbf{r}, t) = c^2(\partial \mathscr{B}_3(\mathbf{r}, t)/\partial x_2 - \partial \mathscr{B}_2(\mathbf{r}, t)/\partial x_3). \qquad (2.2.176)$$

The other field components are treated similarly.

A "solution" of these equations is a commutation rule for $\mathscr{E}$ and $\mathscr{B}$ that ensures that Maxwell's equations are satisfied, e.g., that Eqs. (2.2.175) and (2.2.176) are consistent. One such solution is

$$[\mathscr{E}_j(\mathbf{r}, t), \mathscr{E}_k(\mathbf{r}', t)] = 0, \qquad (2.2.177a)$$
$$[\mathscr{B}_j(\mathbf{r}, t), \mathscr{B}_k(\mathbf{r}', t)] = 0, \qquad (2.2.177b)$$
$$[\mathscr{E}_j(\mathbf{r}, t), \mathscr{B}_k(\mathbf{r}', t)] = (i\hbar/\varepsilon_0)\varepsilon_{jkl} \, \partial \, \delta(\mathbf{r} - \mathbf{r}')/\partial x_l. \qquad (2.2.177c)$$

These are the quantum conditions we seek. They show that the three components of $\mathscr{E}$ (or of $\mathscr{B}$) form a complete, commuting set of observables at every instant, but that $\mathscr{E}$ and $\mathscr{B}$ are not simultaneously measurable. These are *equal time* commutators and do not directly give a commutator such as

$$[\mathscr{E}_j(\mathbf{r}, t), \mathscr{E}_k(\mathbf{r}', t')],$$

where $t \neq t'$. Such commutators are of great importance for understanding the implications of relativity for quantum theory, but they are not needed for our purposes.[50, 51]

The fact that $\mathscr{E}$ and $\mathscr{B}$ do not commute implies uncertainty relations[50] for measurements of $\mathscr{E}$ and $\mathscr{B}$, just as $[x, p_x] = i\hbar$ implies the familiar uncertainty relation for position and momentum. To understand this distinctly nonclassical feature of quantum theory, we consider a very nonclassical result of our earlier analysis, namely, the need to use a *large* charge on a test body in order to measure $\mathscr{E}$ with precision. A test body used to measure a field strength in classical physics can have an arbitrarily small charge, and so it does not perturb the field it is monitoring. In contrast, the test body used in quantum theory *does* perturb the field it is being used to measure, and this perturbation is not exactly calculable because the position and momentum of the test body are not precisely known. This unknown perturbation is proportional to q. Thus, we have the following picture of the measurement of two noncommuting field strengths: Suppose there is a field given, and we measure $\mathscr{E}_x$ and then $\mathscr{B}_y$. The uncertainty in the measurement of $\langle\mathscr{E}_x\rangle$ is $\sim q^{-1}$. This changes $\mathscr{B}_y$ by an unknown perturbation $\sim q$. When we now measure $\langle\mathscr{B}_y\rangle$, our result differs from the value of $\mathscr{B}_y$ prior to the measurement of $\mathscr{E}_x$ by an uncertain amount $\sim q$. The product of these uncertainties is then $\sim (q^{-1})(q) \sim 1$, and is, in principle, independent of the properties of the test body.

The Hamiltonian used above is itself a Hermitian operator representing the energy in the field. Two other important operators are those that represent the linear momentum, **P**, and angular momentum, **J**. Classically, the linear momentum is the space integral of $\varepsilon_0 \mathscr{E} \times \mathscr{B}$. In quantum theory, the appropriate operator is the Hermitian form

$$\mathbf{P} = \tfrac{1}{2}\varepsilon_0 \int d\tau \, [\mathscr{E} \times \mathscr{B} - \mathscr{B} \times \mathscr{E}]; \qquad (2.2.178)$$

for example,

$$P_1 = \tfrac{1}{2}\varepsilon_0 \int d\tau \, [\mathscr{E}_2 \mathscr{B}_3 - \mathscr{E}_3 \mathscr{B}_2 - \mathscr{B}_2 \mathscr{E}_3 + \mathscr{B}_3 \mathscr{E}_2].$$

This operator is Hermitian, while $\mathscr{E} \times \mathscr{B}$ alone is not because $\mathscr{E}$ and $\mathscr{B}$ do not commute.

Angular momentum is represented by the operator

$$\mathbf{J} = \tfrac{1}{2}\varepsilon_0 \int d\tau \, \mathbf{r} \times [\mathscr{E} \times \mathscr{B} - \mathscr{B} \times \mathscr{E}], \qquad (2.2.179)$$

in direct analogy with the classical expression, Eq. (2.2.49). Notice that in this integral **r** is a *parameter*, not an operator like $\mathscr{E}$ or $\mathscr{B}$. This is a general characteristic of the quantum theory of fields as opposed to the quantum theory of particles. In field theory, the field variables themselves are operators, while in particle theory, the coordinates of the particles are operators.

2.2.2.3. Photons. The electric and magnetic fields themselves have occupied our attention up to this point. Classical electromagnetism teaches us that it is useful to introduce an auxiliary field, the vector potential $\mathscr{A}$, and to derive $\mathscr{E}$ and $\mathscr{B}$ from it. The vector potential is even more useful in quantum theory,[53] and we now introduce it into our discussion.

Maxwell's equations for free fields are satisfied if the following operator equations hold:

$$\mathscr{E} = -\dot{\mathscr{A}} \qquad \text{(definition of } \mathscr{E}\text{)}, \qquad (2.2.180a)$$

$$\mathscr{B} = \nabla \times \mathscr{A} \qquad \text{(definition of } \mathscr{B}\text{)}, \qquad (2.2.180b)$$

$$\nabla \cdot \mathscr{A} = 0 \qquad \text{(gauge condition)}, \qquad (2.2.180c)$$

$$\nabla^2 \mathscr{A} - \ddot{\mathscr{A}}/c^2 = 0 \qquad \text{(wave equation)}. \qquad (2.2.180d)$$

---

[53] See Feynman et al.,[20] Chapter 15. This discussion of the Bohm–Aharonov effect should be read together with the article by T. Boyer, *Amer. J. Phys.* **40**, 56 (1972).

Just as in classical theory, it is useful to expand $\mathscr{A}$ into its Fourier components with respect to a discrete basis. Thus, the quantum theory analog of Eq. (2.2.130) can be written as

$$\mathscr{A}(\mathbf{r}, t) = \sum_{\mathbf{k},\lambda} N_\mathbf{k}[a_{\mathbf{k},\lambda}(t)\hat{\mathbf{e}}_\lambda(\mathbf{k})e^{i\mathbf{k}\cdot\mathbf{r}} + a^\dagger_{\mathbf{k},\lambda}(t)\hat{\mathbf{e}}_\lambda^*(\mathbf{k})e^{-i\mathbf{k}\cdot\mathbf{r}}].$$

Here, $N_\mathbf{k}$ is a normalization constant that we will choose in a moment; $a_{\mathbf{k},\lambda}$ and $a^\dagger_{\mathbf{k},\lambda}$ are time dependent operators that replace the classical expansion coefficients $a_{\mathbf{k},\lambda}$ and $a^*_{\mathbf{k},\lambda}$ in Eq. (2.2.130); $\hat{\mathbf{e}}_\lambda(\mathbf{k})$ is a polarization basis vector, the same in both the classical and the quantum expansions. The polarization index λ takes on the values 1, 2 (linear polarization basis) or +, − (helicity basis). The summation is over all modes $(\mathbf{k}, \lambda)$ of the field. We will abbreviate $(\mathbf{k}, \lambda)$ by α, so that, for example,

$$a_{\mathbf{k},\lambda} \leftrightarrow a_\alpha, \quad \text{and} \quad \hat{\mathbf{e}}_\lambda(\mathbf{k}) \to \hat{\mathbf{e}}_\alpha.$$

In order that the gauge condition be satisfied, we require

$$\mathbf{k} \cdot \hat{\mathbf{e}}_\alpha = 0; \tag{2.2.181}$$

to satisfy the wave equation, we require

$$\dot{a}_\alpha(t) = -i\omega_\alpha a_\alpha(t), \tag{2.2.182}$$

where

$$\omega_\alpha = kc.$$

This last condition is an operator equation,

$$\hbar\omega_\alpha a_\alpha(t) = i\hbar\dot{a}_\alpha(t) = [a_\alpha(t), H],$$

which we will solve below.

Thus far, we have simply reexpressed the operators that represent observables of the field. The great utility of the decomposition of the field into its modes is the ease with which certain *states* of the field can be described. The operators $a_\alpha^\dagger$ will turn out to be operators that create photons of type α. To see this in detail, we must first recast the operators $\mathscr{E}$, $\mathscr{B}$, $\mathbf{P}$, and H in terms of the a_α and $a_\alpha^\dagger$. This task is straightforward but tedious. In doing the calculations, we recall that

$$\int d\tau\, e^{i(\mathbf{k}-\mathbf{k}')\cdot\mathbf{r}} = L^3 \delta_{\mathbf{k},\mathbf{k}'}, \quad \hat{\mathbf{e}}_\lambda(\mathbf{k}) \cdot \hat{\mathbf{e}}_{\lambda'}(\mathbf{k}) = \delta_{\lambda,\lambda'}, \tag{2.2.126}$$

and that we are abbreviating $(\mathbf{k}, \lambda)$ by α. Then H, for example, takes the form

$$H = \sum_\alpha N_\alpha^2 \omega_\alpha^2 \varepsilon_0 L^3 (a_\alpha a_\alpha^\dagger + a_\alpha^\dagger a_\alpha).$$

The coefficient of $(a_\alpha a_\alpha^\dagger + a_\alpha^\dagger a_\alpha)$ is conventionally chosen to be $\tfrac{1}{2}\hbar\omega_\alpha$. Thus, the normalization constant N_α is fixed: $N_\alpha = \sqrt{\hbar/2\varepsilon_0 \omega_\alpha L^3}$. The results for

2.2. LIGHT

the other operators are

$$\mathscr{A} = \sum_\alpha \sqrt{\hbar/2\varepsilon_0\omega_\alpha L^3}[a_\alpha(t)\hat{\mathbf{e}}_\alpha e^{i\mathbf{k}\cdot\mathbf{r}} + a_\alpha^\dagger(t)\hat{\mathbf{e}}_\alpha^* e^{-i\mathbf{k}\cdot\mathbf{r}}], \qquad (2.2.183\text{a})$$

$$\mathscr{E} = \sum_\alpha (i\omega_\alpha)\sqrt{\hbar/2\varepsilon_0\omega_\alpha L^3}[a_\alpha(t)\hat{\mathbf{e}}_\alpha e^{i\mathbf{k}\cdot\mathbf{r}} - a_\alpha^\dagger(t)\hat{\mathbf{e}}_\alpha^* e^{-i\mathbf{k}\cdot\mathbf{r}}], \qquad (2.2.183\text{b})$$

$$\mathscr{B} = \sum_\alpha i\sqrt{\hbar/2\varepsilon_0\omega_\alpha L^3}[a_\alpha(t)\mathbf{k}\times\hat{\mathbf{e}}_\alpha e^{i\mathbf{k}\cdot\mathbf{r}} - a_\alpha^\dagger(t)\mathbf{k}\times\hat{\mathbf{e}}_\alpha^* e^{-i\mathbf{k}\cdot\mathbf{r}}], \qquad (2.2.183\text{c})$$

$$\mathbf{P} = \sum_\alpha \hbar\mathbf{k}(a_\alpha^\dagger a_\alpha + a_\alpha a_\alpha^\dagger)/2, \qquad (2.2.183\text{d})$$

$$H = \sum_\alpha \hbar\omega_\alpha(a_\alpha^\dagger a_\alpha + a_\alpha a_\alpha^\dagger)/2. \qquad (2.2.183\text{e})$$

Using this form for H, we can solve the equation of motion for a_α as promised. We have

$$i\hbar\dot{a}_\alpha(t) = \hbar\omega_\alpha a_\alpha(t) = [a_\alpha(t), H] = \sum_\beta \tfrac{1}{2}\hbar\omega_\beta[a_\alpha(t), a_\beta a_\beta^\dagger + a_\beta^\dagger a_\beta].$$

One solution[54] of this equation is

$$[a_\alpha(t), a_\beta^\dagger(t)] = \delta_{\alpha\beta}, \qquad [a_\alpha(t), a_\beta(t)] = [a_\alpha^\dagger(t), a_\beta^\dagger(t)] = 0.$$

These commutation rules can be taken as the basic quantum condition instead of the commutation relations on $\mathscr{E}$ and $\mathscr{B}$. They imply the equation of motion for a_α, and so cause $\mathscr{A}$ to obey the wave equation, and thereby insure that Maxwell's equations hold.

The operators $a_\alpha(t)$ are explicitly time dependent, but according to Eq. (2.2.182) this time dependence is very simple:

$$a_\alpha(t) = a_\alpha(0)e^{-i\omega_\alpha t}. \qquad (2.2.184)$$

The operator $a_\alpha(0)$ will be called simply a_α from now on; it does not depend on time. The commutation rules are

$$[a_\alpha, a_\beta^\dagger] = \delta_{\alpha\beta}, \qquad (2.2.185\text{a})$$

$$[a_\alpha, a_\beta] = [a_\alpha^\dagger, a_\beta^\dagger] = 0. \qquad (2.2.185\text{b})$$

In terms of the operators a_α and $a_\alpha^\dagger$, H and $\mathbf{P}$ take on particularly simple forms:

$$H = \sum_\alpha \hbar\omega_\alpha a_\alpha^\dagger a_\alpha + \underbrace{\sum_\alpha \tfrac{1}{2}\hbar\omega_\alpha}_{\text{(zero-point energy)}}$$

---

[54] This is the standard solution. Comments on related solutions are made by S. Schweber, *Phys. Rev.* **78**, 613 (1950).

and

$$\mathbf{P} = \sum_\alpha \hbar \mathbf{k} a_\alpha^\dagger a_\alpha + \sum_\alpha \tfrac{1}{2}\hbar \mathbf{k}$$

(zero-point momentum).

The zero-point energy provides an additive (although infinite) constant to the energy of the free field; we will discard this constant in our further discussions. This is a rather cavalier treatment of an intriguing term. Interested readers will find a valuable set of references in a note by Boyer.[55]

The Hamiltonian we use is thus

$$H = \sum_\alpha \hbar \omega_\alpha a_\alpha^\dagger a_\alpha. \qquad (2.2.186)$$

The eigenstates of this Hamiltonian can be used to expand any state of the free electromagnetic field, and so we next investigate these eigenstates.[56] First, we note that the eigenvalues of H are necessarily nonnegative. If $H|\psi_E\rangle = E|\psi_E\rangle$, then

$$E = \sum_\alpha \hbar\omega_\alpha \langle \psi_E | a_\alpha^\dagger a_\alpha | \psi_E \rangle = \sum_\alpha \hbar\omega_\alpha \langle \phi_E | \phi_E \rangle,$$

where $|\phi_E\rangle = a_\alpha |\psi_E\rangle$. This is a sum of terms that are not negative, and so $E \geq 0$.

Next, we note that if $H|\psi_E\rangle = E|\psi_E\rangle$, then

$$H(a_\alpha |\psi_E\rangle) = a_\alpha H|\psi_E\rangle + [H, a_\alpha]|\psi_E\rangle$$
$$= E a_\alpha |\psi_E\rangle - \hbar\omega_\alpha a_\alpha |\psi_E\rangle = (E - \hbar\omega_\alpha)(a_\alpha |\psi_E\rangle),$$

where Eq. (2.2.186) has been used to evaluate $[H, a_\alpha]$. By repeatedly applying a_α to $|\psi_E\rangle$, a chain of eigenvalues E, $E - \hbar\omega_\alpha$, $E - 2\hbar\omega_\alpha$, ... is obtained. This chain must terminate, for we have seen that no eigenvalue can be negative. Thus, there is a last state in the chain, say $|\psi_{E0}\rangle$, for which $a_\alpha |\psi_{E0}\rangle = 0$. Then, $E = E0 + n\hbar\omega_\alpha$. The state $|0\rangle$ on which all the a_α give zero, $a_\alpha |0\rangle = 0$ for all α, is called the ground state or the "vacuum." Out of the vacuum state we will build the world of the free field. The vacuum has zero energy, zero momentum, and zero angular momentum:

$$H|0\rangle = 0, \qquad \mathbf{P}|0\rangle = 0, \quad \text{and} \quad \mathbf{J}|0\rangle = 0.$$

We now reverse the process that led to $|0\rangle$. Consider the state $|1_\alpha\rangle = a_\alpha^\dagger |0\rangle$. Using the commutation relations of Eq. (2.2.185), we find that

$$H|1_\alpha\rangle = \hbar\omega_\alpha |1_\alpha\rangle \quad \text{and} \quad \mathbf{P}|1_\alpha\rangle = \hbar\mathbf{k}|1_\alpha\rangle.$$

[55] T. Boyer, *Amer. J. Phys.* **42**, 518 (1974).
[56] This discussion is the standard one. See, e.g., Merzbacher[1] or Gottfried.[49]

This state is not an eigenstate of the angular-momentum operator; it cannot be because $[J_i, P_j] \neq 0$, in general.[†] The angular momentum properties of the state $|1_\alpha\rangle$ are most easily described using the helicity basis, because the helicity quantum numbers $\hbar\lambda_\alpha = \pm\hbar$ are the eigenvalues of the operator $\mathbf{J}\cdot\hat{\mathbf{P}}$, which *does* commute with $\mathbf{P}$. A simple form for this operator can be found starting from Eqs. (2.2.179) and (2.2.183). In the helicity basis,

$$\mathbf{J}\cdot\hat{\mathbf{P}} = \sum_\alpha \hbar\lambda_\alpha a_\alpha^\dagger a_\alpha.$$

Comparison of this equation and (2.2.183) shows at once that $\mathbf{J}\cdot\hat{\mathbf{P}}$ does indeed commute with $\mathbf{P}$. Thus, the state $|1_\alpha\rangle$ has energy $\hbar\omega_\alpha$, momentum $\hbar\mathbf{k}$, and helicity λ_α. It is said to contain one photon with these quantum numbers.

An example of a two-photon state is $|1_\alpha, 1_\beta\rangle = a_\alpha^\dagger a_\beta^\dagger |0\rangle$. The energy, momentum, and helicity for this state are $\hbar(\omega_\alpha + \omega_\beta)$, $\hbar(\mathbf{k}_\alpha + \mathbf{k}_\beta)$, and $\lambda_\alpha + \lambda_\beta$, respectively. Because $[a_\alpha^\dagger, a_\beta^\dagger] = 0$, the states $|1_\alpha, 1_\beta\rangle$ and $|1_\beta, 1_\alpha\rangle$ are identical; this two-photon state is *symmetric* under exchange of the photons. Thus, photons are *bosons*. Multiphoton states are obtained in the same way:

$$|1_\alpha, 1_\beta, 1_\gamma, \ldots\rangle = a_\alpha^\dagger a_\beta^\dagger a_\gamma^\dagger \cdots |0\rangle,$$

where $\alpha \neq \beta \neq \gamma \neq \cdots$. The quantum numbers are additive, just as in the two-photon case, and the states are symmetric under any exchange of photons.

Because photons are bosons, states can exist in which a single mode contains many photons. These states are common in laser operation and at low frequencies, e.g., radio waves. Such a state is proportional to $(a_\alpha^\dagger)^{n_\alpha}|0\rangle$. It has energy $n_\alpha\hbar\omega_\alpha$, momentum $n_\alpha\hbar\mathbf{k}_\alpha$, and helicity $n_\alpha\lambda_\alpha$. However, this state is not normalized as written. The normalized state is readily shown to be

$$|n_\alpha\rangle = [(a_\alpha^\dagger)^{n_\alpha}/\sqrt{n_\alpha!}]|0\rangle$$

by using the commutator

$$[a_\alpha, (a_\alpha^\dagger)^{n_\alpha}] = n_\alpha(a_\alpha^\dagger)^{n_\alpha-1}, \qquad (2.2.187)$$

which can be proved by induction. Thus, a general normalized state is

$$|n_\alpha, n_\beta, \ldots\rangle = [(a_\alpha^\dagger)^{n_\alpha}(a_\beta^\dagger)^{n_\beta} \quad /\sqrt{n_\alpha!n_\beta!\cdots}]|0\rangle. \qquad (2.2.188)$$

The description of the states of an electromagnetic field in terms of the vectors $|\{n_\alpha\}\rangle = |n_\alpha, n_\beta, \ldots\rangle$ is called the occupation number representation. The number n_α is the number of photons occupying the mode α in the

---

[57] See Gottfried,[49] Chapter 8.

[†] For some purposes, particularly in gamma-ray spectroscopy, it is convenient to choose $\mathbf{J}\cdot\mathbf{J} = J^2$ to be diagonal instead of $\mathbf{P}$. Such "spherical photons" are discussed by Gottfried.[57]

energy eigenstate $|\{n_\alpha\}\rangle$. The operator $a_\alpha^\dagger a_\alpha$ is useful in this connection, for it simply counts the number of photons in mode α. For example,

$$a_\beta^\dagger a_\beta |n_\alpha, n_\beta, n_\gamma, \ldots\rangle = n_\beta |n_\alpha, n_\beta, n_\gamma, \ldots\rangle. \quad (2.2.189a)$$

The actions of the operators a_α and $a_\alpha^\dagger$ on these states will be useful later. Their actions are

$$a_\beta |n_\alpha, n_\beta, n_\gamma, \ldots\rangle = \sqrt{n_\beta} |n_\alpha, n_\beta - 1, n_\gamma, \ldots\rangle \quad (2.2.189b)$$

and

$$a_\beta^\dagger |n_\alpha, n_\beta, n_\gamma, \ldots\rangle = \sqrt{n_\beta + 1} |n_\alpha, n_\beta + 1, n_\gamma, \ldots\rangle. \quad (2.2.189c)$$

These equations are easily derived from the definition of the normalized states in Eq. (2.2.188) and the commutator of Eq. (2.2.187). The operator a_β is seen to decrease the number of photons of type β by one, while $a_\beta^\dagger$ increases their number by one. Thus, a_β is called a destruction operator, or an annihilation operator, while $a_\beta^\dagger$ is called a creation operator.

We now can put all of the above together and write down the matrix elements of the vector potential between two states in the occupation number representation. We use the notation $|\{n_\alpha\} \pm 1_{\mathbf{k},\lambda}\rangle$ to indicate that the state $|\{n_\alpha\} + 1_{\mathbf{k},\lambda}\rangle(|\{n_\alpha\} - 1_{\mathbf{k},\lambda}\rangle)$ contains one photon more (less) in the $(\mathbf{k}, \lambda)$ mode than does $|\{n_\alpha\}\rangle$. The matrix element at $t = 0$ is

$$\langle\{n_\beta'\}|\mathscr{A}(\mathbf{r}, 0)|\{n_\alpha\}\rangle$$
$$= \sum_{\mathbf{k},\lambda} \sqrt{\hbar/2\varepsilon_0 \omega_k L^3} \{\sqrt{n_{\mathbf{k},\lambda}}\ \hat{\mathbf{e}}_\lambda(\mathbf{k})\ e^{i\mathbf{k}\cdot\mathbf{r}} \langle\{n_\beta'\}|\{n_\alpha\} - 1_{\mathbf{k},\lambda}\rangle$$
$$+ \sqrt{n_{\mathbf{k},\lambda} + 1}\ \hat{\mathbf{e}}_\lambda^*(\mathbf{k}) e^{-i\mathbf{k}\cdot\mathbf{r}} \langle\{n_\beta'\}|\{n_\alpha\} + 1_{\mathbf{k},\lambda}\rangle\}. \quad (2.2.190)$$

This is the matrix element that will be needed in Chapter 2.3. To obtain the matrix element for any other time t, make the replacements

$$\sqrt{n_{\mathbf{k},\lambda}} \to \sqrt{n_{\mathbf{k},\lambda}} e^{-i\omega_k t}, \quad \sqrt{n_{\mathbf{k},\lambda} + 1} \to \sqrt{n_{\mathbf{k},\lambda} + 1} e^{+i\omega_k t}.$$

The operators we have used in this chapter are explicitly time dependent because we have chosen to work in the Heisenberg picture. We did this to make the correspondence with classical electromagnetism clearer; the operators $\mathscr{E}$ and $\mathscr{B}$ obey Maxwell's equations. In Chapter 2.3, however, perturbation theory is developed in a standard manner utilizing the probably more familiar Schrödinger picture. The operators in the Schrödinger picture are exactly the ones used so far, but *evaluated always at $t = 0$*. Readers unfamiliar with the correspondence between the two pictures may profitably consult a number of texts.[1] Here we will just provide a short "dictionary" relating the two pictures (Table III).

2.2. LIGHT

TABLE III. Dictionary

Object	Schrödinger	Heisenberg	Relation
Vector potential	$\mathscr{A}_S(\mathbf{r})$	$\mathscr{A}_H(\mathbf{r}, t)$	$\mathscr{A}_S(\mathbf{r}) = \mathscr{A}_H(\mathbf{r}, 0)$
State vector	$|\psi_S(t)\rangle$	$|\psi_H\rangle$	$|\psi_S(0)\rangle = |\psi_H\rangle$
n-Photon states (basis vectors)	$|n\rangle = (n!)^{-1/2}(a^\dagger)^n|0\rangle$	$|n, t\rangle = (n!)^{-1/2}(a^\dagger(t))^n|0\rangle$	$|n\rangle = e^{-in\omega t}|n, t\rangle$ $= |n, 0\rangle$

The last line shown in Table III is sometimes confusing. The *basis vectors* in the Schrödinger picture are constructed from the time independent operators a_α. A state vector $|\psi_S(t)\rangle$ is time dependent because its coefficients are time dependent:

$$|\psi_S(t)\rangle = \sum_n c_n(t)|n\rangle.$$

The *basis vectors* in the Heisenberg picture are time dependent, constructed from the operators $a^\dagger(t) = e^{i\omega t}a^\dagger(0)$. The state vector $|\psi_H\rangle$ is therefore independent of time:

$$|\psi_H\rangle = \sum_n c_n(t)|n, t\rangle = \sum_n (c_n(0)e^{-in\omega t})(e^{+in\omega t}|n, 0\rangle) = \sum_n c_n(0)|n\rangle = |\psi_S(0)\rangle.$$

2.2.2.4. *Intensity and Polarization of a Light Beam.* The classical description of a light beam includes specification of intensity, polarization, and some measure of coherence. To carry this description into quantum theory, we need to reexpress these ideas in the photon language of the preceding section. For the notions of intensity and polarization, this reexpression is fairly straightforward. Discussion of coherence, however, is most easily done in terms of the "coherent-state representation" rather than the occupation number representation. We will defer discussion of coherence until the next section, in which the coherent states are introduced.

Intensity is the energy transmitted across a unit area per unit time. What is measured, though, is a time average of this quantity. This average is what we want to convert into a Hermitian operator.

We start with the classical Poynting vector,

$$\mathbf{S} = c^2\varepsilon_0 \mathscr{E} \times \mathscr{B},$$

and insert the quantum operators $\mathscr{E}$ and $\mathscr{B}$. We make the replacement

$$\mathscr{E} \times \mathscr{B} \to \tfrac{1}{2}(\mathscr{E} \times \mathscr{B} - \mathscr{B} \times \mathscr{E})$$

so that the expression is Hermitian (see the discussion after Eq. (2.2.178)). We then average this over time. This physically appropriate time averaging has an important consequence for the mathematics. The quantum form of **S** is

$$\mathbf{S} = \sum_{\alpha,\alpha'} (\hbar c \sqrt{\omega_\alpha \omega_{\alpha'}}/2L^3)\{-\hat{\mathbf{e}}_\alpha \times (\hat{\mathbf{k}}_{\alpha'} \times \hat{\mathbf{e}}_{\alpha'})a_\alpha a_{\alpha'} e^{i(\mathbf{k}_\alpha + \mathbf{k}_{\alpha'})\cdot\mathbf{r}} e^{i(\omega_\alpha + \omega_{\alpha'})t}$$

$$+ \hat{\mathbf{e}}_\alpha^* \times (\hat{\mathbf{k}}_{\alpha'} \times \hat{\mathbf{e}}_{\alpha'})a_\alpha^\dagger a_{\alpha'} e^{-i(\mathbf{k}_\alpha - \mathbf{k}_{\alpha'})\cdot\mathbf{r}} e^{i(\omega_\alpha - \omega_{\alpha'})t}$$

$$+ \text{Hermitian conjugate}\}. \tag{2.2.191}$$

Averaging over time eliminates the terms in $a_\alpha a_{\alpha'}$ and $a_\alpha^\dagger a_{\alpha'}^\dagger$ because ω_α and $\omega_{\alpha'}$ are both positive, and so the time average of $e^{+i(\omega_\alpha + \omega_{\alpha'})t}$ is zero. The averaging also forces the terms $a_\alpha^\dagger a_{\alpha'}$ in the sum to have the same frequency, $\omega_\alpha = \omega_{\alpha'}$. The result for the time-averaged operator is

$$\langle \mathbf{S} \rangle_{\text{time}} = \sum_{\alpha,\alpha'} (\hbar \omega_\alpha c/2L^3)[\hat{\mathbf{e}}_\alpha^* \times (\hat{\mathbf{k}}_{\alpha'} \times \hat{\mathbf{e}}_{\alpha'})a_\alpha^\dagger a_{\alpha'} e^{-i(\mathbf{k}_\alpha - \mathbf{k}_{\alpha'})\cdot\mathbf{r}}$$

$$+ \text{Hermitian conjugate}]\delta(\omega_\alpha, \omega_{\alpha'}).$$

We now compute the expected value of this operator for a state describing a collimated, polarized light beam.

The proper description of a real beam of light is difficult. It is dependent on a multitude of physical processes in the source. Fortunately, intensity and polarization measurements depend on only a relatively small number of parameters that characterize the beam, and so our task is possible. We compute the intensity for two cases, one very special and one quite general, obtaining the same form for both cases.

A rather special case is a beam described by a single vector $|\psi\rangle$, a "pure state." We expand $|\psi\rangle$ in the occupation number representation:

$$|\psi\rangle = \sum_\beta c(\{n_\beta\})|\{n_\beta\}\rangle. \tag{2.2.192}$$

Here, the occupied modes included in the sum all have $\mathbf{k}_\beta$ within a solid angle $\Delta\Omega_{\hat{n}_0}$ about a particular direction $\hat{\mathbf{n}}_0$, and they all have polarization λ. The probability that the set of occupation numbers is $\{n_\beta\}$ in the state $|\psi\rangle$ is $|c(\{n_\beta\})|^2$. The expected value of $\langle \mathbf{S} \rangle_{\text{time}}$ in this state $|\psi\rangle$ is called $\mathbf{I}_\psi$:

$$\mathbf{I}_\psi = \sum_\beta |c(\{n_\beta\})|^2 n_\beta (\hbar\omega_\beta c/L^3)\hat{\mathbf{n}}_0$$

If the beam is reasonably monochromatic, the occupied modes all have ω_β within a range $\Delta\omega$ about a particular frequency ω. This form for $\mathbf{I}_\psi$ can be turned into something more familiar by using the prescription, valid for large L^3,

$$L^{-3}\sum_\mathbf{k} \to (2\pi)^{-3}\int d^3k.$$

2.2. LIGHT

This expresses the fact that, for large L, the volume of k space associated with each discrete k value is $(2\pi/L)^3$. Using this, we obtain

$$\mathbf{I}_\psi = \int d^3k (2\pi)^{-3} L^3 |c(\{n_\beta\})|^2 n_{\mathbf{k},\lambda}(\hbar\omega_\mathbf{k} c/L^3)\hat{\mathbf{n}}_0$$

$$= \Delta\Omega_{\hat{n}_0} (L/2\pi)^3 k^2 \,\Delta k \, \bar{n}_\psi(\mathbf{k}, \lambda)(\hbar\omega c/L^3)\hat{\mathbf{n}}_0,$$

where $\bar{n}_\psi(\mathbf{k}, \lambda)$ is the mean number of photons in the mode $(\mathbf{k}, \lambda)$ in the state $|\psi\rangle$. In terms of frequency, $\omega = kc$, this is

$$\mathbf{I}_\psi = \underbrace{[(L/2\pi)^3(\omega^2/c^3)\,\Delta\Omega_{\hat{n}_0}\,\Delta\omega]}_{\text{number of occupied modes}} \underbrace{[(\hbar\omega c/L^3)\hat{\mathbf{n}}_0]}_{\substack{\text{intensity}\\\text{per photon}}} \underbrace{\bar{n}_\psi(\omega, \lambda)}_{\substack{\text{mean number}\\\text{of photons}}}. \quad (2.2.193)$$

Usually, a beam of light is not describable by a single pure state $|\psi\rangle$ as in Eq. (2.2.192). Instead, it is a mixture of many different such $|\psi\rangle$, each with some probability p_ψ. Then, for each one of the states $|\psi\rangle$, the preceding result gives an intensity $\mathbf{I}_\psi$, and so the intensity of the beam, $\mathbf{I}_{\text{beam}}$, is

$$\mathbf{I}_{\text{beam}} = \sum_\psi p_\psi \mathbf{I}_\psi. \quad (2.2.194)$$

But Eq. (2.2.193) only involves ψ in the term $\bar{n}_\psi(\omega, \lambda)$, and so we can define, for the whole beam, a mean number of photons in the $(\mathbf{k}, \lambda)$ mode:

$$\bar{n}(\omega, \lambda) = \sum_\psi p_\psi \bar{n}_\psi(\omega, \lambda).$$

Then,

$$\mathbf{I}_{\text{beam}} = [(L/2\pi)^3(\omega^2/c^3)\,\Delta\Omega_{\hat{n}_0}\,\Delta\omega][(\hbar\omega c/L^3)\hat{\mathbf{n}}_0]\bar{n}(\omega, \lambda) \quad (2.2.195\text{a})$$

$$\equiv \mathbf{I}(\omega)\,\Delta\omega\,\Delta\Omega_{\hat{n}_0}. \quad (2.2.195\text{b})$$

The significance of this result is simply that only the mean number of photons per mode is needed to specify the intensity. (We will see in Section 2.2.2.5 that the intensity we have just found is the simplest of a set of functions characterizing a beam. These functions, correlation functions, do require more information for their specification.)

The polarization state of the beam can be described by Stokes's parameters, which are obtained by several intensity measurements (see Section 2.2.1.6.), just as if the beam were specified by classical concepts. A simple example is worked out in 2.4.

2.2.2.5. *Coherence.* The preceding paragraphs have been concerned with quantizing classical electromagnetic theory. It is certainly possible to continue in this vein for a discussion of coherence, but it is interesting to invert the viewpoint, to wholeheartedly adopt the quantum view as the primary one, to ask questions about photons instead of field strengths. So, in this paragraph, we will not "measure intensity," but we will instead "count photons." We thus switch our image of detection apparatus from an alcohol soaked cloth plus prism (see Chapter 1.1) to a solid-state detector plus photomultiplier tube (see the discussions in other chapters of these volumes). For the rest of this section, we consider only one mode; the frequency is ω, the polarization vector is $\hat{\mathbf{e}}_\lambda(\mathbf{k}) = \hat{\mathbf{e}}_x$. The mode index is therefore dropped to simplify the notation.

The simplest model of a photon detector is probably an atom that can be excited from state α to state α' by a photon of the radiation field we wish to study. To discuss this properly, we need to anticipate a bit of the material in Chapter 2.3. The information we need is this: the probability of a transition from state $|i_{\text{field}}\rangle|\alpha_{\text{atom}}\rangle$ to the state $|f_{\text{field}}\rangle|\alpha'_{\text{atom}}\rangle$ is proportional to the square of the matrix element

$$\langle \alpha'|x|\alpha\rangle\langle f|\mathscr{E}(\mathbf{r}, t)|i\rangle.$$

This is the dipole approximation. Because a photon is absorbed, we deal with the part of $\mathscr{E}(\mathbf{r}, t)$ that contains the annihilation operator. This is called $\mathscr{E}^{(+)}(\mathbf{r}, t)$. The probability of absorbing (counting) a photon from state $|i\rangle$ is proportional to the sum of

$$|\langle f|\mathscr{E}^{(+)}(\mathbf{r}, t)|i\rangle|^2$$

over all states $|f\rangle$ that can be connected to $|i\rangle$ by $\mathscr{E}^{(+)}$. (Our imagined detector is perfect; its effects do not vary for the different final states $|f\rangle$.) But

$$\sum_f |\langle f|\mathscr{E}^{(+)}(\mathbf{r}, t)|i\rangle|^2 = \sum_f \langle i|\mathscr{E}^{(-)}(\mathbf{r}, t)|f\rangle\langle f|\mathscr{E}^{(+)}(\mathbf{r}, t)|i\rangle$$

$$= \langle i|\mathscr{E}^{(-)}(\mathbf{r}, t)\mathscr{E}^{(+)}(\mathbf{r}, t)|i\rangle,$$

where $\mathscr{E}^{(-)} = \mathscr{E}^{(+)\dagger}$. This is the case if the radiation field is described by the single vector $|i\rangle$. If, instead, the state is described by a mixture of many such vectors, each with probability p_i, then the counting rate is proportional to

$$\sum_i p_i\langle i|\mathscr{E}^{(-)}(\mathbf{r}, t)\mathscr{E}^{(+)}(\mathbf{r}, t)|i\rangle \equiv \langle G^{(1)}(\mathbf{r}, t)\rangle. \quad (2.2.196)$$

For example, if the radiation field is a mixture of the occupation number states, then

$$\langle G^{(1)}(\mathbf{r}, t)\rangle = (\hbar\omega/2L^3\varepsilon_0)\sum_n p_n n = \bar{n}(\hbar\omega/2L^3\varepsilon_0).$$

Comparison of this result with Eq. (2.2.195) for the intensity of a beam shows that $\langle G^{(1)}\rangle$ is proportional to the intensity. We begin to see the relation between photon counting and intensity measurements.

We can now ask more complicated questions, such as "What is the probability of detecting one photon at $\mathbf{r}_1, t_1$ *and* one photon at $\mathbf{r}_2, t_2$?" Repeating the argument leading to Eq. (2.2.196), we see that this is proportional to

$$\sum_{f,i} p_i |\langle f|\mathscr{E}^{(+)}(2)\mathscr{E}^{(+)}(1)|i\rangle|^2 = \sum_i p_i \langle i|\mathscr{E}^{(-)}(1)\mathscr{E}^{(-)}(2)\mathscr{E}^{(+)}(2)\mathscr{E}^{(+)}(1)|i\rangle$$

$$\equiv \langle G^{(2)}(1,2)\rangle,$$

where we have abbreviated $\mathscr{E}(\mathbf{r}_1, t_1)$ by $\mathscr{E}(1)$. The "coherence" of a state is specified by the values of the correlation functions $\langle G^{(n)}(1, 2, \ldots, n)\rangle$. For example, if the initial state is the one photon state $|1\rangle$, then

$$0 = \langle G^{(2)}\rangle = \langle G^{(3)}\rangle = \cdots.$$

Thus, a monochromatic, polarized beam containing only one photon is incoherent in the sense that the expected value of all but one of the correlation functions vanishes.

We can ask what the correlation functions would be in a classical coherent state. Such a state is, for example,

$$\mathscr{E}(\mathbf{r}, t) = \mathscr{E}e^{i\mathbf{k}\cdot\mathbf{r} - i\omega t} + \mathscr{E}^* e^{-i\mathbf{k}\cdot\mathbf{r} + i\omega t} = \mathscr{E}^{(+)} + \mathscr{E}^{(-)}.$$

The part called $\mathscr{E}^{(+)}$ has the same time dependence, $e^{-i\omega t}$, that the part of the quantum operator that contains the annihilation operator does. The correlation functions for this classical coherent state then factor in the form

$$\langle G^{(n)}\rangle = (\mathscr{E}^*)^n (\mathscr{E})^n. \tag{2.2.197}$$

During the early 1960's, Glauber addressed the question of whether there are quantum states whose correlation functions exhibit the same sort of behavior as those for classical coherent states. The answer is yes, and he constructed and used such quantum coherent states in several investigations, beginning in 1963.[58] We refer the reader to these papers for complete discussions of these states. Here we will simply introduce them.

The quantum states whose correlation functions factor as do the classical ones in Eq. (2.2.197) are eigenstates of the quantum operator $\mathscr{E}^{(+)}$, or, equivalently, of the annihilation operator a. We label by $|\eta\rangle$ such a state whose eigenvalue is the complex number η, and we expand it in the occupation number representation. Thus,

$$a|\eta\rangle = \eta|\eta\rangle \quad \text{and} \quad |\eta\rangle = \sum_{n=0}^{\infty} c_n |n\rangle.$$

Because the operator a is not Hermitian, η need not be real, and the states $|\eta\rangle$ and $|\rho\rangle$, with $\eta \neq \rho$, need not be orthogonal. Using Eq. (2.2.187), one

---

[58] R. J. Glauber, (a) *Phys. Rev.* **130**, 2529 (1963); (b) **131**, 2766, (1963). These references are reprinted together with a large number of other interesting papers on coherence and quantum optics, in L. Mandel and E. Wolf (eds.), "Selected Papers on Coherence and Fluctuations of Light." Dover, New York, 1970.

finds that the coefficients c_n are

$$c_n = c_0(n!)^{-1/2}\eta^n,$$

and normalization of $|\eta\rangle$ requires that

$$c_0 = \exp(-\tfrac{1}{2}|\eta|^2).$$

Thus,

$$|\eta\rangle = \exp(-\tfrac{1}{2}|\eta|^2) \sum_{n=0}^{\infty} (\eta^n/\sqrt{n!})|n\rangle. \tag{2.2.198}$$

For these states, we have

$$\langle G^{(n)} \rangle \sim (\eta^*)^n(\eta)^n,$$

the same factorization seen in the classical case in Eq. (2.2.197). Furthermore, the expected value of $\mathscr{E}$ in these states is

$$\langle \eta|\mathscr{E}|\eta\rangle = \sqrt{\hbar\omega/2L^3\varepsilon_0}(i\eta e^{i\mathbf{k}\cdot\mathbf{r}-i\omega t} - i\eta^* e^{-i\mathbf{k}\cdot\mathbf{r}+i\omega t}). \tag{2.2.199}$$

This expectation value looks just like a classical electric field with the complex amplitude $\mathscr{E} = i\eta\sqrt{\hbar\omega/2L^3\varepsilon_0}$. Thus, these states are well adapted for considering the classical limit where $\hbar \to 0$, but $\hbar\eta^2$ stays finite.

These quantum coherent states are not orthogonal. We have

$$\langle \eta|\rho\rangle = \sum_{n,m} \exp(-\tfrac{1}{2}|\eta|^2 - \tfrac{1}{2}|\rho|^2)(n!m!)^{-1/2}(\eta^*)^n(\rho)^m \langle n|m\rangle$$

$$= \exp(-\tfrac{1}{2}|\eta|^2 - \tfrac{1}{2}|\rho|^2) \sum_n (\eta^*\rho)^n/n!$$

$$= \exp(-\tfrac{1}{2}|\eta|^2 - \tfrac{1}{2}|\rho|^2 + \eta^*\rho), \tag{2.2.200a}$$

and so

$$|\langle \eta|\rho\rangle|^2 = \exp(-|\eta - \rho|^2). \tag{2.2.200b}$$

If η is very different from ρ, then $|\eta\rangle$ and $|\rho\rangle$ are nearly orthogonal.

Finally, the coherent states are useful because they *are* complete; in fact,[58]

$$\mathbf{1} = 1/\pi \int d^2\eta \, |\eta\rangle \langle \eta|, \tag{2.2.201a}$$

where

$$d^2\eta = d(\operatorname{Re}\eta)\, d(\operatorname{Im}\eta). \tag{2.2.201b}$$

Indeed, these states are *overcomplete*—each is expandable in terms of all the others. Using Eqs. (2.2.200) and (2.2.201), we have

$$|\eta\rangle = 1/\pi \int d^2\rho \exp(-\tfrac{1}{2}|\eta|^2 - \tfrac{1}{2}|\rho|^2 + \eta\rho^*)|\rho\rangle.$$

The coherent states form an alternative to the occupation number states as a set of basis vectors for the description of states of the electromagnetic field. They are exactly the fields radiated by a classical current distribution[59] and are frequently used in discussions of laser fields.[44, 45] In a quite different context, they appear in a very interesting discussion of a wide class of uncertainty relations.[60]

2.3. Interaction of Light and Matter

2.3.1. Time-Dependent Perturbation Theory.

2.3.1.1. Time-Dependent Basis States.
The time-dependent Schrödinger equation

$$H|\Psi(t)\rangle = i\hbar\, \partial/\partial t|\Psi(t)\rangle \tag{2.3.1}$$

permits, in principle, the calculation of the state of a system $|\Psi(t)\rangle$ at any time if the state is known at some instant, $|\Psi(t_0)\rangle$, and the explicit time dependence of the Hamiltonian is known as well as its functional dependence on the coordinates and conjugate momenta for the system.[61] This equation also holds if the Hamiltonian has no explicit time dependence. If, for example,

$$H_0|u_n\rangle = E_n|u_n\rangle \tag{2.3.2}$$

is the Schrödinger equation for such a time-independent Hamiltonian, with eigenfunctions $|u_n\rangle$ and corresponding energy eigenvalues E_n, it is easy to see that the set of functions $e^{-iE_n t/\hbar}|u_n\rangle$ are solutions to Eq. (2.3.1):

$$H_0 e^{-iE_n t/\hbar}|u_n\rangle = i\hbar\,(\partial e^{-iE_n t/\hbar}/\partial t)|u_n\rangle. \tag{2.3.3}$$

These state functions, $e^{-iE_n t/\hbar}|u_n\rangle$, are obviously time-dependent descriptions of stationary states since the spatial description and eigenvalues do not change in time. The set of functions $e^{-iE_n t/\hbar}|u_n\rangle$ is complete and can be made orthonormal, so that

$$\langle u_k|e^{iE_k t/\hbar}e^{-iE_n t/\hbar}|u_n\rangle = \langle u_k|u_n\rangle e^{i(E_k-E_n)t/\hbar} = \delta_{k,n}. \tag{2.3.4}$$

Thus, they provide a convenient basis set of functions. One way of looking at the time dependence of a general state of the system is to expand the function $|\Psi(t)\rangle$ as a linear combination of the set of functions $e^{-iE_n t/\hbar}|u_n\rangle$, the expansion coefficients being dependent on time:

$$|\Psi(t)\rangle = \sum_n c_n(t) e^{-iE_n t/\hbar}|u_n\rangle. \tag{2.3.5}$$

The summation in Eq. (2.3.5) is understood to be a summation over the

---

[59] Merzbacher,[1] Chapter 22; see also Glauber.[38b]
[60] P. Carruthers and M. M. Nieto, *Rev. Mod. Phys.* **40**, 411 (1968).
[61] Merzbacher,[1] p. 450; Schiff,[1] p. 280.

eigenfunctions with discrete eigenvalues and integration over the eigenfunctions having continuous eigenvalues. The condition that $|\Psi(t)\rangle$ be normalized then leads to a requirement on the $c_n(t)$'s that

$$\sum_n c_n^*(t)c_n(t) = \sum_n |c_n(t)|^2 = 1. \tag{2.3.6}$$

Equation (2.3.6) permits the interpretation of $|c_n(t)|^2$ as the probability that at time t the system is in the state $|u_n\rangle$, and thus the changing state of the system in time can be followed by having a knowledge of the expansion coefficients $c_n(t)$.

2.3.1.2. *Transition Probabilities.* Let us write the time-dependent Hamiltonian H as a sum of a time-independent part H_0 with known eigenfunctions and eigenvalues as given by Eq. (2.3.2), and a part H' which may depend on time, so that Eq. (2.3.1) can be written as

$$\{H_0 + H'\}|\Psi(t)\rangle = i\hbar \, \partial/\partial t |\Psi(t)\rangle. \tag{2.3.7}$$

This is a form of the Hamiltonian which is useful in many situations of interest, in particular, the interaction of particle systems with electromagnetic radiation.

Substituting the expansion in terms of eigenfunctions of H_0 given in Eq. (2.3.5) for $|\Psi(t)\rangle$, we can write

$$\{H_0 + H'\} \sum_n c_n(t) e^{-iE_n t/\hbar} |u_n\rangle = i\hbar \, \partial \left(\sum_n c_n(t) e^{-iE_n t/\hbar} \right) \Big/ \partial t |u_n\rangle. \tag{2.3.8}$$

Operating on this equation from the left with $\langle u_k| e^{iE_k t/\hbar}$ we get

$$c_k(t)E_k + \sum_n c_n(t)\langle u_k|H'|u_n\rangle e^{i(E_k - E_n)t/\hbar} = i\hbar \, \partial c_k(t)/\partial t + c_k(t)E_k. \tag{2.3.9}$$

Define $\omega_{kn} = (E_k - E_n)/\hbar$ and rearrange Eq. (2.3.9) to obtain the equation of motion for the expansion coefficients:

$$i\hbar \, \partial c_k(t)/\partial t = \sum_n c_n(t)\langle u_k|H'|u_n\rangle e^{i\omega_{kn}t}. \tag{2.3.10}$$

These equations are a system of coupled first-order linear differential equations; the order of this system is equal to the number of functions in the basis set. This system of equations is difficult to solve in general; however, approximate solutions can be carried out for several cases which are important in the interpretation of physical phenomena and measurements.

Let us assume that the system starts in a definite one of the states of H_0, $|u_a\rangle$; that is $|c_a(0)| = 1$, and all other expansion coefficients are zero: $c_b(0) = 0$ for $b \neq a$. If it turns out that the rate of change of the c_b's is so

2.3. INTERACTION OF LIGHT AND MATTER

small that only $c_a(t)$ is appreciably different from zero during the time interval of interest, and that $|c_a(t)| \cong 1$ during this time interval, then Eqs. (2.3.10) are uncoupled and the resulting equations,

$$i\hbar\, dc_b(t)/dt = \langle u_b|H'|u_a\rangle e^{i\omega_{ba}t}, \qquad (2.3.11)$$

can be integrated directly to give

$$c_b(t) = (i\hbar)^{-1} \int_0^t \langle u_b|H'|u_a\rangle e^{i\omega_{ba}t'}\, dt' \qquad (2.3.12)$$

as an approximate solution, valid as long as all the $c_b(t)$'s remain very small.[†] Under these circumstances, the probability that the system, having started in state $|u_a\rangle$ at time 0, is in state $|u_b\rangle$ at time t is $P_{ba}(t) = |c_b(t)|^2$. We say that $P_{ba}(t)$ is the probability that the system has made a transition from state $|u_a\rangle$ to state $|u_b\rangle$ in time t.

The integration of Eq. (2.3.12) depends on the form of the interaction Hamiltonian matrix elements. The system that we will consider will be a material system of charged particles in a radiation field. Initially, we will assume that the particles are spinless. The Hamiltonian operator for a free radiation field was given in Eq. (2.2.186) as

$$H_R = \sum_{\alpha=\{\mathbf{k},\lambda\}} a_\alpha^\dagger a_\alpha \hbar\omega_\alpha, \qquad (2.3.13)$$

the sum being over the modes of the normalization volume and two polarizations for each propagation vector. The classical Hamiltonian for a system of charged, spinless particles interacting with a radiation field was given in Eq. (2.2.170). We have

$$H = H_R + \sum_j (2m_j)^{-1}(\mathbf{p}_j - q_j\mathscr{A}(\mathbf{r}_j))^2 + \sum_{j\neq i}\sum_i \tfrac{1}{2}(q_iq_j/4\pi\varepsilon_0 r_{ij}). \qquad (2.3.14)$$

---

[62] Heitler,[4] pp. 163–174.
[63] Merzbacher,[1] p. 481.
[64] Breen[37]; H. Griem, "Plasma Spectroscopy," p. 31. McGraw-Hill, New York, 1964.

[†] Heitler[62] has discussed a somewhat more general perturbation theory. He uses this technique in treatments of resonance fluorescence and line breadth. A simple approach to line breadth is to assume that $c_a(t)$ decreases exponentially with time,[63] so that the integral in Eq. (2.3.12) has an additional factor $e^{-\gamma t}$. If the particle system interacts only with the radiation field, the damping constant for $c_a(t)$ can be found in terms of the transition matrix elements from state $|u_a\rangle$ to all other states, and is called the natural damping constant. If there are other time-dependent terms due to interactions between particle systems, then the damping constant is larger. The additional width of the transition is called collisional or statistical broadening, depending on the nature of the interaction.[64]

Here, the electromagnetic field is in the Coulomb gauge. Equation (2.3.14) may be expanded to read

$$H = H_R + \left\{ \sum_j p_j^2/2m_j + \sum_{j \ne i} \sum_i \tfrac{1}{2}(q_i q_j/4\pi\varepsilon_0 r_{ij}) \right\}$$
$$- \sum_j (q_j/2m_j)(\mathbf{p}_j \cdot \mathscr{A}(\mathbf{r}_j) + \mathscr{A}(\mathbf{r}_j) \cdot \mathbf{p}_j) + \sum_j (q_j/2m_j)\mathscr{A}(\mathbf{r}_j) \cdot \mathscr{A}(\mathbf{r}_j). \tag{2.3.15}$$

The last term in Eq. (2.3.15) is of interest since its quadratic dependence on the vector potential describes processes that involve two photons, two-photon absorption and emission, and the scattering of a photon from a material system. (Two-photon processes also arise from the $\mathbf{p} \cdot \mathscr{A}$ term in second order.) The normal diamagnetism of an atomic system can also be calculated from this term. Treatments of photon scattering are outlined by Heitler[65] and Sakurai,[66] while calculation of two-photon absorption and emission has been reviewed by Peticolas[67] and the calculation of the diamagnetism of an atomic system is given by Park.[68] Since the $\mathscr{A} \cdot \mathscr{A}$ term is not significant for the single-photon processes we want to describe, it will be omitted in the following discussion. Since the vector potential is in the Coulomb gauge, the term involving $\mathbf{p}_j \cdot \mathscr{A}(\mathbf{r}_j)$ in Eq. (2.3.15) can be written as $\mathscr{A}(\mathbf{r}_j) \cdot \mathbf{p}_j$. This is because the quantum-mechanical operator for $\mathbf{p}_j$ is $-i\hbar\nabla_j$, and in the Coulomb gauge $\nabla_j \cdot \mathscr{A}(\mathbf{r}_j)$ is zero, and so the momentum operator will commute with $\mathscr{A}(\mathbf{r}_j)$; that is, $\nabla_j \cdot \{\mathscr{A}(\mathbf{r}_j) f(\mathbf{r}_j)\} = \mathscr{A}(\mathbf{r}_j) \cdot \nabla_j f(\mathbf{r}_j)$. Thus, we can write the total Hamiltonian for the radiation field plus material system as

$$H = H_R + H_M + H_I,$$

where

$$H_M = \sum_j \mathbf{p}_j^2/2m_j + \sum_{j \ne i} \sum_i \tfrac{1}{2}(q_i q_j/4\pi\varepsilon_0 r_{ij}), \tag{2.3.16}$$
$$H_I = -\sum_j (q_j/m_j)\mathscr{A}(\mathbf{r}_j) \cdot \mathbf{p}_j,$$

and H_R is given in Eq. (2.3.13). This Hamiltonian is for spinless charged particles. It is important to consider the modifications of the interaction

---

[65] Heitler,[4] p. 190.

[66] J. J. Sakurai, "Advanced Quantum Mechanics," p. 47. Addison-Wesley, Reading, Massachusetts, 1967.

[67] W. L. Peticolas, *Annu. Rev. Phys. Chem.* **18**, 233 (1967); W. L. Peticolas, R. Norris, and K. E. Rieckhoff, *J. Chem. Phys.* **42**, 4164 (1965).

[68] D. Park, "Introduction to the Quantum Theory," 2nd ed., pp. 251, 280. McGraw-Hill, New York, 1974; C. Kittel, "Introduction to Solid State Physics," p. 749. Wiley, New York, 1971.

2.3. INTERACTION OF LIGHT AND MATTER

Hamiltonian that occur when the particles have intrinsic magnetic moments. The considerations involved are not of extreme difficulty, but they do require more space than we have available to be developed properly. The reader is urged to consult, for example, the excellent discussion by Slater.[69] We will consider the nature of the spin interaction terms later in order to investigate their influence on the transition rates.

We will use $H_R + H_M$ as the unperturbed Hamiltonian for the system, and use the time-dependent Schrödinger-picture eigenstates of this Hamiltonian as a basis in terms of which to expand the time-dependent state of the system. These Schrödinger states will be direct products of the form

$$e^{-i\Sigma_\alpha n_\alpha \omega_\alpha t}|\{n_\alpha\}\rangle e^{-iE_a't/\hbar}|u_a'\rangle = e^{-i(\Sigma_\alpha n_\alpha \omega_\alpha + (E_a'/\hbar))t}|\{n_\alpha\}, u_a'\rangle, \quad (2.3.17)$$

where $|u_a'\rangle$ satisfies $H_M|u_a'\rangle = E_a'|u_a'\rangle$, and $|\{n_\alpha\}\rangle$ satisfies Eq. (2.2.189).

The form of the interaction Hamiltonian H_I can be obtained from Eq. (2.2.183) for the vector potential, where a_α and $a_\alpha^\dagger$ are, respectively, annihilation and creation operators for the electromagnetic field, and $\mathbf{p}_j$ is the usual quantum-mechanical momentum operator. Thus, the matrix element of the interaction term appearing in Eq. (2.3.12) is

$$\langle u_b|H_I|u_a\rangle e^{i\omega_{ba}t}$$
$$= \langle u_b', \{n_\alpha(b)\}|e^{i(\Sigma_\alpha n_\alpha(b)\omega_\alpha + (E_b'/\hbar))t}$$
$$\times (\sum_j (q_j/m_j)\mathscr{A}(\mathbf{r}_j)\cdot\mathbf{p}_j)e^{-i(\Sigma_\alpha n_\alpha(a)\omega_\alpha + (E_a'/\hbar))t}|u_a', \{n_\alpha(a)\}\rangle, \quad (2.3.18)$$

with

$$\mathscr{A}(\mathbf{r}_j) = \sum_\alpha \sqrt{\hbar/L^3 2\omega_\alpha \varepsilon_0}\{a_\alpha \hat{\mathbf{e}}_\alpha e^{i\mathbf{k}_\alpha\cdot\mathbf{r}_j} + a_\alpha^\dagger \hat{\mathbf{e}}_\alpha^* e^{-i\mathbf{k}_\alpha\cdot\mathbf{r}_j}\}. \quad (2.3.19)$$

We can write

$$\mathscr{A}(\mathbf{r}_j)\cdot\mathbf{p}_j = \sum_\alpha \sqrt{\hbar/L^3 2\omega_\alpha \varepsilon_0}\{a_\alpha \hat{\mathbf{e}}_\alpha\cdot\mathbf{p}_j e^{i\mathbf{k}_\alpha\cdot\mathbf{r}_j} + a_\alpha^\dagger \hat{\mathbf{e}}_\alpha^*\cdot\mathbf{p}_j e^{-i\mathbf{k}_\alpha\cdot\mathbf{r}_j}\}, \quad (2.3.20)$$

and so the interaction matrix element can be written as

$$\langle u_b|H_I|u_a\rangle e^{i\omega_{ba}t}$$
$$= \sum_\alpha \sqrt{\hbar/L^3 2\omega_\alpha \varepsilon_0}\sum_j (q_j/m_j)\{\langle\{n_\alpha(b)\}|a_\alpha|\{n_\alpha(a)\}\rangle\langle u_b'|\hat{\mathbf{e}}_\alpha\cdot\mathbf{p}_j e^{i\mathbf{k}_\alpha\cdot\mathbf{r}_j}|u_a'\rangle$$
$$+ \langle\{n_\alpha(b)\}|a_\alpha^\dagger|\{n_\alpha(a)\}\rangle\langle u_b'|\hat{\mathbf{e}}_\alpha^*\cdot\mathbf{p}_j e^{-i\mathbf{k}_\alpha\cdot\mathbf{r}_j}|u_a'\rangle\}$$
$$\times \exp\{i[((E_b' - E_a')/\hbar) + \sum_\alpha (n_\alpha(b) - n_\alpha(a))\omega_\alpha]t\}. \quad (2.3.21)$$

---
[69] Slater,[47] pp. 161–200.

The two terms in this matrix element cannot be effective simultaneously. By properties of the annihilation and creation operators given in Eq. (2.2.189), if the term involving a_α is to be nonzero, then the final state must differ from the initial state only in having one less α photon for one particular α; if the term involving $a_\alpha^\dagger$ is to be nonzero, the final field state must differ from the initial state only in having one more α photon for some particular α. Either one or the other of these conditions may hold, but not both. Therefore, we can simplify Eq. (2.3.21) by considering emission and absorption separately. We have then for emission

$$\langle u_b|H_1|u_a\rangle e^{i\omega_{ba}t} = \sqrt{\hbar(n_\alpha + 1)/L^3 2\omega_\alpha \varepsilon_0} \sum_j (q_j/m_j)$$

$$\times \langle u_b'|\hat{\mathbf{e}}_\alpha^* \cdot \mathbf{p}_j e^{-i\mathbf{k}_\alpha \cdot \mathbf{r}_j}|u_a'\rangle e^{i([(E_{b'}-E_{a'})/\hbar]+\omega_\alpha)t},$$
(2.3.22a)

and for absorption

$$\langle u_b|H_1|u_a\rangle e^{i\omega_{ba}t} = \sqrt{\hbar n_\alpha/L^3 2\omega_\alpha \varepsilon_0} \sum_j (q_j/m_j)$$

$$\times \langle u_b'|\hat{\mathbf{e}}_\alpha \cdot \mathbf{p}_j e^{i\mathbf{k}_\alpha \cdot \mathbf{r}_j}|u_a'\rangle e^{i([(E_{b'}-E_{a'})/\hbar]-\omega_\alpha)t}.$$
(2.3.22b)

At this point we make an approximation that leads us to the results for the most important interaction between a system of charged particles and an electromagnetic field, the electric-dipole approximation. We have, $|\mathbf{k}_\alpha| = 2\pi/\lambda_\alpha$, so $|\mathbf{k}_\alpha \cdot \mathbf{r}_j| = (2\pi r_j/\lambda_\alpha) \cos(\mathbf{k}_\alpha, \mathbf{r}_j)$. For radiation of much longer wavelength than the dimensions of the particle system, $r/\lambda \ll 1$,[†] we can expand the exponential functions in the form

$$e^{i\mathbf{k}_\alpha \cdot \mathbf{r}_j} = 1 + i\mathbf{k}_\alpha \cdot \mathbf{r}_j - \tfrac{1}{2}(\mathbf{k}_\alpha \cdot \mathbf{r}_j)^2 + \cdots.$$

For now, we will keep only the constant term; we will later remark on the influence of the other terms in this expansion. With this approximation, we can write

$$\langle u_b'|\hat{\mathbf{e}}_\alpha \cdot \mathbf{p}_j e^{i\mathbf{k}_\alpha \cdot \mathbf{r}_j}|u_a'\rangle \cong \langle u_b'|\hat{\mathbf{e}}_\alpha \cdot \mathbf{p}_j|u_a'\rangle. \quad (2.3.23)$$

We now use a quantum-mechanical result relating the particle position and momentum operators and a Hamiltonian operator of the form of H_M[70] [Eq. (2.3.16)]:

$$\mathbf{p}_j = (im_j/\hbar)(H_M \mathbf{r}_j - \mathbf{r}_j H_M). \quad (2.3.24)$$

[70] Merzbacher,[1] p. 465.

[†] For example, if the particle system has size $r = 3$ Å, the approximation $e^{i\mathbf{k}\cdot\mathbf{r}} \cong 1$ is good to better than 1% for radiation of any wavelength longer than $\lambda = 300$ Å.

2.3. INTERACTION OF LIGHT AND MATTER

Using Eq. (2.3.24), we can write Eq. (2.3.23) as

$$\langle u_b'|\hat{\mathbf{e}}_\alpha \cdot \mathbf{p}_j|u_a'\rangle = im_j((E_b' - E_a')/\hbar)\langle u_b'|\hat{\mathbf{e}}_\alpha \cdot \mathbf{r}_j|u_a'\rangle. \quad (2.3.25)$$

When we substitute this into Eqs. (2.3.22), we obtain for emission

$$\langle u_b|H_1|u_a\rangle e^{i\omega_{ba}t} = i\sqrt{\hbar(n_\alpha + 1)/L^3 2\omega_\alpha \varepsilon_0}((E_b' - E_a')/\hbar)$$

$$\times \left\langle u_b'\middle|\hat{\mathbf{e}}_\alpha^* \cdot \sum_j q_j \mathbf{r}_j\middle|u_a'\right\rangle e^{i([(E_b' - E_a')/\hbar] + \omega_\alpha)t}$$

(2.3.26a)

and for absorption

$$\langle u_b|H_1|u_a\rangle e^{i\omega_{ba}t} = i\sqrt{\hbar n_\alpha/L^3 2\omega_\alpha \varepsilon_0}((E_b' - E_a')/\hbar)$$

$$\times \left\langle u_b'\middle|\hat{\mathbf{e}}_\alpha \cdot \sum_j q_j \mathbf{r}_j\middle|u_a'\right\rangle e^{i([(E_b' - E_a')/\hbar] - \omega_\alpha)t}.$$

(2.3.26b)

Equations (2.3.26a) and (2.3.26b) give the matrix elements that must be integrated in Eq. (2.3.12) to find $c_b(t)$, from which the transition probability can be calculated. We define $\mathbf{Q}_1 = \sum_j q_j \mathbf{r}_j$, the electric dipole moment operator for the particle system. For emission we have

$$c_b(t) = (i\hbar)^{-1} \int_0^t i\sqrt{\hbar(n_\alpha + 1)/L^3 2\omega_\alpha \varepsilon_0}((E_b' - E_a')/\hbar)$$

$$\times \langle u_b'|\hat{\mathbf{e}}_\alpha^* \cdot \mathbf{Q}_1|u_a'\rangle e^{i([(E_b' - E_a')/\hbar] + \omega_\alpha)t'} dt'$$

$$= \sqrt{(n_\alpha + 1)/\hbar L^3 2\omega_\alpha \varepsilon_0}((E_b' - E_a')/\hbar)$$

$$\times \langle u_b'|\hat{\mathbf{e}}_\alpha^* \cdot \mathbf{Q}_1|u_a'\rangle \int_0^t e^{i([(E_b' - E_a')/\hbar] + \omega_\alpha)t'} dt'. \quad (2.3.27)$$

Now,

$$\int_0^t e^{i([(E_b' - E_a')/\hbar] + \omega_\alpha)t'} dt'$$

$$= [-i/([(E_b' - E_a')/\hbar] + \omega_\alpha)]\{e^{i([(E_b' - E_a')/\hbar] + \omega_\alpha)t} - 1\},$$

(2.3.28)

and so

$$c_b(t) = -i\sqrt{(n_\alpha + 1)/\hbar L^3 2\omega_\alpha \varepsilon_0}((E_b' - E_a')/\hbar)\langle u_b'|\hat{\mathbf{e}}_\alpha^* \cdot \mathbf{Q}_1|u_a'\rangle$$

$$\times [(\exp[i([(E_b' - E_a')/\hbar] + \omega_\alpha)t] - 1)/([(E_b' - E_a')/\hbar] + \omega_\alpha)]$$

(2.3.29)

From Eq. (2.3.29), we can write the transition probability for emission as

$$P_{ba}(t) = |c_b(t)|^2 = [(n_\alpha + 1)/\hbar L^3 2\omega_\alpha \varepsilon_0]((E_b' - E_a')/\hbar)^2 |\langle u_b'|\hat{\mathbf{e}}_\eta^* \cdot \mathbf{Q}_1|u_a'\rangle|^2$$

$$\times [4\sin^2 \tfrac{1}{2}([(E_b' - E_a')/\hbar] + \omega_\alpha)t/([(E_b' - E_a')/\hbar] + \omega_\alpha)^2].$$

(2.3.30)

A similar treatment for the case of absorption gives

$$P_{ba}(t) = (n_\alpha/\hbar L^3 2\omega_\alpha \varepsilon_0)((E_b' - E_a')/\hbar)^2 |\langle u_b'|\hat{\mathbf{e}}_\alpha \cdot \mathbf{Q}_1|u_a'\rangle|^2$$
$$\times [4\sin^2 \tfrac{1}{2}([(E_b' - E_a')/\hbar] - \omega_\alpha)t / ([(E_b' - E_a')/\hbar] - \omega_\alpha)^2]. \quad (2.3.31)$$

We can separate the transition probability for emission into two parts; one is called induced emission and depends on n_α and thus on the radiation field intensity, while the other is called spontaneous emission and is independent of the radiation field intensity. Equations (2.3.30) and (2.3.31) show that the probability of induced emission from a to b is the same as the probability of absorption from b to a ($E_a' > E_b'$ in both cases). This equality was found by Einstein using thermodynamic arguments prior to the development of quantum mechanics.

The computation of a transition rate from the transition probability of Eq. (2.3.30) or (2.3.31) requires finding the expected number of transitions in some time interval for the transition of interest. This is done by multiplying the transition probability $P_{ba}((E_b' - E_a')/\hbar, \omega_\alpha, t)$ by the density for the initial matter state and the density for the final matter state, integrating over the initial and final matter state energies, and summing over the states of the radiation field. This gives an expression of the form

$$\int_{E_b'} \int_{E_a'} \sum_\alpha P_{ba}(\omega'(b,a), \omega_\alpha, t) \rho(E_b') \rho(E_a') \, dE_a' \, dE_b',$$

where we define $\omega'(b,a) = (E_b' - E_a')/\hbar$. The probability that state b has energy between E_b' and $E_b' + dE_b'$ is $\rho(E_b') \, dE_b'$. Because $E_b' = E_a' + \hbar\omega'(b,a)$, by a change of variables in the integrations over E_b' and E_a' we can write the above expression as

$$\int_{\omega'(b,a)} \sum_\alpha P_{ba}(\omega'(b,a), \omega_\alpha, t) \left\{ \int_{E_a'} \hbar \rho(E_a' + \hbar\omega'(b,a)) \rho(E_a') \, dE_a' \right\} d\omega'(b,a).$$

We call the term in braces $F(\omega'(b,a))$, the line-shape function for the transition. Note that $\int_{-\infty}^{\infty} F(\omega) \, d\omega = 1$. We can replace the sum over radiation-field modes by an integral over radiation frequencies and directions by using the prescription $\sum_\alpha \to \int \rho(\omega_\alpha) \, d\omega_\alpha \, d\Omega$, where $\rho(\omega_\alpha)$ is the density of field states as a function of frequency. We can then write the expression for the computation of the expected number of transitions by time t as

$$\int_{\omega'(b,a)} \int_{\omega_\alpha} \int_{\Delta\Omega} P_{ba}(\omega'(b,a), \omega_\alpha, t) F(\omega'(b,a)) \rho(\omega_\alpha) \, d\Omega \, d\omega_\alpha \, d\omega'(b,a).$$

This integral is called $w_{b,a}t$. We shall see below that in many cases, $w_{b,a}$ is time-independent. It is then called the transition rate from state a to state b.

2.3. INTERACTION OF LIGHT AND MATTER

The quantities that determine the transition rate are, therefore, the transition probability, $P_{ba}(\omega'(b, a), \omega_\alpha, t)$, the line-shape function, $F(\omega'(b, a))$, and the state density of the radiation field, $\rho(\omega_\alpha)$. The form of the transition probability is given by Eqs. (2.3.30) and (2.3.31). We see that, for long times, P_{ba} is a function sharply peaked at $\omega_\alpha = \pm \omega'(b, a)$. (See, for example, the diagram in Schiff,[1] p. 284.)

The line-shape function is generally sharply peaked for some value of $\omega'(b, a)$. The specific form of this function is determined by interactions of the material system with its surroundings and by higher-order interactions with the radiation field. The effects of these interactions are called collisional damping and radiation damping, respectively. These terms are discussed at length by Heitler[62] and Breen.[64] The density of radiation states may be a sharply peaked function of ω_α, as in a laser, or a slowly varying function of ω_α, as in a thermal source.

There are several cases of particular interest in the evaluation of transition rates:

(a) Absorption, induced emission, and spontaneous emission involving discrete states for the matter and a continuous distribution of frequencies for the radiation field. An example of this case is absorption of light from a nonmonochromatic beam by a dilute gas. For this case, the line-shape function is essentially a delta function, and integration over $\omega'(b, a)$ replaces the variable $\omega'(b, a)$ by a definite value $\omega(b, a)$. The radiation-field state density is a slowly varying function of ω_α compared with the function $P_{ba}(\omega(b, a), \omega_\alpha, t)$, which is sharply peaked for large t. Thus, the radiation-field state density in the integrand may be approximated by its value at $\omega_\alpha = \omega(b, a)$ and removed from the integral. The remaining integral can be evaluated directly.

(b) Absorption and induced emission involving discrete states for the radiation field, and a broadened line-shape function for the matter. An example of this case is the interaction of a laser beam with a dense gas. In this case, the density of radiation-field states is sharply peaked, and so integration over ω_α replaces the variable ω_α by a definite value $\tilde{\omega}_\alpha$ wherever it appears in the integrand. Since the line-shape factor is a slowly varying function of $\omega'(b, a)$ compared with the function $P_{ba}(\omega'(b, a), \tilde{\omega}_\alpha, t)$ for large t, it can be approximated by its value at $\omega'(b, a) = \tilde{\omega}_\alpha$ and removed from the integral. The remaining integral can be evaluated directly.

(c) Absorption, induced emission, and spontaneous emission involving a broadened line-shape function and a continuous distribution of frequencies for the radiation field. An example of this case is the absorption of infrared radiation by molecules in a dense gas. The transition probability $P_{ba}(\omega'(b, a), \omega_\alpha, t)$ is a sharply peaked function for large t; thus, the integration over ω_α replaces the radiation-field state density by its value at

$\omega_\alpha = \omega'(b, a)$. The integration over $\omega'(b, a)$ is then a convolution of the line-shape factor with the radiation-field state density, both of which must be known in order to evaluate the resulting integral.

Cases in which the transition probability P_{ba} is not sharply peaked imply excitation or deexcitation over short time periods. Under these circumstances, the transition rate is a function of time; this is not at present a case of interest in spectroscopy. For analysis of such a case, see Wu et al.[71]

Let us first consider, as an example of case (a), the transition probability for spontaneous emission by a particle system with discrete, well-defined states. The spontaneous emission part of Eq. (2.3.30) is

$$P_{ba}(t) = (\hbar L^3 2\omega_\alpha \varepsilon_0)^{-1} \omega'(b, a)^2 |\langle u_b'|\hat{\mathbf{e}}_\alpha^* \cdot \mathbf{Q}_1|u_a'\rangle|^2$$
$$\times [4 \sin^2 \tfrac{1}{2}(\omega'(b, a) + \omega_\alpha)t/(\omega'(b, a) + \omega_\alpha)^2]. \quad (2.3.32)$$

The line-shape factor is essentially a delta function, and integration over $\omega'(b, a)$ simply replaces $\omega'(b, a)$ by the definite value $\omega(b, a)$. The emitted photon may be in any mode of the radiation field, and so we must sum Eq. (2.3.32) over all values of α to obtain the total transition probability. As we have noted above, we can replace the sum over modes by an appropriate integral over frequencies. The number of modes associated with an angular frequency interval $d\omega$ with propagation direction in the solid angle interval $d\Omega_\mathbf{k}$, and having a definite polarization or helicity, is given by [see Eq. (2.2.193)]

$$\rho(\omega, \lambda) \, d\omega \, d\Omega_\mathbf{k} = (L^3 \omega^2/(2\pi)^3 c^3) \, d\omega \, d\Omega_\mathbf{k}. \quad (2.3.33)$$

Using Eq. (2.3.33) in Eq. (2.3.32), we can write, for the probability of emission by time t,

$$w_{b,a} t = \int_{\Delta\Omega} d\Omega_\mathbf{k} \int_0^\infty (L^3 \omega^2/(2\pi)^3 c^3)(\hbar L^3 2\omega\varepsilon_0)^{-1} \omega(b, a)^2 |\langle u_b'|\hat{\mathbf{e}}_\alpha^* \cdot \mathbf{Q}_1|u_a'\rangle|^2$$
$$\times [4 \sin^2 \tfrac{1}{2}(\omega(b, a) + \omega)t/(\omega(b, a) + \omega)^2] \, d\omega$$
$$= [(2\pi)^2 4\pi\varepsilon_0 c^3 \hbar]^{-1} \int_{\Delta\Omega} d\Omega_\mathbf{k} |\langle u_b'|\hat{\mathbf{e}}_\alpha^* \cdot \mathbf{Q}_1|u_a'\rangle|^2$$
$$\times \int_0^\infty \omega \, \omega(b, a)^2 [4 \sin^2 \tfrac{1}{2}(\omega(b, a) + \omega)t/(\omega(b, a) + \omega)^2] \, d\omega. \quad (2.3.34)$$

The integral over frequency can be written as

$$\int_{-\omega(a,b)t/2}^\infty \omega \, \omega(a, b)^2 [(\sin^2 \zeta)/\zeta^2] 2t \, d\zeta,$$

[71] C. S. Wu, Y. K. Lee, N. Benczer-Koller, and P. Simons, *Phys. Rev. Lett.* **5**, 432–435 (1960).

2.3. INTERACTION OF LIGHT AND MATTER

where

$$\zeta = (-\omega(a, b) + \omega)t/2 \quad \text{and} \quad \omega(a, b) = ((E_a - E_b)/\hbar) > 0.$$

The function $(\sin^2 \zeta)/\zeta^2$ is peaked at $\zeta = 0$ such that, as time goes on, the only contributions to the integral come from a small region around $\omega = \omega(a, b)$. We can, therefore, replace ω by $\omega(a, b)$ and remove it from the integral, obtaining

$$2t\omega(a, b)^3 \int_{-\omega(a,b)t/2}^{\infty} [(\sin^2 \zeta)/\zeta^2]\, d\zeta.$$

In this integral, the lower limit can be extended to $-\infty$ with practically no loss of accuracy since almost all the contributions to the integral come from a small range around $\zeta = 0$. The integral $\int_{-\infty}^{\infty} (\sin^2 \zeta)/\zeta^2\, d\zeta$ has the value π, so we can write

$$\int_0^{\infty} \omega\, \omega(b, a)^2 [4 \sin^2 \tfrac{1}{2}(\omega(b, a) + \omega)t/(\omega(b, a) + \omega)^2]\, d\omega = 2\pi t\omega(a, b)^3. \tag{2.3.35}$$

In many applications of interest, there are many similar particle systems oriented randomly with respect to one another. In this case, we can write, for some particular orientation of the particle system,

$$|\langle u_b'|\hat{\mathbf{e}}_\alpha^* \cdot \mathbf{Q}_1|u_a'\rangle|^2 = \cos^2 \eta |\langle u_b'|\mathbf{Q}_1|u_a'\rangle|^2,$$

where η is the angle between $\hat{\mathbf{e}}_\alpha^*$ and $\mathbf{Q}_1$, and

$$|\langle u_b'|\mathbf{Q}_1|u_a'\rangle|^2 \equiv |\langle u_b'|Q_{1x}|u_a'\rangle|^2 + |\langle u_b'|Q_{1y}|u_a'\rangle|^2 + |\langle u_b'|Q_{1z}|u_a'\rangle|^2.$$

For a given $\hat{\mathbf{e}}_\alpha^*$-direction, an average over the random orientations of the particle systems gives $\langle \cos^2 \eta \rangle_{\text{av}} = \tfrac{1}{3}$, and so we can replace $|\langle u_b'|\hat{\mathbf{e}}_\alpha^* \cdot \mathbf{Q}_1|u_a'\rangle|^2$ by $\tfrac{1}{3}|\langle u_b'|\mathbf{Q}_1|u_a'\rangle|^2$ when dealing with a large, randomly oriented collection of particle systems. To compute the total decay rate, we must include radiation into all directions. Thus, the integral over $d\Omega_k$ is

$$\int_{4\pi} |\langle u_b'|\hat{\mathbf{e}}_\alpha^* \cdot \mathbf{Q}_1|u_a'\rangle|^2\, d\Omega_k = (4\pi/3)|\langle u_b'|\mathbf{Q}_1|u_a'\rangle|^2.$$

We can write, for the probability of emission from a randomly oriented particle system,

$$w_{b,a} t = [4\omega(a, b)^3/3\hbar 4\pi\varepsilon_0 c^3]|\langle u_b'|\mathbf{Q}_1|u_a'\rangle|^2 t. \tag{2.3.36}$$

The transition probability per unit time in this case is

$$w_{b,a} = [4\omega(a, b)^3/3\hbar 4\pi\varepsilon_0 c^3]|\langle u_b'|\mathbf{Q}_1|u_a'\rangle|^2. \tag{2.3.37}$$

Let us next consider, as a second example of case (a), the absorption of radiation by a particle system with discrete states from a radiation beam with an intensity $I(\omega)$, where $I(\omega)$ is the energy flux $[(J/m^2)/\text{sec}]$ per unit angular frequency interval of photons with a given polarization and propagating within $d\Omega_k$.

The sum over all values of α is replaced in the now familiar way by an integration over angular frequency weighted by the density of field states in the beam. This gives

$$w_{b,a}t = \int_0^\infty (L^3\omega_\alpha^2/(2\pi)^3c^3)(\bar{n}_\alpha/\hbar L^3 2\omega_\alpha \varepsilon_0)\omega(b,a)^2$$
$$\times [4\sin^2 \tfrac{1}{2}(\omega(b,a) - \omega_\alpha)t/(\omega(b,a) - \omega_\alpha)^2]\,d\omega_\alpha$$
$$\times \int_{\Delta\Omega} d\Omega_k\,|\langle u_b'|\hat{\mathbf{e}}_\alpha \cdot \mathbf{Q}_1|u_a'\rangle|^2. \qquad (2.3.38)$$

We use Eq. (2.2.195) to obtain

$$\bar{n}_\alpha = (2\pi)^3 c^2 I(\omega_\alpha)/\hbar\omega_\alpha^3; \qquad (2.3.39)$$

Eq. (2.3.38) then becomes

$$w_{b,a}t = (2\pi/\hbar^2 c 4\pi\varepsilon_0) \int_{\Delta\Omega} d\Omega_k\,|\langle u_b'|\hat{\mathbf{e}}_\alpha \cdot \mathbf{Q}_1|u_a'\rangle|^2$$
$$\times \int_0^\infty (I(\omega_\alpha)/\omega_\alpha^2)\omega(b,a)^2[4\sin^2 \tfrac{1}{2}(\omega(b,a) - \omega_\alpha)t/(\omega(b,a) - \omega_\alpha)^2]\,d\omega_\alpha$$
$$= [4\pi^2 t I(\omega(b,a))/\hbar^2 c 4\pi\varepsilon_0] \int_{\Delta\Omega} d\Omega_k\,|\langle u_b'|\hat{\mathbf{e}}_\alpha \cdot \mathbf{Q}_1|u_a'\rangle|^2, \qquad (2.3.40)$$

where we have used the same arguments that led to Eq. (2.3.35). This last integral, if averaged over random orientations of the molecules, can be replaced by

$$\int_{\Delta\Omega} d\Omega_k\,|\langle u_b'|\hat{\mathbf{e}}_\alpha \cdot \mathbf{Q}_1|u_a'\rangle|^2 = \tfrac{1}{3}|\langle u_b'|\mathbf{Q}_1|u_a'\rangle|^2\,\Delta\Omega,$$

so that the transition probability per unit time for absorption by a randomly oriented particle system is

$$w_{b,a} = [4\pi^2 I(\omega(b,a))\,\Delta\Omega/3\hbar^2 c 4\pi\varepsilon_0]|\langle u_b'|\mathbf{Q}_1|u_a'\rangle|^2, \qquad (2.3.41)$$

with

$$\omega(b,a) = (E_b - E_a)/\hbar > 0.$$

The transition probability per unit time for induced emission can be calculated in the same way as that for absorption. The result is

$$w_{b,a} = [4\pi^2 I(\omega(a,b))\,\Delta\Omega/3\hbar^2 c 4\pi\varepsilon_0]|\langle u_b'|\mathbf{Q}_1|u_a'\rangle|^2, \qquad (2.3.42)$$

where

$$\omega(a,b) = (E_a - E_b)/\hbar > 0.$$

We can compare the relative importance of induced and spontaneous emission. Considering emission of a particular polarization into a solid

2.3. INTERACTION OF LIGHT AND MATTER

angle $\Delta\Omega$, we have

$$\frac{w_{b,a}(\text{induced emission into }\Delta\Omega\text{ steradians})}{w_{b,a}(\text{spontaneous emission into }\Delta\Omega\text{ steradians; one polarization})}$$

$$= \frac{4\pi^2 I(\omega(b,a))\,\Delta\Omega\,|\langle u_b'|\mathbf{Q}_1|u_a'\rangle|^2/3\hbar^2 4\pi\varepsilon_0 c}{(\Delta\Omega/4\pi)\frac{1}{2}4\omega^3(b,a)|\langle u_b'|\mathbf{Q}_1|u_a'\rangle|^2/3\hbar 4\pi\varepsilon_0 c^3}$$

$$= \frac{2(4\pi^3)I(\omega(b,a))c^2}{\hbar\omega(b,a)^3}. \tag{2.3.43}$$

From Eq. (2.3.39) we can write

$$\frac{w_{b,a}(\text{induced emission})}{w_{b,a}(\text{spontaneous emission})} = \bar{n}_\alpha,$$

where $\bar{n}_\alpha$ is the average number of α photons in the beam. For a radiation field in thermal equilibrium, $\bar{n}_\alpha = (e^{\hbar\omega_\alpha/kT} - 1)^{-1}$ (Planck's law), where $\hbar\omega_\alpha$ is the photon energy.[72] For this situation, induced emission is important only for those wavelengths or temperatures such that $\hbar\omega_\alpha/kT \ll 1$. Thus, at ordinary temperatures, induced emission is important in the microwave region of the spectrum.

In discussing the absorption of radiation, we frequently want to compute an absorption coefficient. We can obtain this in a straightforward manner from Eqs. (2.3.41) and (2.3.42). For a transition between states $|u_a'\rangle$ and $|u_b'\rangle$, the net transition probability is the difference between absorption and induced emission. Spontaneous emission occurs randomly into 4π sr and thus does not provide an important contribution to the intensity of a narrowly directed beam. We can determine the change in intensity of a beam of intensity $I(\omega)$ per unit frequency passing through a thickness Δx of absorber as follows.

The loss of intensity per unit frequency interval, $\Delta I(\omega_\alpha)\,\Delta\Omega$, equals the number of absorbers per unit area, $N_a\,\Delta x$, where N_a is the number of absorbers per unit volume, times the transition rate per absorber per unit frequency interval, times the energy of the absorbed photon, $\hbar\omega_\alpha$. The transition rate per absorber for a transition from state a to state b has been shown to be

$$w_{b,a} = \int_{\omega_\alpha} [t^{-1} \int_{\omega'(b,a)} \int_{\Delta\Omega} P_{ba}(\omega'(b,a),\omega_\alpha,t) F(\omega'(b,a))\rho(\omega_\alpha)\,d\omega'(b,a)\,d\Omega]\,d\omega_\alpha. \tag{2.3.44}$$

We note that we previously carried out the integration over $\omega'(b,a)$, treating the line-shape function $\Gamma(\omega'(b,a))$ as a delta function centered on frequency $\omega(b,a)$. Now, however, we allow F to have a more general form and refrain

---

[72] Slater,[47] pp. 133–135.

from integrating over $\omega'(b, a)$ at this point. The expression in the square brackets in Eq. (2.3.44) is the transition rate per absorber per unit frequency interval of the radiation, $w_{b,a}(\omega_\alpha)$. We can evaluate $w_{b,a}(\omega_\alpha)$ by now carrying out the integration over $\omega'(b, a)$. This requires P_{ba} from Eq. (2.3.31), $\rho(\omega_\alpha)$ and $I(\omega_\alpha)$ from Eqs. (2.3.33) and (2.3.39), and use of the fact that $(\sin^2 \zeta)/\zeta^2$ is peaked at $\zeta = 0$, as discussed just before Eq. (2.3.35). This gives

$$w_{b,a}(\omega_\alpha) = t^{-1} \int_{\omega'(b,a)} \int_{\Delta\Omega} (\bar{n}_\alpha/2\hbar L^3 \varepsilon_0 \omega_\alpha) \omega'(b,a)^2 |\langle u_b'|\hat{\mathbf{e}}_\alpha \cdot \mathbf{Q}_1|u_a'\rangle|^2$$
$$\times [4 \sin^2 \tfrac{1}{2}(\omega'(b,a) - \omega_\alpha)t/(\omega'(b,a) - \omega_\alpha)^2]$$
$$\times F(\omega'(b,a))\rho(\omega_\alpha) \, d\Omega \, d\omega'(b,a)$$
$$= I(\omega_\alpha) 4\pi^2/4\pi\varepsilon_0 \hbar c \int_{\Delta\Omega} |\langle u_b'|\hat{\mathbf{e}}_\alpha \cdot \mathbf{Q}_1|u_a'\rangle|^2 \, d\Omega \, F(\omega_\alpha)/h. \quad (2.3.45)$$

Note that $w_{b,a}(\omega_\alpha)$ is a rate per unit frequency, and so is dimensionless. The change in intensity per unit frequency in passing through the matter because of absorption is

$$\Delta I(\omega_\alpha) \, \Delta\Omega = -N_a \, \Delta x \, w_{b,a}(\omega_\alpha) \hbar \omega_\alpha.$$

Likewise, the change in intensity per unit frequency in passing through the matter because of induced emission is

$$\Delta I(\omega_\alpha) \, \Delta\Omega = +N_b \, \Delta x \, w_{a,b}(\omega_\alpha) \hbar \omega_\alpha.$$

The discussion after Eq. (2.3.31) shows that $w_{a,b}(\omega_\alpha) = w_{b,a}(\omega_\alpha)$. This can also be seen directly from Eq. (2.3.45). The net change in intensity per unit frequency is then

$$\Delta I(\omega_\alpha) \, \Delta\Omega = -(N_a - N_b) \, \Delta x \, w_{b,a}(\omega_\alpha) \hbar \omega_\alpha$$
$$= -(N_a - N_b) I(\omega_\alpha)(4\pi^2/4\pi\varepsilon_0 \hbar c)$$
$$\times \int_{\Delta\Omega} |\langle u_b'|\hat{\mathbf{e}}_\alpha \cdot \mathbf{Q}_1|u_a'\rangle|^2 \, d\Omega \, F(\omega_\alpha) \omega_\alpha \, \Delta x$$
$$\equiv -k_{ba}(\omega_\alpha) I(\omega_\alpha) \, \Delta\Omega \, \Delta x.$$

Thus,

$$I(\omega_\alpha) = I_0(\omega_\alpha) e^{-k_{ba}(\omega_\alpha)x}.$$

The absorption coefficient $k_{ba}(\omega_\alpha)$ has the dimensions of (length)$^{-1}$ and can be measured in an absorption experiment with an instrument of high resolution. (The effect of the instrument on the observed frequency dependence of $k(\omega)$ is discussed, for example, by Houghton and Smith.[73]) The integrated

---

[73] J. T. Houghton and S. D. Smith, "Infrared Physics," pp. 51–52. Oxford Univ. Press, London and New York, 1966.

2.3. INTERACTION OF LIGHT AND MATTER

absorption coefficient, or line strength, is

$$k_{ba} = \int_{\text{line}} k_{ba}(\omega)\,d\omega.$$

Because $F(\omega)$ is sharply peaked about $\omega = \omega(b, a)$, and its integral is normalized to unity, we have

$$k_{ba} = (N_a - N_b)(4\pi^2\omega(b, a)/4\pi\varepsilon_0\hbar c)(1/\Delta\Omega)\int_{\Delta\Omega}|\langle u_b'|\hat{\mathbf{e}}_\alpha \cdot \mathbf{Q}_1|u_a'\rangle|^2\,d\Omega.$$

Thus,

$$k_{ba}(\omega_\alpha) = k_{ba}[\omega_\alpha/\omega(b, a)]F(\omega_\alpha).$$

Using the same method to average k_{ba} over the random orientation of the absorbers as was used to obtain Eq. (2.3.36), we obtain, finally,

$$k_{ba} = (N_a - N_b)[4\pi^2\omega(b, a)/3(4\pi\varepsilon_0\hbar c)]|\langle u_b'|\mathbf{Q}_1|u_a'\rangle|^2. \quad (2.3.46)$$

The contribution of N_b to k_{ba} can be of importance in cases in which the upper state has an appreciable population, and is dominant in cases of population inversion, as in experiments involving optical pumping. The contribution of N_b is usually negligible if the energy levels are primarily thermally populated and the upper level is more than a few hundred wavenumbers above state a. In many cases, state a is the ground state and the levels are thermally populated. Then, N_a is approximately N, the number of particle systems per unit volume, and so $k_{ba}(\omega)$ in fact depends only on frequency and is written simply $k(\omega)$.

Another approach to the absorption coefficient has been used in studies of relaxation phenomena.[74] This approach is especially useful in situations involving random motions of the particle systems under consideration; for example, the orientations of molecules in a gas. The result[75] of this analysis is that the absorption coefficient $k(\omega)$ can be written

$$k(\omega) = [4\pi N/6(4\pi\varepsilon_0\hbar c)]\int_{-\infty}^{\infty} e^{-i\omega t}\langle \mathbf{Q}_1(0) \cdot \mathbf{Q}_1(t)\rangle_{\text{av}}\,dt, \quad (2.3.47)$$

where $\langle \mathbf{Q}_1(0) \cdot \mathbf{Q}_1(t)\rangle_{\text{av}}$ is the autocorrelation function for the dipole moment, averaged over the states available to the particle system.

2.3.2. The Multipole Expansion

The dipole-moment matrix element $\langle u_b'|\mathbf{Q}_1|u_a'\rangle$ is zero in some cases because of symmetry properties of the states a and b. In these cases, we say that the transition is forbidden under a dipole transition because the

---

[74] R. G. Gordon, *Advan. Magn. Reson.* **3**, 1–42 (1968).
[75] R. Kubo, in "Lectures in Theoretical Physics" (W. E. Brittin and L. G. Dunham, eds.). Wiley (Interscience), New York, 1959.

transition rate computed above is zero. The higher-order terms in the expansion of $e^{i\mathbf{k}\cdot\mathbf{r}}$ then become important, and lead to transitions mediated by higher multipole interactions.

The systematic discussion of these multipole interactions is carried out in several places.[76–78] Here, we discuss only the multipoles next in importance to the electric-dipole moment. These are the magnetic-dipole and electric-quadrupole moments.

We have so far neglected the magnetic interactions of charged particles because we have worked in the dipole approximation. As we shall see below, even particles without intrinsic magnetic moments have magnetic interactions coming from the $\mathbf{p}\cdot\mathscr{A}$ term in the Hamiltonian. If the particles have, in addition, a magnetic moment, then there are additional terms in the interaction Hamiltonian. The forms of these terms are probably best obtained from a relativistic treatment of quantum mechanics, that is, the Dirac equation.[79] However, the result of this treatment is that in the nonrelativistic limit the most important of these terms is exactly the term one would expect from classical considerations, namely $-\mathbf{M}_1\cdot\mathscr{B}$, where $\mathbf{M}_1$ is the magnetic moment of the particle and $\mathscr{B}$ is the magnetic field strength. Thus, the interaction Hamiltonian we consider now is

$$H_1 = -(q/m)\mathscr{A}\cdot\mathbf{p} - \mathbf{M}_1\cdot\mathscr{B} = -(q/m)\mathscr{A}\cdot\mathbf{p} + \mu g \mathbf{S}\cdot(\nabla\times\mathscr{A}). \tag{2.3.48}$$

This equation is written for one particle of charge q, mass m, gyromagnetic ratio μg, and spin angular momentum $\mathbf{S}$. For an electron, $\mu = e/2m_e$ is one Bohr magneton, and the g-factor, g, is 2. For protons and neutrons, $\mu = e/2m_p$ is one nuclear magneton, and the g-factors are $g_p = 5.6$ and $g_n = -3.8$. (Sometimes the spin-dependent term in Eq. (2.3.48) is written in terms of the Pauli matrices $\boldsymbol{\sigma}$ instead of $\mathbf{S} = \frac{1}{2}\hbar\boldsymbol{\sigma}$. Then, the g-factors are one-half as large, being 1, 2.8, and -1.9 for electron, proton, and neutron, respectively.) The g-factor for an electron can be computed by using relativistic quantum electrodynamics,[80] but there is, at present, no fundamental theory from which to derive g_p and g_n unambiguously.

We can see the effect of higher-order terms in the expansion of $\mathscr{A}$ by looking at the terms in H_1 giving rise to photon emission from an atom. The

---

[76] M. E. Rose, "Multipole Fields." Wiley, New York, 1955.

[77] B. W. Shore and D. H. Menzel, "Principles of Atomic Spectra," p. 440. Wiley, New York, 1968.

[78] H. J. Rose and D. M. Brink, *Rev. Mod. Phys.* **39**, 306 (1967).

[79] J. C. Slater, "Quantum Theory of Atomic Structure," Vol. 2, pp. 167–180. McGraw-Hill, New York, 1960.

[80] Bjorken and Drell,[49] Chapter 8.

2.3. INTERACTION OF LIGHT AND MATTER

part of H_I governing emission of a photon from an atom into the mode α is

$$H_{I,\alpha} = \sqrt{\hbar/2L^3\omega_\alpha\varepsilon_0} \sum_{\text{electrons}} \{(e/m)e^{-i\mathbf{k}_\alpha \cdot \mathbf{r}_j}\hat{\mathbf{e}}_\alpha^* \cdot \mathbf{p}_j$$
$$+ (e/2m)g\mathbf{S}_j \cdot (-i\mathbf{k}_\alpha \times \hat{\mathbf{e}}_\alpha^*)e^{-i\mathbf{k}_\alpha \cdot \mathbf{r}_j}\}, \quad (2.3.49)$$

where $\mathbf{p}_j$, $\mathbf{r}_j$, and $\mathbf{S}_j$ are the momentum, position, and spin operators for the jth electron. Consider now a single electron, drop the subscript j for the moment, and choose a coordinate system. We take $\mathbf{k}_\alpha$ along the x-axis and $\hat{\mathbf{e}}_\alpha$ along the y-axis (so we are considering linearly polarized radiation). Then, the first term in braces in Eq. (2.3.49) is

$$(e/m)e^{-ik_\alpha x}p_y = (e/m)p_y - (ie\omega_\alpha/mc)xp_y + \cdots,$$

while the second term is

$$-(ie\omega_\alpha/2mc)gS_z + \cdots,$$

where the terms we have dropped are proportional to ω^2, ω^3, and higher powers of ω. The term $(e/m)p_y$ is the electric-dipole term treated earlier, while the remaining terms,

$$-(ie\omega_\alpha/2mc)(2xp_y + gS_z), \quad (2.3.50)$$

contain the electric-quadrupole and magnetic-dipole contributions. To see this, we write

$$(2xp_y + gS_z) = (xp_y + yp_x + xp_y - yp_x + gS_z) = (xp_y + yp_x) + (L_z + gS_z). \quad (2.3.51)$$

Using $\mathbf{p} = (im/\hbar)[H, \mathbf{r}]$, we can express Eq. (2.3.51) as

$$(2xp_y + gS_z) = (im/\hbar)(Hxy - xyH) + (L_z + gS_z). \quad (2.3.52)$$

The term in xy is the contribution of the single electron under consideration to the electric-quadrupole moment of the atom, while the term $L_z + gS_z$ is its contribution to the magnetic-dipole moment.

We now return to Eq. (2.3.49) and do the sum over electrons, keeping only the terms through order ω in the braces. This gives the expression

$$H_{I,\alpha} = \sqrt{\hbar/2L^3\omega_\alpha\varepsilon_0} \Big((e/m)\sum_j p_{y,j} + (e/2m)(-i\omega_\alpha/c)\{(im/\hbar)[H, \sum_j x_j y_j]$$
$$+ \sum_j (L_{z,j} + gS_{z,j})\}\Big). \quad (2.3.53)$$

Now, $-e\sum_j x_j y_j$ is the xy-component of the electric-quadrupole-moment tensor $\mathbf{Q}_2$ of the atom, and $-(e/2m)\sum_j (L_{z,j} + gS_{z,j})$ is the z-component of the magnetic-dipole moment $\mathbf{M}_1$. Thus, the matrix element of Eq. (2.3.53) between initial and final atomic states $|u_a'\rangle$ and $|u_b'\rangle$, respectively, is

$$\langle u_b'|H_{I,\alpha}|u_a'\rangle$$
$$= \sqrt{\hbar/2L^3\omega_\alpha\varepsilon_0}\,\{i\omega(b,a)\langle u_b'|Q_{1,y}|u_a'\rangle$$
$$- (\omega_\alpha\omega(b,a)/2c)\langle u_b'|Q_{2,xy}|u_a'\rangle + (i\omega_\alpha/c)\langle u_b'|M_{1,z}|u_a'\rangle\}. \quad (2.3.54)$$

This is the result in the particular coordinate system we have used. But this matrix element must be invariant under rotations, and so we write (2.3.54) in invariant form, using

$$Q_{2,xy} = \hat{\mathbf{k}}_\alpha \cdot \mathbf{Q}_2 \cdot \hat{\mathbf{e}}_\alpha^* \quad \text{and} \quad M_{1,z} = \mathbf{M}_1 \cdot (\hat{\mathbf{k}}_\alpha \times \hat{\mathbf{e}}_\alpha^*).$$

We define $\mathbf{Q}_2$ to be the traceless part of $-e\sum_{\text{electrons}} \mathbf{rr}$. Sometimes $\mathbf{Q}_2$ is defined to include the trace.[81] This is irrelevant for our pruposes because the trace is $\{-e\sum_{\text{electrons}} r^2\}\mathbf{1}$, and this does not contribute to radiation because

$$\hat{\mathbf{k}}_\alpha \cdot \mathbf{1} \cdot \hat{\mathbf{e}}_\alpha^* = \hat{\mathbf{k}}_\alpha \cdot \hat{\mathbf{e}}_\alpha^* = 0.$$

There is no monopole radiation. Thus,

$$\langle u_b'|H_{I,\alpha}|u_a'\rangle = \sqrt{\hbar/2L^3\omega_\alpha\varepsilon_0}\,\{i\omega(b,a)\langle u_b'|\mathbf{Q}_1|u_a'\rangle \cdot \hat{\mathbf{e}}_\alpha^*$$
$$+ (i\omega_\alpha/c)\langle u_b'|\mathbf{M}_1|u_a'\rangle \cdot (\hat{\mathbf{k}}_\alpha \times \hat{\mathbf{e}}_\alpha^*)$$
$$- (\omega_\alpha\omega(b,a)/2c)\hat{\mathbf{k}}_\alpha \cdot \langle u_b'|\mathbf{Q}_2|u_a'\rangle \cdot \hat{\mathbf{e}}_\alpha^* + \cdots\}. \quad (2.3.55)$$

If the electric-dipole matrix element is not zero, then the remaining terms are negligible. This is the case worked out in preceding sections. If it does vanish, then the transition may be a mixture of the magnetic-dipole and electric-quadrupole types. The analysis of this general case is carried out by Rose and Brink.[78] For many cases of interest, only one or the other of these terms is not zero. Then we can use the nonzero part of Eq. (2.3.55) in the development of time-dependent perturbation theory. The discussion leading to Eq. (2.3.35) shows that ω_α may be replaced by $\omega(b,a)$. Then, the transition rate for emission into all directions (4π sr) for the magnetic-dipole ($M1$) case is

$$w_{b,a}(M1) = [4\omega(b,a)/3(4\pi\varepsilon_0\hbar c)c^2](\omega(b,a)/c)^2|\langle u_b'|\mathbf{M}_1|u_a'\rangle|^2, \quad (2.3.56)$$

---

[81] E. U. Condon and G. H. Shortley, "The Theory of Atomic Spectra," p. 94. Cambridge Univ. Press, London and New York, 1935; Corson and Lorrain,[2] p. 69; R. B. Leighton, "Principles of Modern Physics," p. 470. McGraw-Hill, New York, 1959.

2.3. INTERACTION OF LIGHT AND MATTER

while for the electric-quadrupole (E2) case it is

$$w_{b,a}(E2) = [4\omega(b, a)/10(4\pi\varepsilon_0 \hbar c)c^2](\omega^2(b, a)/2c)^2$$
$$\times \sum_{i,j} \langle u_b'|Q_{2,ij}|u_a'\rangle^*\langle u_b'|Q_{2,ij}|u_a'\rangle. \quad (2.3.57)$$

These equations are the analogs of Eq. (2.3.37), which gave the equivalent rate for electric-dipole radiation.

Higher multipole interactions may be of interest in those circumstances when the E1, M1, and E2 matrix elements all vanish. These multipoles are most effectively treated by using spherical tensor methods. The interested reader is referred to Shore and Menzel[77] and Rose and Brink[78] for details.

2.3.3. Selection Rules

The transition rate from state a to state b depends on the matrix elements $\langle u_b'|\mathcal{M}|u_a'\rangle$, where $\mathcal{M}$ is the relevant multipole moment introduced above. This matrix element is strongly limited by general considerations of symmetry. These limitations are partially expressed as *selection rules*, rules which require that certain matrix elements vanish. A basic reference on applications of symmetry to quantum mechanics is the book by Wigner.[5] A useful introduction to this topic is given by Merzbacher.[1] Here, we outline the simplest consequences of some familiar symmetries.

An operator $\mathcal{O}$ acting on a quantum-mechanical system is a symmetry operator if the matrix elements $|\langle\psi|\phi\rangle|$ and $|\langle\mathcal{O}\psi|\mathcal{O}\phi\rangle|$ are equal. A consequence of this equality is Wigner's Theorem,[82] which states that the operator $\mathcal{O}$ may be taken to be a unitary operator unless time reversal is involved. We will not consider time reversal, for it does not lead to selection rules. Consequences of time-reversal symmetry are discussed by Merzbacher,[1] Wigner,[82] and Wick.[83]

The symmetry operations we shall discuss here are inversion and rotation of coordinates. Other important symmetries are exchange of identical particles (this symmetry is expressed in the Pauli principle) and the approximate symmetry of certain atomic and nuclear configurations under unitary, orthogonal, and symplectic groups (these are discussed by Hammermesh[5]).

The selection rules following from symmetry under inversion are the most easily discussed. If the Hamiltonian of a system is invariant under coordinate inversion, then its eigenvectors may be chosen so that they either remain unchanged on inversion or they are multiplied by -1. We say the states represented by these eigenvectors have even or odd parity, respectively.

---

[82] Wigner,[5] p. 175.
[83] G. C. Wick, *Annu. Rev. Nucl. Sci.* **8**, 1–48 (1968).

Operators corresponding to physical quantities, and especially those corresponding to the various multipoles discussed in the preceding section, can also be classified by their behavior under coordinate inversion. For example, the operator $\mathbf{Q}_1$ changes sign under inversion, while $\mathbf{Q}_2$ does not. In the former case, the operator is said to have odd parity, in the latter case, even parity. It is convenient to denote the parity of an operator or a state by $(-)^\pi$, where π is even (odd) for even (odd) parity states and operators.

Matrix elements of operators with definite parity between states having definite parity are strongly constrained by this symmetry under inversion of coordinates. If inversion is a symmetry, and if the parity of $|\psi_i\rangle$ is $(-)^{\pi_i}$, that of $|\psi_f\rangle$ is $(-)^{\pi_f}$, and that of $\mathcal{M}$ is $(-)^\pi$, then the matrix element $\langle\psi_f|\mathcal{M}|\psi_i\rangle$ must equal the matrix element after inversion, namely $(-)^{\pi_i+\pi+\pi_f}\langle\psi_f|\mathcal{M}|\psi_i\rangle$. Thus, unless $\pi_i + \pi + \pi_f$ is even, the matrix element must vanish.

The parities of the multipole-moment operators introduced in the preceding section are as follows:

$$E1: \quad \pi \text{ odd}$$
$$M1: \quad \pi \text{ even}$$
$$E2: \quad \pi \text{ even}$$
$$\vdots$$
$$E(L): \quad (-)^\pi = (-)^L$$
$$M(L): \quad (-)^\pi = (-)^{L+1}.$$

Thus, $E1$ transitions must connect states having opposite parities, while $M1$ and $E2$ transitions connect states having the same parity.

The selection rules arising as a consequence of rotational symmetry are somewhat more involved than those following from inversion symmetry. We first note that the multipoles introduced in the preceding sections behave in the following way under rotations. The operators $\mathbf{Q}_1$ and $\mathbf{M}_1$ are vectors; thus, they transform under rotation like the spherical harmonics[84] Y_1^m. The operator $\mathbf{Q}_2$ transforms like Y_2^m. In general, if a set of operators transforms under rotations like the Y_l^m for fixed l, then the operators are the components of a spherical tensor of rank l. We denote the spherical components of such a tensor by $\mathcal{O}_l^m$. The relations between these spherical components and the Cartesian components used previously can be found in Shore and Menzel.[85]

We denote angular-momentum eigenvectors by $|LM\rangle$; $L_{op}^2|LM\rangle = \hbar^2 L(L+1)|LM\rangle$ and $L_{z,op}|LM\rangle = \hbar M|LM\rangle$. These eigenvectors can be expressed in a coordinate basis as $\langle\hat{\mathbf{r}}|LM\rangle \propto Y_L^M(\theta, \phi)$. The operator $\mathcal{O}_l^m$ acting on $|LM\rangle$ can be seen by the addition theorem for spherical harmonics

---

[84] Merzbacher,[1] p. 188.
[85] Shore and Menzel,[77] p. 164.

2.3. INTERACTION OF LIGHT AND MATTER

to yield[86]

$$\mathcal{O}_l^m |LM\rangle = \sum_{\tilde{l}=|L-l|}^{L+l} C_{\tilde{l}} |\tilde{l}, M+m\rangle. \quad (2.3.58)$$

Evaluation of

$$\langle L'M' | \mathcal{O}_l^m | LM \rangle = \sum_{\tilde{l}=|L-l|}^{L+l} C_{\tilde{l}} \langle L'M' | \tilde{l}, M+m \rangle$$

shows, because of the orthogonality of the angular-momentum eigenvectors, that unless L' is equal to one of the set $|L - l| \leq L' \leq L + l$, the matrix element will be zero. This is the triangle selection rule on angular momentum; the matrix element is zero unless the three numbers L, l, and L' form a closed triangle. The triangle selection rule may be generalized to include half-integral angular momenta. When we allow for the possibility that the angular momenta may be integral or half-integral, we will call the quantum numbers J and J'. The matrix element $\langle J'M' | \mathcal{O}_l^m | JM \rangle$ vanishes unless the three numbers J, l, and J' can form the sides of a closed triangle, and J and J' must be both integral or both half-integral. Further generalization is possible, but not pertinent for electromagnetic transitions.

The triangle selection rule has many applications. For dipole transitions (electric and magnetic), it requires that $J' = J$ or $J \pm 1$. Further, $J = 0 \to J' = 0$ transitions are forbidden because 0, 1, and 0 cannot form a triangle. Quadrupole transitions obey the selection rule $J' = J$, $J \pm 1$ or $J \pm 2$, and $J = 0 \to J' = 0$ transitions are forbidden, as are $J = \frac{1}{2} \to J' = \frac{1}{2}$ transitions.

Rotational symmetry has more precise implications than the selection rules discussed above. Consider states $|n, j, m\rangle$, where j and m are angular-momentum quantum numbers and n represents all other labels needed to specify the state. Then matrix elements of a spherical-tensor operator factor in the following way:

$$\langle n_2 j_2 m_2 | \mathcal{O}_l^m | n_1 j_1 m_1 \rangle = C_{m_1 m m_2}^{j_1 l j_2} \langle n_2 j_2 \| \mathcal{O}_l \| n_1 j_1 \rangle. \quad (2.3.59)$$

This is the Wigner–Eckart theorem.[87] The dependence on the magnetic quantum numbers resides in the coefficients $C_{m_1 m m_2}^{j_1 l j_2}$. These are called Clebsch–Gordan coefficients, vector addition coefficients, or Wigner coefficients, and are tabulated in many places.[88] The other factor in Eq. (2.3.59) is called a reduced matrix element and is independent of the magnetic quantum numbers. There are several definitions of the reduced matrix element in common use. They differ by phases and factors of $\sqrt{2j_2 + 1}$.

---

[86] Merzbacher,[1] p. 188; Jackson,[8] p. 67.
[87] Merzbacher,[1] p. 401; Wigner,[5] p. 245.
[88] A. R. Edmonds, "Angular Momentum in Quantum Mechanics," Princeton Univ. Press, Princeton, New Jesey, 1957; Condon and Shortley,[81] pp. 73–78; Shore and Menzel,[77] p. 268.

The utility of the Wigner–Eckart theorem lies in the fact that after calculating a single matrix element $\langle n_2 j_2 m_2 | \mathcal{O}_l^m | n_1 j_1 m_1 \rangle$ and extracting the reduced matrix element, all the other matrix elements in Eq. (2.3.59) involving the many combinations of m_1, m, and m_2 may be found from tabulations of the $C_{m_1 m m_2}^{j_1 l j_2}$. This is sometimes expressed by saying that the factor $C_{m_1 m m_2}^{j_1 l j_2}$ is geometrical, while the reduced matrix element contains "the physics."

The Clebsh–Gordan coefficients also play the role of angular-momentum coupling coefficients. That is, if two systems are combined, the angular momentum of the composite system is obtained from the angular momenta of the subsystems by

$$|(j_1 j_2) JM\rangle = \sum_{m_1, m_2} C_{m_1 m_2 M}^{j_1 j_2 J} |j_1 m_1\rangle |j_2 m_2\rangle. \qquad (2.3.60)$$

The properties of the $C_{m_1 m_2 M}^{j_1 j_2 J}$ follow from Eq. (2.3.60). For example, they are normalized:

$$\sum_{m_1, m_2} C_{m_1 m_2 M}^{j_1 j_2 J} C_{m_1 m_2 M'}^{j_1 j_2 J'} = \delta_{J, J'} \delta_{M, M'}. \qquad (2.3.61)$$

Other properties of these coefficients may be studied, for example, in Merzbacher,[1] Hamermesh,[5] and Edmonds.[88]

Transitions are sometimes observed that apparently violate the selection rules for a given material system. The apparent violation of selection rules usually indicates only that the state labeling is not completely correct, and that the states contain admixtures of other states with different labels. Such admixtures can come about because of neglected terms in the Hamiltonian that violate the symmetry on which the state labels are based. Interesting cases of this kind in atomic spectra were discussed by Condon and Shortley[89] and more recently by Marrus.[90]

2.4. Applications

2.4.1. Atomic and Nuclear Decay Rates

2.4.1.1. *Angular Distributions.* The spontaneous decay of a hydrogenic atom from the 2p to the 1s state furnishes an instructive application of the ideas developed in Sections 2.1–2.3. First we find the form of the matrix element, then we find the angular distributions and the Stokes parameters of the emitted radiation without actually computing any matrix elements, and finally we evaluate the lifetime. Only this last step requires any detailed knowledge of the wavefunctions. In more complex situations, this must usu-

---

[89] Condon and Shortley,[81] p. 282.
[90] R. Marrus, *Nucl. Instrum. Methods* **110**, 333–342 (1973).

ally be done numerically or in a very approximate way. It is remarkable how much information can be obtained without need of this last step.

The decay 2p → 1s has $\Delta l = 1$, and the parity changes. Thus, it is an E1 allowed process (see Section 2.3.3). The matrix element is, therefore,

$$\langle u_b|H_1|u_a\rangle = (e/m)\langle 1s; \mathbf{k}, \lambda|\mathscr{A}\cdot\mathbf{p}|2p, m\rangle$$
$$= \sqrt{\hbar/2L^3\varepsilon_0\omega}(-i\omega e)\hat{\mathbf{e}}_\lambda^*(\mathbf{k})\cdot\langle 1s|\mathbf{r}|2p, m\rangle. \quad (2.4.1)$$

The rate at which emission of a photon into the $(\mathbf{k}, \lambda)$-mode occurs is thus proportional to the square of the absolute value of

$$\hat{\mathbf{e}}_\lambda^*(\mathbf{k})\cdot\langle 1s|\mathbf{r}|2p, m\rangle,$$

and this varies with the direction of $\mathbf{k}$. This rate depends on the magnetic quantum numbers $m = 0, \pm 1$, and so measurement of the angular distribution of emitted radiation provides a way of studying the processes that populate the different magnetic substates.

Measuring an angular distribution implies setting up a coordinate system. So far, we have only one axis in the problem, the axis of quantization for the magnetic quantum number of the 2p state. This may be the axis of an external field, or the axis of the beam in an experiment involving an accelerator. If a second axis is also defined by the experimental arrangement (for example, the normal to a foil target), then a complete coordinate system can be established. We shall discuss here only the single axis case, and choose this preferred axis to be the quantization axis for the 2p states. Angular distributions must, therefore, be cylindrically symmetric about this axis, and are functions only of θ, the angle between $\mathbf{k}$ and the chosen z-axis. We denote an angular distribution by $I_\lambda^{(m)}(\theta)$, where m is the magnetic quantum number of the 2p state, and λ denotes the polarization type.

These angular distributions are easily computed. The z-axis is fixed, but we may choose the rest of the coordinate system as we wish. A simple choice is to let $\mathbf{k}$ lie in the yz-plane. Then,

$$\mathbf{k} = (\omega/c)(0, \sin\theta, \cos\theta). \quad (2.4.2)$$

The polarization vectors must be perpendicular to $\mathbf{k}$; a convenient choice for $\hat{\mathbf{e}}_1$ is

$$\hat{\mathbf{e}}_1 = (1, 0, 0) = \hat{\mathbf{e}}_x. \quad (2.4.3)$$

Then,

$$\hat{\mathbf{e}}_2 = (0, \cos\theta, -\sin\theta). \quad (2.4.4)$$

We first consider linear polarizations. Suppose the atom is definitely in the magnetic substate $m = +1$. The angular distribution of x-polarized radiation is then proportional to $|\hat{\mathbf{e}}_x\cdot\langle 1s|\mathbf{r}|2p, +1\rangle|^2$:

$$I_x^{(1)}(\theta) \propto |\langle 1s|x|2p, +1\rangle|^2. \quad (2.4.5)$$

For radiation polarized perpendicularly to $\hat{\mathbf{e}}_x$ and $\mathbf{k}$, we have

$$I_\perp^{(1)}(\theta) \propto |\hat{\mathbf{e}}_2 \cdot \langle 1s|\mathbf{r}|2p, +1\rangle|^2$$
$$= |\cos\theta \langle 1s|y|2p, +1\rangle - \sin\theta \langle 1s|z|2p, +1\rangle|^2$$
$$= \cos^2\theta |\langle 1s|y|2p, +1\rangle|^2 \tag{2.4.6}$$

because $\langle 1s|z|2p, +1\rangle = 0$. Now, $|\langle 1s|y|2p, +1\rangle| = |\langle 1s|x|2p, +1\rangle|$ because we can rotate x into y without changing the z-component of angular momentum. Thus, the angular distribution of radiation without regard to polarization is

$$I^{(1)}(\theta) \propto (1 + \cos^2\theta)|\langle 1s|x|2p, +1\rangle|^2. \tag{2.4.7}$$

The angular distributions of radiation emitted during the decay of the states with magnetic quantum numbers $m = 0$ and $m = -1$ are computed in the same way. The results for $m = -1$ are identical to those just obtained after replacing $|\langle 1s|x|2p, +1\rangle|$ by $|\langle 1s|x|2p, -1\rangle|$. The results for $m = 0$ are

$$I_x^{(0)}(\theta) = 0, \tag{2.4.8a}$$
$$I_\perp^{(0)}(\theta) \propto \sin^2\theta |\langle 1s|z|2p, 0\rangle|^2. \tag{2.4.8b}$$

The angular distribution of radiation emitted from a collection of atoms thus depends on the fraction of the atoms that are in the various magnetic substates. Let w_m be the fraction of atoms having magnetic quantum number m along the chosen z-axis. Thus,

$$w_1 + w_0 + w_{-1} = 1.$$

Then the observed unpolarized angular distribution is

$$I(\theta) = w_1 I^{(1)}(\theta) + w_0 I^{(0)}(\theta) + w_{-1} I^{(-1)}(\theta)$$
$$\propto (w_1|\langle 1s|x|2p, +1\rangle|^2 + w_{-1}|\langle 1s|x|2p, -1\rangle|^2)(1 + \cos^2\theta)$$
$$+ w_0|\langle 1s|z|2p, 0\rangle|^2 \sin^2\theta. \tag{2.4.9}$$

Symmetry arguments can simplify calculations such as this. For example, on reflection in the xy-plane, the state $|2p, +1\rangle$ goes into $|2p, -1\rangle$, while the coordinate x stays the same. Thus, $|\langle 1s|x|2p, +1\rangle| = |\langle 1s|x|2p, -1\rangle|$. Further, if the atoms are randomly oriented so that $w_m = \frac{1}{3}$ for all m, then the radiation must be isotropic, for no axis is singled out. This means that

$$\tfrac{2}{3}|\langle 1s|x|2p, +1\rangle|^2(1 + \cos^2\theta) + \tfrac{1}{3}|\langle 1s|z|2p, 0\rangle|^2 \sin^2\theta$$

is independent of θ. Thus, we learn that

$$|\langle 1s|z|2p, 0\rangle|^2 = 2|\langle 1s|x|2p, +1\rangle|^2.$$

2.4. APPLICATIONS

Putting these facts together, we obtain

$$I(\theta) \propto (w_1 + w_{-1})(1 + \cos^2 \theta) + 2w_0 \sin^2 \theta$$
$$\propto 1 + [(1 - 3w_0)/(1 + w_0)] \cos^2 \theta. \quad (2.4.10)$$

This shows the measurement of the angular distribution of radiation without regard to polarization can be used to determine the fraction of the atoms having magnetic quantum number $m = 0$. The coefficient of $\cos^2 \theta$ in Eq. (2.4.10) is sometimes called the "polarization" of the collection of atoms, but it is more properly termed an alignment parameter. It determines only w_0, and hence the sum $(w_1 + w_{-1})$, but it does not, for example, distinguish between collections with $w_1 = 1$ and those with $w_{-1} = 1$.

Calculation of the angular distribution of radiation considering circular polarizations is no harder than the preceding computation and yields additional information. Instead of $\hat{\mathbf{e}}_1$ and $\hat{\mathbf{e}}_2$, we use

$$\hat{\mathbf{e}}_{\pm} = (1/\sqrt{2})(\hat{\mathbf{e}}_1 \pm i\hat{\mathbf{e}}_2). \quad (2.4.11)$$

The angular distributions of positive- and negative-helicity photons are

$$I_{\pm}(\theta) \propto w_1 \tfrac{1}{4}(1 \pm \cos \theta)^2 + w_{-1} \tfrac{1}{4}(1 \mp \cos \theta)^2 + w_0 \tfrac{1}{2} \sin^2 \theta$$
$$= \tfrac{1}{4}(1 + w_0) \pm \tfrac{1}{2}(w_1 - w_{-1}) \cos \theta + \tfrac{1}{4}(1 - 3w_0) \cos^2 \theta$$
$$\propto 1 \pm [2(w_1 - w_{-1})/(1 + w_0)] \cos \theta + [(1 - 3w_0)/(1 + w_0)] \cos^2 \theta.$$
$$(2.4.12)$$

The population parameters w_m may thus be inferred from measurements of circularly polarized angular distributions.

Measurement of the Stokes parameters defined in Section 2.2.1.6 is another way to obtain information about population parameters. Instead of measuring an angular distribution of intensity, we measure the intensity transmitted through a variety of polarizers for a fixed direction of $\mathbf{k}$. To illustrate this, we compute $\langle S(\alpha) \rangle$ defined in Eq. (2.2.83) for the collection of atoms with population parameters w_m. We must pick some direction for $\mathbf{k}$; let us choose $\theta = \pi/2$. Then, $\hat{\mathbf{e}}_1 = \hat{\mathbf{e}}_x$ and $\hat{\mathbf{e}}_2 = -\hat{\mathbf{e}}_z$. Following the discussion leading to Eq. (2.2.83), we compute the intensity polarized at angle α to $\hat{\mathbf{e}}_1$. This polarization vector is $\hat{\mathbf{e}}_\alpha = +\cos \alpha \hat{\mathbf{e}}_1 + \sin \alpha \hat{\mathbf{e}}_2$. The intensity $|\langle S(\alpha) \rangle|$ then satisfies

$$|\langle S(\alpha) \rangle| \propto w_1 |\langle 1s|x|2p, +1 \rangle \cos \alpha|^2 + w_{-1} |\langle 1s|x|2p, -1 \rangle \cos \alpha|^2$$
$$+ w_0 |\langle 1s|z|2p, 0 \rangle \sin \alpha|^2$$
$$\propto (w_1 + w_{-1}) \cos^2 \alpha + 2w_0 \sin^2 \alpha$$
$$\propto 1 - (\tfrac{3}{2}w_0/(1 + \tfrac{1}{2}w_0)) \cos 2\alpha.$$

Thus, the ratio of the Stokes parameters M/I as a function of α can be used to find w_0. This is an alternative to measuring the angular distribution.

It is interesting that the Stokes parameter C is zero for a collection of atoms characterized only by population parameters w_m. What would a nonzero value for C mean? To see this, we compute $\langle \mathbf{S}(\alpha) \rangle$ again, but this time for a special collection of atoms, all the atoms having the wavefunction $a|2p, +1\rangle + b|2p, 0\rangle$, $|a|^2 + |b|^2 = 1$. These atoms exhibit some coherence between the different magnetic substates, for the relative phase of a and b is fixed, not just the probabilities $w_1 = |a|^2$ and $w_0 = |b|^2$. Now,

$$|\langle \mathbf{S}(\alpha) \rangle| \propto |\cos \alpha \langle 1s|x|2p, +1\rangle a - \sin \alpha \langle 1s|z|2p, 0\rangle b|^2.$$

Thus, the Stokes parameter C is proportional to

$$\mathrm{Re}\{ab^*\langle 1s|x|2p, +1\rangle \langle 1s|z|2p, 0\rangle^*\}.$$

A nonzero value for C thus implies a more general collection of atoms where the relative phases of the several $|2p, m\rangle$ states are not random. More can be learned about the collection than simply the population fractions w_m. A systematic discussion of these points using the density matrix formalism can be found in Fano and Macek,[91] and a recent application can be found in Berry et al.[92]

A direct calculation using Eq. (2.2.89) shows that the Stokes parameter ratio $\mathcal{S}/I$ satisfies

$$\mathcal{S}/I \propto (I_+ - I_-)/(I_+ + I_-),$$

and thus can be used to infer relative populations of the states $|2p, +1\rangle$ and $|2p, -1\rangle$.

Angular distributions of electric-dipole radiation from states with total angular momentum greater than 1 may be computed in much the same way as above. If a transition involves higher multipoles, and especially if two multipoles both contribute (e.g., $E2$ and $M1$), then the angular distribution of radiation is best calculated by the more systematic methods discussed by Rose and Brink.[78] They are primarily concerned with nuclear problems, but their formalism is general.

2.4.1.2. *Decay Rates*. Angular distributions of radiation depend primarily on the spins and parities of the states involved in the transition, for these determine the multipolarity of the transition. Computation of a total decay rate requires more detailed knowledge. Let us again consider the $2p \to 1s$ transition in hydrogen. The lifetimes of the different magnetic substates are equal because of rotational invariance, and so we may concentrate on any one of them. The $m = 0$ state is easy to handle.

[91] U. Fano and J. Macek, *Rev. Mod. Phys.* **45**, 533 (1973).
[92] H. G. Berry, L. J. Curtis, D. G. Ellis, and R. M. Schectman, *Phys. Rev. Lett.* **32**, 751 (1974).

2.4. APPLICATIONS

The transition rate [following from Eq. (2.3.34)] is the integral over solid angle of the product of the square of the matrix element in Eq. (2.4.1), the density of states of the radiation field, and a factor $2\pi/\hbar^2$:

$$w = (2\pi/\hbar^2) \int_{\Delta\Omega} \underbrace{\{(L^3 \, d\Omega_k/(2\pi)^3)(\omega^2/c^3)\}}_{\substack{\text{density of states} \\ \rho(\omega) \, d\Omega_k}} \underbrace{\{(\hbar/2L^3\omega\varepsilon_0)\omega^2 e^2 |\hat{e}_\lambda^*(\mathbf{k}) \cdot \langle 1s|\mathbf{r}|2p, 0\rangle|^2\}}_{\substack{\text{square of} \\ \langle u_b|H_1|u_a \rangle}} \quad (2.4.13)$$

To get the total decay rate, we sum over all polarizations that can be emitted (for the state with $m = 0$, only z-polarized light is possible) and integrate over all directions of emission. Thus, using Eq. (2.4.8b), we obtain

$$w = (e^2/4\pi\varepsilon_0 \hbar c)(\omega/2\pi c^2)\omega^2 \int d\Omega_k \sin^2\theta |\langle 1s|z|2p, 0\rangle|^2$$

$$= \tfrac{4}{3}(e^2/4\pi\varepsilon_0 \hbar c)(\omega^3/c^2)|\langle 1s|z|2p, 0\rangle|^2. \quad (2.2.14)$$

The matrix element must be calculated by using the wave functions for hydrogen:

$$\psi_{1s}(\mathbf{r}) = (\pi a^3)^{-1/2} e^{-r/a} \quad (2.4.15)$$

and

$$\psi_{2p,0}(\mathbf{r}) = \tfrac{1}{8}\sqrt{2}(\pi a^3)^{-1/2}(r/a) e^{-r/2a} \cos\theta, \quad (2.4.16)$$

where a is the Bohr radius. The result[93] is that for the 2p → 1s transition in hydrogen, $w^{-1} = \tau = 1.6 \times 10^{-9}$ sec.

Usually the correct wave functions are not known, and so the matrix element in Eq. (2.4.13) (and its analogs for higher multipoles) cannot be evaluated exactly. Thus, estimates of the matrix elements are useful. A standard set of estimates used in nuclear physics is due to Weisskopf; the estimates are called Weisskopf units.[94] The transition is assumed to involve a single nucleon; the radial wavefunction of the nucleon needed to evaluate the matrix element is taken to be constant within the nucleus and zero outside; the initial angular momentum (including spin) of the nucleon is taken to be $l + \tfrac{1}{2}$, where l is the multipolarity of the transitions, and the final angular momentum of the nucleon is taken to be $\tfrac{1}{2}$. The matrix element can then be calculated. For E1 decays this gives

$$w(\text{Weisskopf, E1}) = (e^2/4\pi\varepsilon_0 \hbar c)(\omega R/2c)^2 \omega,$$

where $R = 1.2 A^{1/3} \times 10^{-15}$ m (A is the mass number).

Similar Weisskopf units exist for M1, E2, and higher multipole transitions. Nuclear transition rates range from a small fraction of a Weisskopf unit to on the order of 100 Weisskopf units. These high rates are presumably due to collective transitions in which many nucleons take part coherently.

---

[93] H. A. Bethe and E. E. Salpeter, "Quantum Mechanics of One- and Two-Electron Atoms." Academic Press, New York, 1957.

[94] A. Bohr and B. R. Mottelson, "Nuclear Structure," p. 387. Benjamin, New York, 1969.

2.4.2 Molecular Transitions

Vibrational–rotational transitions in diatomic molecules furnish another example of an application of the theory of transition rates developed in Section 2.3. We shall briefly review the quantum mechanics of diatomic molecules, compute the relevant transition matrix elements, and use these to find integrated absorption coefficients.

2.4.2.1. *Molecular States.* The wavefunction for a general molecule involves coordinates of the electrons and the atomic nuclei. We shall see that this wavefunction is approximately separable into an electronic part and a part containing only nuclear coordinates. This latter part, the wavefunction for the nuclear motion, obeys a Schrödinger equation whose potential energy term is determined by the electronic configuration.

The coordinates needed to discuss the molecular problem are shown in Fig. 3. The coordinate system XYZ is an inertial coordinate system. The

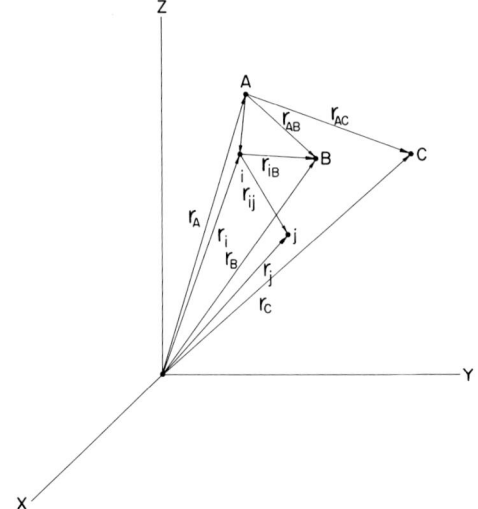

FIG. 3. Coordinates for the description of a molecule.

nuclei are designated by capital letters A, B, C, etc., and the electrons are designated by lower case letters i, j, k, etc. The Hamiltonian of this molecular system can then be written as [compare Eq. (2.3.16)]

$$H_M = -\sum_A (\hbar^2/2M_A) \nabla_A^2 - (\hbar^2/2m) \sum_i \nabla_i^2 + \tfrac{1}{2} \sum_{A \neq B} \sum_B (Z_A Z_B e^2/4\pi\varepsilon_0 r_{AB})$$
$$- \sum_A \sum_i (Z_A e^2/4\pi\varepsilon_0 r_{Ai}) + \tfrac{1}{2} \sum_{i \neq j} \sum_j (e^2/4\pi\varepsilon_0 r_{ij}). \quad (2.4.17)$$

This Hamiltonian includes only the kinetic energy of the particles and the Coulomb interactions between the particles. Other interactions are smaller in magnitude and can be treated as perturbations on this Hamiltonian.[95]

Let us define the electronic Hamiltonian:

$$H_e = -(\hbar^2/2m)\sum_i \nabla_i^2 + \tfrac{1}{2}\sum_{i\neq j}\sum_j (e^2/4\pi\varepsilon_0 r_{ij}) - \sum_A \sum_i (Z_A e^2/4\pi\varepsilon_0 r_{Ai}). \quad (2.4.18)$$

The eigenvalues and eigenfunctions of this Hamiltonian may be obtained by treating the nuclear coordinates $\{r_N\}$ as fixed parameters. The resulting eigenvalue equation is

$$H_e(\{r_i\};\{r_N\})\psi_e(\{r_i\};\{r_N\}) = E_e(\{r_N\})\psi_e(\{r_i\};\{r_N\}). \quad (2.4.19)$$

Passing over the considerable difficulties that are involved in actually solving Eq. (2.4.19), we use the solutions formally. We propose to write the eigenfunctions of the full molecular Hamiltonian in the form

$$\psi = \psi_e(\{r_i\};\{r_N\})\phi_n(\{r_N\}), \quad (2.4.20)$$

where $\phi_n(\{r_N\})$ is a function of the nuclear coordinates alone. With this form for the wavefunction, we can write the full Schrödinger equation with total energy E' as

$$\left\{H_e(\{r_i\};\{r_N\}) + \sum_A [-(\hbar^2/2M_A)\nabla_A^2 + \tfrac{1}{2}\sum_{B\neq A}(Z_A Z_B e^2/4\pi\varepsilon_0 r_{AB})]\right\}$$
$$\times \psi_e(\{r_i\};\{r_N\})\phi_n(\{r_N\}) = E'\psi_e(\{r_i\};\{r_N\})\phi_n(\{r_N\}).$$

This can be rewritten, by using Eq. (2.4.19), as

$$\left\{\sum_A [-(\hbar^2/2M_A)\nabla_A^2 + \tfrac{1}{2}\sum_{B\neq A}(Z_A Z_B e^2/4\pi\varepsilon_0 r_{AB})] + E_e(\{r_N\})\right\}$$
$$\times \psi_e(\{r_i\};\{r_N\})\phi_n(\{r_N\})$$
$$= E'\phi_n(\{r_N\})\psi_e(\{r_i\};\{r_N\}) + \left\{\sum_A (\hbar^2/2M_A)(\phi_n(\{r_N\})\nabla_A^2 \psi_e(\{r_i\};\{r_N\})\right.$$
$$\left. + \nabla_A \phi_n(\{r_N\})\cdot \nabla_A \psi_e(\{r_i\};\{r_N\}))\right\}. \quad (2.4.21)$$

If the term in braces on the right-hand side of Eq. (2.4.21) is negligible, then this equation involves only nuclear coordinates, the electronic part being

---

[95] H. H. Nielsen, *Rev. Mod. Phys.* **23**, 90–136 (1951).

hidden in $E_e(\{\mathbf{r}_N\})$. Neglecting the term in braces gives the Born–Oppenheimer approximation,[96] which is discussed in numerous texts.[97] An effective potential for the nuclear motion is defined by

$$V_n(\{\mathbf{r}_N\}) \equiv \tfrac{1}{2} \sum_A \sum_{B \neq A} (Z_A Z_B e^2/4\pi\varepsilon_0 r_{AB}) + E_e(\{\mathbf{r}_N\}). \qquad (2.4.22)$$

The first term is the Coulomb repulsion of the several nuclei; the second term is the electronic contribution to the internuclear potential. Molecular bound states can occur only if this term is sufficiently negative to make $V_n(\{\mathbf{r}_N\})$ attractive for some range of $\{\mathbf{r}_N\}$.

The approximate equation for the nuclear wavefunction $\phi_n(\{\mathbf{r}_N\})$ is then

$$\left[\sum_A -(\hbar^2/2M_A)\,\boldsymbol{\nabla}_A^{\,2} + V_n(\{\mathbf{r}_N\})\right]\phi_n(\{\mathbf{r}_N\}) = E'\phi_n(\{\mathbf{r}_N\}). \qquad (2.4.23)$$

The potential $V_n(\{\mathbf{r}_N\})$ is different for each electronic state; we will consider $V_n(\{\mathbf{r}_N\})$ only for the electronic ground state.

2.4.2.2. *Diatomic Molecule Rotation–Vibration States.* We now specialize Eq. (2.4.23) to the case of a diatomic molecule. The nuclear coordinates $\{\mathbf{r}_N\}$ are then $\mathbf{r}_1$ and $\mathbf{r}_2$, referred to an inertial coordinate system. The potential V_n depends only on the internuclear separation $|\mathbf{r}_1 - \mathbf{r}_2| \equiv |\mathbf{r}| \equiv r$. The dependence of Eq. (2.4.23) on the center of mass coordinate $\mathbf{R} = M_1\mathbf{r}_1 + M_2\mathbf{r}_2$ can be separated by using

$$-(\hbar^2/2M_1)\,\boldsymbol{\nabla}_1^{\,2} - (\hbar^2/2M_2)\,\boldsymbol{\nabla}_2^{\,2} = -(\hbar^2/2(M_1+M_2))\,\boldsymbol{\nabla}_R^{\,2} - (\hbar^2/2\mu)\,\boldsymbol{\nabla}_r^{\,2}, \qquad (2.4.24)$$

where $\mu = M_1 M_2/(M_1 + M_2)$ is the reduced mass.

Equation (2.4.23) then becomes

$$[-(\hbar^2/2(M_1+M_2))\,\boldsymbol{\nabla}_R^{\,2} - (\hbar^2/2\mu)\,\boldsymbol{\nabla}_r^{\,2} + V_n(r)]\phi_n(\mathbf{R},\mathbf{r}) = E'\phi_n(\mathbf{R},\mathbf{r}).$$

Thus, the wavefunction $\phi_n(\mathbf{R},\mathbf{r})$ has the form

$$\phi_n(\mathbf{R},\mathbf{r}) = e^{i\mathbf{K}\cdot\mathbf{R}}\psi_n(\mathbf{r}), \qquad (2.4.25)$$

where $\psi_n(\mathbf{r})$ is the wavefunction for the internal motion and satisfies

$$[-(\hbar^2/2\mu)\,\boldsymbol{\nabla}_r^{\,2} + V_n(r)]\psi_n(\mathbf{r}) = E\psi_n(\mathbf{r}). \qquad (2.4.26)$$

Here, $E = E' - (\hbar^2 K^2/2(M_1 + M_2))$ is the internal energy.

[96] M. Born and J. R. Oppenheimer, *Ann. Phys.* **84**, 457 (1927).
[97] J. C. Slater, "Quantum Theory of Molecules and Solids," p. 9. McGraw-Hill, New York, 1963; M. Born and K. Huang, "Dynamical Theory of Crystal Lattices," p. 166. Oxford Univ. Press, London and New York, 1954.

Equation (2.4.26) separates into a radial part and an angular part because $V(r)$ depends only on r. We denote the spherical coordinates of **r** by (r, θ, ϕ) and use[98]

$$\mathbf{V}_r^2 = (1/r^2)(\partial/\partial r)r^2(\partial/\partial r) - (L^2/r^2), \quad (2.4.27)$$

where

$$L^2 = -\hbar^2[(1/\sin\theta)(\partial/\partial\theta)\sin\theta(\partial/\partial\theta) + (1/\sin^2\theta)(\partial^2/\partial\phi^2)].$$

Then,

$$\psi_n(\mathbf{r}) = ru_J(r)Y_J^M(\theta, \phi),$$

where

$$L^2 Y_J^M(\theta, \phi) = \hbar^2 J(J+1) Y_J^M(\theta, \phi),$$

$$L_z Y_J^M(\theta, \phi) = -i\hbar(\partial/\partial\phi) Y_J^M(\theta, \phi) = \hbar M Y_J^M(\theta, \phi),$$

and $u_J(r)$ satisfies

$$[-(\hbar^2/2\mu)(d^2/dr^2) + (J(J+1)\hbar^2/2\mu r^2) + V(r)]u_J(r) = E_J(r)u_J(r). \quad (2.4.28)$$

We expand $V(r)$ about the equilibrium internuclear separation r_0, the point at which the net internuclear force vanishes; that is, $(\partial V/\partial r)_{r_0} = 0$. Then, putting $r = r_0 + x$, we have

$$V(r) = V(r_0) + (1/2!)x^2(\partial^2 V/\partial r^2)_{r_0} + (1/3!)x^3(\partial^3 V/\partial r^3)_{r_0} + \cdots$$

$$= V(r_0) + \tfrac{1}{2}kx^2 + ax^3 + \cdots. \quad (2.4.29)$$

Note that x can be positive or negative. We choose the zero of energy so that $V(r_0) = 0$. We next make the expansion

$$(J(J+1)\hbar^2/2\mu r^2) = (J(J+1)\hbar^2/2\mu r_0^2) - (2x/r_0)(J(J+1)\hbar^2/2\mu r_0^2) + \cdots. \quad (2.4.30)$$

The equation for $u(r)$ can now be written in terms of $F_J(x) = u_J(r_0 + x)$. We have

$$[-(\hbar^2/2\mu)(d^2/dx^2) + \tfrac{1}{2}kx^2 + (J(J+1)\hbar^2/2\mu r_0^2)$$
$$+ (-2x/r_0)(J(J+1)\hbar^2/2\mu r_0^2) + ax^3 + \cdots]F_J(x) = EF_J(x). \quad (2.4.31)$$

The term linear in x is the leading rotation–vibration coupling term. Since $x \ll r_0$, it is small. The term in x^3 is the first of an infinite series of terms representing the anharmonicity of $V(r)$. They are also small. As a first approximation, we neglect rotation–vibration coupling and anharmonic

[98] Merzbacher,[1] pp. 176–180.

terms. We then have that $F_J(x)$ obeys the equation

$$(-(\hbar^2/2\mu)(d^2/dx^2) + \tfrac{1}{2}kx^2)F_J(x) = (E - (J(J + 1)\hbar^2/2\mu r_0^2))F_J(x). \tag{2.4.32}$$

This is the equation for a simple harmonic oscillator. The differential operator is independent of J, and so the J index on F is superfluous in this approximation. The boundary condition is that F vanish as x goes to $+\infty$. Since the nuclear motion is such that $x \ll r_0$, $F(x)$ also is small for x very negative, and we make negligible error in choosing the boundary conditions $F(x) \to 0$ as $x \to \pm \infty$. In this case, the solutions of Eq. (2.4.32) are[99]

$$F(x) = (\alpha/\sqrt{\pi}\, 2^v v!)^{1/2} e^{-\alpha x^2/2} H_v(\alpha x), \tag{2.4.33}$$

where $\alpha = \sqrt{\mu k/\hbar}$, H_v is the Hermite polynomial of order v, and $v = 0, 1, 2, 3, \ldots$. The eigenvalues of Eq. (2.4.32) are

$$(E - (J(J + 1)\hbar^2/2\mu r_0^2)) = h\nu(v + \tfrac{1}{2}),$$

where $\nu = (2\pi)^{-1}\sqrt{k/\mu}$ is the oscillator frequency. In molecular problems, frequency is often expressed in wavenumbers; $\tilde{\nu} = \nu/c$ is the frequency in wavenumbers. The quantity $\hbar^2/2\mu r_0^2$ is written hcB, where $B = h/8\pi^2\mu r_0^2 c$ is called the rotator constant. Using these definitions, the energy spectrum of a diatomic molecule is, in lowest order,

$$E_{vJ} = hcBJ(J + 1) + hc\tilde{\nu}(v + \tfrac{1}{2}). \tag{2.4.34}$$

The corresponding wavefunctions are

$$\psi_{vJM}(\mathbf{r}) = rF_v(r - r_0)Y_J^M(\theta, \phi) \equiv \langle \mathbf{r}|vJM\rangle, \tag{2.4.35}$$

where F_v is given by Eq. (2.4.33).

The corrections to Eq. (2.4.34) produced by rotation–vibration coupling and the anharmonic terms can be computed in first- and second-order perturbation theory.[95] The result is that higher vibrational levels have their energies lowered slightly from the values given by Eq. (2.4.34), and the rotator constant becomes a function of v and J. These terms also affect the state vectors, adding to $|vJM\rangle$ small admixtures of states $|v'JM\rangle$ with $v' \neq v$.

2.4.2.3. *Absorption Coefficients.* The energy spectrum of diatomic molecules having been determined, we now consider transitions between states. Line strengths for diatomic molecules can be computed from Eq. (2.3.46) if we know the electric-dipole matrix elements. A calculation of these matrix elements from first principles is not feasible, so we make a model of the situation. The dipole moment of a diatomic molecule can only be oriented

---

[99] Merzbacher,[1] p. 61.

along **r**, so we introduce the expansion

$$\mathbf{Q}_1 = \hat{\mathbf{e}}_r Q_1^{(0)} + \hat{\mathbf{e}}_r x(\partial Q_1/\partial r)_{r_0} + \cdots = \hat{\mathbf{e}}_r Q_1^{(0)} + \hat{\mathbf{e}}_r x q_e + \cdots. \tag{2.4.36}$$

The first term is the permanent dipole moment of the molecule; the second term may be thought of as an effective charge q_e oscillating with the internuclear separation. We treat $Q_1^{(0)}$ and q_e as unknown parameters to be determined by experiments, and we shall compute line strengths k_{ba} in terms of these parameters.

An important result of Section 2.3.1.2 that we shall use is Eq. (2.3.46) relating the line strength to dipole moment matrix elements:

$$k_{ba} = (N_a - N_b)[4\pi^2\omega(b, a)/3(4\pi\varepsilon_0 \hbar c)]|\langle u_b'|\mathbf{Q}_1|u_a'\rangle|^2. \tag{2.3.46}$$

The matter states $|u'\rangle$ are the $|vJM\rangle$ introduced above. For each transition $|v_a J_a M_a\rangle \to |v_b J_b M_b\rangle$ there is a corresponding line strength, but no single one of these is observed because ordinarily all the $2J + 1$ magnetic substates are degenerate in the initial state and in the final state. (Exceptions to this situation are experiments involving molecular Zeeman effects.)[100] Thus, to find the total rate from a particular state $|v_a J_a M_a\rangle$ to all the $2J_b + 1$ states $|v_b J_b M_b\rangle$, we must sum Eq. (2.3.46) over all the values of M_b. Furthermore, a particular molecule is equally likely to be in any one of the $2J_a + 1$ states $|v_a J_a M_a\rangle$, and so we should average the transition rate over the values of M_a. Thus, the line strength that is usually observed is represented by

$$k_{ba} = (2J_a + 1)^{-1} \sum_{M_a, M_b} k(v_b J_b M_b, v_a J_a M_a)$$

$$= (N_a - N_b)[4\pi^2\omega(b, a)/3(4\pi\varepsilon_0 \hbar c)]$$

$$\left\{(2J_a + 1)^{-1} \sum_{M_a, M_b} |\langle v_b J_b M_b|\mathbf{Q}_1|v_a J_a M_a\rangle|^2\right\}. \tag{2.4.37}$$

We now consider the expression in braces in Eq. (2.4.37). The matrix element itself splits into two factors. Using $\mathbf{Q}_1$ from Eq. (2.4.36), we find

$$\langle v_b J_b M_b|\mathbf{Q}_1|v_a J_a M_a\rangle = [Q_1^{(0)}\langle v_b|v_a\rangle + q_e\langle v_b|x|v_a\rangle]\langle J_b M_b|\hat{\mathbf{e}}_r|J_a M_a\rangle. \tag{2.4.38}$$

The vibrational factor can be computed in a variety of ways[101]; the result is

$$Q_1^{(0)}\delta_{v_a,v_b} + q_e\sqrt{\hbar/2\omega\mu}(\sqrt{v_b}\,\delta_{v_b,v_a+1} + \sqrt{v_a}\,\delta_{v_b,v_a-1}), \tag{2.4.39}$$

---

[100] C. H. Townes and A. L. Schawlow, "Microwave Spectroscopy," Chapter 11. McGraw-Hill, New York, 1955.
[101] Merzbacher,[1] Chapters 5 and 15.

where $\hbar\omega = hc\tilde{\nu}$ is the energy of the vibrational transition $v_a \to v_b$. Thus, there are vibrational selection rules; $v_b = v_a$ (first term) and $v_b = v_a \pm 1$ (second term). The rotational factor also yields selection rules. The states $|JM\rangle$ have parity $(-)^J$, while $\hat{\mathbf{e}}_r$ is a spherical tensor of order 1 and has odd parity. Thus, rotational invariance implies $J_b = J_a$ or $J_a \pm 1$, and parity eliminates $J_b = J_a$. Therefore, possible transitions are

$$\Delta v = 0, \quad \Delta J = \pm 1 \quad \text{(pure rotation)},$$
$$\Delta v = \pm 1, \quad \Delta J = \pm 1 \quad \text{(rotation–vibration)} \tag{2.4.40}$$

The rotation–vibration transitions in absorption ($\Delta v = +1$) are split into two branches: $\Delta J = +1$ is the R-branch and $\Delta J = -1$ is the P-branch.

The rotational factor can be computed in several ways. Here we just assemble the components, which are

$$|\langle J_b M_b| \cos\theta |J_a M_a\rangle|^2 = ((2J_a + 1)/(2J_b + 1))(C_{000}^{J_a 1 J_b})^2 (C_{M_a 0 M_b}^{J_a 1 J_b})^2 \tag{2.4.41a}$$

and

$$|\langle J_b M_b| \sin\theta\, e^{\pm i\phi} |J_a M_a\rangle|^2 = ((2J_a + 1)/(2J_b + 1))(C_{000}^{J_a 1 J_b})^2 (C_{M_a \pm 1 M_b}^{J_a 1 J_b})^2. \tag{2.4.41b}$$

These results can be derived by use of the relation $\cos\theta = \sqrt{4\pi/3}\, Y_1^0(\theta, \phi)$ and the integral[102]

$$\int d\Omega\, Y_{J_b}^{M_b *}(\Omega) Y_1^m(\Omega) Y_{J_a}^{M_a}(\Omega) = \sqrt{3/4\pi}\sqrt{(2J_a + 1)/(2J_b + 1)}\, C_{000}^{J_a 1 J_b} C_{M_a m M_b}^{J_a 1 J_b}.$$

The summation and averaging then give

$$(2J_a + 1)^{-1} \sum_{M_a, M_b} |\langle J_b M_b|\hat{\mathbf{e}}_r|J_a M_a\rangle|^2$$
$$= (2J_a + 1)^{-1}((2J_a + 1)/(2J_b + 1))(C_{000}^{J_a 1 J_b})^2 \sum_{M_a, M_b, m}(C_{M_a m M_b}^{J_a 1 J_b})^2$$
$$= (C_{000}^{J_a 1 J_b})^2, \tag{2.4.42}$$

where we have used

$$\sum_{M_a, M_b, m}(C_{M_a m M_b}^{J_a 1 J_b})^2 = (2J_b + 1)$$

The values of $C_{000}^{J_a 1 J_b}$ are

$$C_{000}^{J 1 J+1} = \sqrt{(J+1)/(2J+1)} \tag{2.4.43a}$$

and

$$C_{000}^{J 1 J-1} = -\sqrt{J/(2J+1)}. \tag{2.4.43b}$$

---

[102] Merzbacher,[1] p. 396.

The line strength of Eq. (2.4.37) can now be computed for the allowed transitions $\Delta v = 0, \Delta J = +1$ and $\Delta v = +1, \Delta J = \pm 1$.

(a) $\Delta v = 0, \Delta J = +1$. Purely rotational transitions have $\omega(b, a) = 4\pi cB(J + 1)$, and so

$$k_{ba} = (N_a - N_b)[16\pi^3 cB(J + 1)/3(4\pi\varepsilon_0 hc)]|Q_1^{(0)}|^2 ((J + 1)/(2J + 1)). \quad (2.4.44)$$

The populations N_a and N_b can be related to N, the number of molecules per unit volume, if the levels are populated thermally. At low temperatures, almost all the molecules have $v = 0$, so that

$$N_J \approx N(2J + 1)(hcB/kT)e^{-J(J+1)hcB/kT}, \quad (2.4.45)$$

where k is Boltzmann's constant and T is the absolute temperature. Using Eq. (2.4.45) in Eq. (2.4.44) gives, for low temperatures,

$$k_{J+1,J} = (32\pi^4 c/3k(4\pi\varepsilon_0))(NB^2/T)(J + 1)^2|Q_1^{(0)}|^2 e^{-J(J+1)hcB/kT}. \quad (2.4.46)$$

(b) $\Delta v = +1, \Delta J = \pm 1$. Rotational–vibrational line strengths are obtained by combining N_J from Eq. (2.4.45), the frequency $\omega(b, a)$ from Eq. (2.4.34), the vibrational factor of the matrix element from Eq. (2.4.39) and the rotational factor of the matrix element from Eq. (2.4.42). The results are, for low temperatures,

P-branch: $$k_{J-1,J} = (2\pi^2 h/3(4\pi\varepsilon_0))(NB/kT)(J|q_e|^2/\mu)e^{-J(J+1)hcB/kT} \quad (2.4.47)$$

and

R-branch: $$k_{J+1,J} = (2\pi^2 h/3(4\pi\varepsilon_0))(NB/kT)((J + 1)|q_e|^2/\mu)e^{-J(J+1)hcB/kT}. \quad (2.4.48)$$

At high temperatures, states for which $v \neq 0$ may be significantly populated. Rotation–vibration bands originating from those states are called "hot bands." The partition function must then include vibrational degrees of freedom so that N_a and N_b are related to N in a more complicated way.

2.5. Conclusion

The theory discussed in this chapter is, in its relativistic form, the most impressively confirmed of all the theories that claim to be truly fundamental. Furthermore, its range of applicability is enormous, ranging from the emission of gamma radiation in experiments at ultrarelativistic energies, to the atomic processes through which we see, to dielectric phenomena involving cooperative action of many molecules. Because the theory of the electromagnetic interaction is so well confirmed, it seems certain that more and more subtle features of the description of matter will be probed by spectroscopic methods. Many currently interesting spectroscopic applications are discussed in subsequent chapters of this book.

3. NUCLEAR AND ATOMIC SPECTROSCOPY

3.1. Gamma-Ray Region*

3.1.1. Energy and Intensity

Gamma-ray photons represent the shortest wavelength region of the electromagnetic spectrum, from about 10^{-9} cm on down. They are produced in deexcitation of nuclear states and in bremsstrahlung. Discussion here will be restricted to gamma rays from nuclear decays, representing the wavelength region 10^{-9}–10^{-11} cm or the energy range of about 10 keV to 10 MeV.

Gamma rays are the dominant mode of decay for nuclear levels up to an excitation energy at which nucleon emission can occur (about 8 MeV for medium and heavy nuclei). Thus, gamma-ray spectroscopy provides an important means for studying the lower nuclear levels. Nuclear levels can also decay by beta decay, emission of alpha particles, conversion electrons, or pair production. These alternate decay modes will be discussed below as they relate to gamma emission.

The decay of a state of initial angular momentum (or "spin," referring to the nucleus as a whole) and parity $J_a^{\pi_a}$ to a final state $J_b^{\pi_b}$ can occur if the angular momentum L and parity π carried off by the electromagnetic field satisfy the relations

$$\mathbf{J}_a = \mathbf{L} + \mathbf{J}_b, \qquad (3.1.1)$$

$$\pi_a = \pi_b \pi. \qquad (3.1.2)$$

The electromagnetic radiation field has multipoles of $L = 1, 2, 3, \ldots$ (dipole, quadrupole, octupole, ...). There is no monopole radiation ($L = 0$), so that decay of one zero-spin state to another zero-spin state cannot occur by single photon emission.

The parity of the multipole radiation is different for electric and magnetic radiation of the same multipole, and is given by

$$\pi_e = (-)^L, \qquad (3.1.3)$$

$$\pi_m = (-)^{L+1}. \qquad (3.1.4)$$

---

*Chapter 3.1 is by James C. Legg and Gregory G. Seaman.

Since nuclear states have definite parity, electromagnetic transitions between two nuclear states cannot have both electric and magnetic multipoles of the same angular momentum. The transitions with even parity (no nuclear state parity change) are $M1, E2, M3, E4, \ldots$, and the transitions with odd parity are $E1, M2, E3, M4, \ldots$.

The transition probability for a given multipole radiation has been presented by Blatt and Weisskopf[1] for the case of a single nucleon changing state. This "single-particle estimate," or "Weisskopf rate," is very approximate, and in some cases is about a factor of 100 different from the actual rate because many nucleons contribute to the radiation. The calculation is carried out by assuming the nucleus to be a spinless, infinite potential well, and the nucleon makes a transition from an initial state of angular momentum $L + \frac{1}{2}$ to a final state of angular momentum $\frac{1}{2}$. The wave function is taken to be constant throughout the nucleus, which has radius R.

For electric transitions, the estimated transition probability is

$$\tau^{-1}(EL) = \frac{2(L+1)}{L[(2L+1)!!]^2} \left(\frac{3}{L+3}\right)^2 \frac{e^2}{\hbar c} \left(\frac{\omega R}{c}\right)^{2L} \omega \quad [\text{sec}^{-1}]. \tag{3.1.5}$$

For magnetic transitions, the contribution of the magnetic moment of the nucleon of mass M is added in, obtaining

$$\tau^{-1}(ML) = \frac{20(L+1)}{L[(2L+1)!!]^2} \left(\frac{3}{L+3}\right)^2 \frac{e^2}{\hbar c} \left(\frac{\hbar}{McR}\right)^2 \left(\frac{\omega R}{c}\right)^{2L} \omega \quad [\text{sec}^{-1}]. \tag{3.1.6}$$

The $(\omega R/c)^{2L}$ term is the ratio of the nuclear radius to the reduced wavelength $(R/\lambdabar)^{2L}$. This term is less than unity even for such an extreme case as a uranium nucleus ($R \sim 8 \times 10^{-13}$ cm) and a 10-MeV gamma ray ($\lambdabar = 2 \times 10^{-12}$ cm). The L-dependent multiplicative factor decreases strongly with increasing L, and so the entire transition probability falls rapidly with increasing multipolarity.

In atoms, photon transitions between electronic states are $E1$ transitions, while magnetic and higher-order multipoles are "forbidden" because the electronic states are destroyed in collisions before the slower decay modes can occur. (These forbidden transitions, such as the $2s_{1/2} \rightarrow 1s_{1/2}$ $M1$ transition, can be of sufficient intensity in high Z atoms to be seen,[2] or can occur in highly ionized atoms.) However, nuclear states that can only decay by gamma emission are not destroyed by other processes, so that all multipolarities allowed by the spin sequences are observed.

---

[1] J. M. Blatt and V. F. Weisskopf, "Theoretical Nuclear Physics," p. 623. Wiley, New York. 1952.

[2] H. R. Rosner and C. P. Bhalla, *Z. Phys.* **231**, 347 (1970).

3.1. GAMMA-RAY REGION

The relative intensity of magnetic radiation to electric radiation of the same multipolarity is roughly proportional to $(v/c)^2$, or about 0.05.[3] This can be explained on the basis of the current density being smaller than the charge density,[3] or from an uncertainty principle argument for dipole fields.[1,4] Thus, the radiation expected from an $L = 1$ multipole field will be almost pure $E1$ if there is a parity change, since the $M2$ radiation will be reduced by the $(v/c)^2$ factor as well as from the higher multipolarity. However, $M1$ radiation will usually be accompanied by $E2$ radiation. Completely pure multipole radiation occurs if J_a or J_b is zero. The various possible combinations are expressed in Table I, with particular examples given in Table II.

TABLE I. Dominant Multipole Radiation

$\|J_a - J_b\|$	No parity change	Parity change
0,[a] 1	$M1, E2^b$	$E1$
2	$E2$	$M2, E3^b$
3	$M3, E4^b$	$E3$
4	$E4$	$M4, E5^b$

[a] If $J_a = J_b \neq 0$.
[b] Only if allowed by the vector relation $\mathbf{J}_L = \mathbf{L} + \mathbf{J}_b$.

TABLE II. Examples of Multipole Radiation

J_a^π	$J_b^{\pi_b}$	Radiation
$\frac{1}{2}^+$	$\frac{1}{2}^+$	$M1$
$\frac{1}{2}^+$	$\frac{1}{2}^-$	$E1$
$\frac{3}{2}^+$	$\frac{1}{2}^+$	$M1, E2$
1^+	0^+	$M1$
1^-	0^+	$E1$
2^+	0^+	$E2$
2^+	2^+	$M1, E2$
3^-	0^+	$E3$

The radiation pattern for "mixed" transitions depends strongly upon the relative intensities of the two decay modes. As will be discussed below, the relative intensity of $L + 1$ radiation to L radiation, defined as $\delta = \langle |L+1|\rangle/\langle |L|\rangle$, can be extracted from the angular distribution of the radiation pattern if the decaying state is aligned in a known way.

[3] D. H. Wilkinson, in "Nuclear Spectroscopy" (F. Ajzenberg-Selove, ed.), Part B, p. 855. Academic Press, New York, 1960.
[4] P. Marmier and E. Sheldon, "Physics of Nuclei and Particles," Vol. I, p. 418 ff. Academic Press, New York, 1969.

Finally, the Weisskopf estimates of the transition probabilities can be expressed more compactly if the nuclear radius is approximated by $R = r_0 A^{1/3}$, where $r_0 = 1.2 \times 10^{-13}$ cm. If the gamma-ray energy is given in MeV, then the radiative widths are[5] as shown in Table III.

TABLE III. Weisskopf Transition Probabilities (E_γ in MeV)

$\Gamma_W(E1) = 6.8 \times 10^{-2} A^{2/3} E_\gamma^3$ eV	$\Gamma(M1) = 2.1 \times 10^{-2} E_\gamma^3$ eV
$\Gamma_W(E2) = 4.9 \times 10^{-8} A^{4/3} E_\gamma^5$ eV	$\Gamma(M2) = 1.5 \times 10^{-8} A^{2/3} E_\gamma^5$ eV
$\Gamma_W(E3) = 2.3 \times 10^{-14} A^2 E_\gamma^7$ eV	$\Gamma(M3) = 6.8 \times 10^{-15} A^{4/3} E_\gamma^7$ eV
$\Gamma_W(E4) = 6.8 \times 10^{-21} A^{8/3} E_\gamma^9$ eV	$\Gamma(M4) = 2.1 \times 10^{-21} A^2 E_\gamma^9$ eV
$\Gamma_W(E5) = 1.6 \times 10^{-27} A^{10/3} E_\gamma^{11}$ eV	$\Gamma(M5) = 4.9 \times 10^{-28} A^{8/3} E_\gamma^{11}$ eV

As an example of the level widths that result from these estimates, a 1-MeV $E1$ gamma transition in ^{40}Ca has a width of 0.8 eV, or a lifetime of 8.2×10^{-16} sec. For $E2$ radiation, the width is 6.7×10^{-6} eV, and has a lifetime of 9.8×10^{-11} sec. The corresponding $M1$ radiation has a width of 2.1×10^{-2} eV and a lifetime of 3.1×10^{-13} sec.

For low-energy gamma-ray transitions between states with widely differing spins, the lifetimes can be extremely long. For lifetimes longer than about 0.01 sec, the long-lived nuclear state is known as an isomeric state. These are usually $E3$, $M3$, and higher-multipole transitions, although a few $E1$ and $M2$ cases are known for very low-energy transitions. Several examples[6] are shown in Fig. 1.

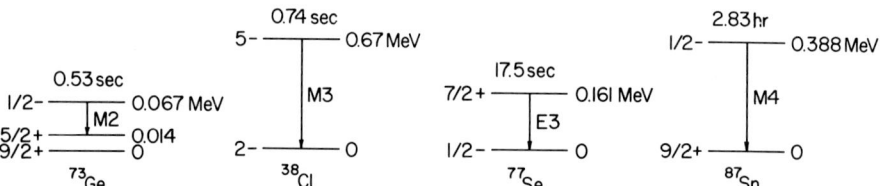

FIG. 1. Examples of isomeric transitions in four nuclei. The long lifetimes are the result of both large multipolarity and low-energy gamma-ray transitions.

3.1.1.1. Alternate Decay Modes of Nuclear States.

As noted above, a single photon transition is strictly forbidden for 0^+ to 0^+ nuclear states. However, if the energy of the state is greater than 1.2 MeV, the 0^+ state may decay by emission of an electron–positron pair. This process, called $E0$

---

[5] D. H. Wilkinson, in "Nuclear Spectroscopy" (F. Ajzenberg-Selove, ed.), Part B, p. 858. Academic Press, New York, 1960.

[6] C. M. Lederer, J. M. Hollander, and I. Perlman, "Table of Isotopes," 6th ed. Wiley, New York, 1967.

decay, is especially important for those nuclei in which the first excited state is a 0^+ state because there are then no intermediate levels through which decay can proceed. Only three nuclei fall in this category: ^{16}O, ^{40}Ca, and ^{90}Zr. Experiments have been made to look for 0^+ to 0^+ transitions by double gamma decay, but no conclusive evidence yet exists for this process. A recent study that includes a summary of earlier work is that by Beardsworth et al.[7] Pair emission may compete with gamma-ray decay, such as the 4.28-MeV 0^+ level in ^{48}Ca, which decays 23% by pair emission to the 0^+ ground state and 77% by E2 gamma decay to the 3.83-MeV 2^+ state.[8]

A more common decay process that competes with any multipolarity gamma decay is emission of conversion electrons. The probability of this process is proportional to the density of electrons at the nucleus, and thus is more important for higher-Z nuclei. Further, since the process competes with gamma decay, levels in which the lifetime is long are more likely to decay by electron emission. Therefore, for low-energy or high-multipolarity transitions, the conversion process will be large and may be the dominant decay mode. For example, the 0.187-MeV $M1$ transition to the ground state in ^{203}Pb has about 1.4 times as many conversion electrons as gamma rays.[6]

The energy of the conversion electron emitted is equal to the gamma-ray energy minus the binding energy of the electron, $E_e = E_\gamma - E_B$. Several energies of electrons can occur since the $K, L, M, \ldots$ electrons have finite densities at the nucleus. For example, in ^{212}Pb conversion electrons up to the P shell are observed.[9] Of course, if the transition energy is less than the binding energy of a particular electron shell, then those electrons will not be emitted.

The ratio of probability for conversion electrons to the probability for gamma rays is defined as the conversion coefficient α, and is the sum of all the electron shells involved, $\alpha = \alpha_K + \alpha_L + \cdots$. Thus, the total number of transitions for a level that decays by gamma emission is $N_\gamma(1 + \alpha)$, where N_γ is the number of gamma rays. Extensive tabulations of calculated conversion coefficients have been produced for point nuclei[10] and for finite size nuclei.[11] The effects of nuclear size are important for heavy nuclei and high-energy gamma rays. The values of K-conversion coefficients for two nuclei as a function of multipolarity and gamma-ray energy are given in Fig. 2.

[7] E. Beardsworth, R. Hensler, J. W. Tape, N. Benczer-Koller, W. Darcey, and J. R. MacDonald, *Phys. Rev. C* **8**, 216 (1973).

[8] N. Benczer-Koller, G. G. Seaman, M. C. Bertin, J. W. Tape, and J. R. MacDonald, *Phys. Rev. C* **2**, 1037 (1970).

[9] A. Flammersfeld, *Z. Phys.* **114**, 227 (1939).

[10] M. E. Rose, G. H. Goertzel, B. I. Spinrad, J. Harr, and P. Strong, *Phys. Rev.* **83**, 79 (1951).

[11] L. A. Sliv and I. M. Band, Rep. 57 ICC K1, 57 ICC K2, Univ. of Illinois (1956); *in* "Alpha-, Beta-, and Gamma-Ray Spectroscopy" (K. Siegbahn, ed.), p. 1639. North-Holland Publ., Amsterdam, 1965.

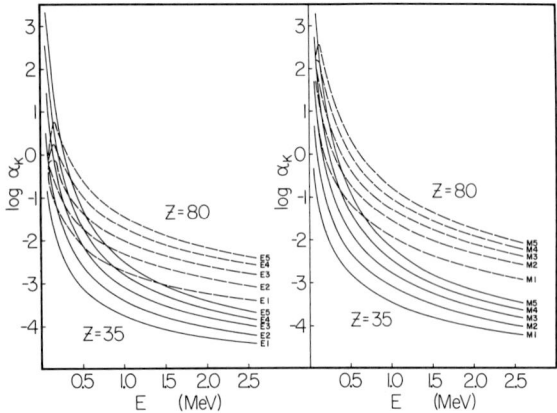

FIG. 2. Conversion coefficients for K electrons for electric (left) and magnetic (right) transitions as a function of transition energy for nuclei of $Z = 35$ and $Z = 80$. The conversion coefficients depend strongly upon multipolarity, so that a measurement of the conversion coefficient for a given transition can serve to determine the multipolarity. The values are taken from L. A. Sliv and I. M. Band, Rep. 57 ICC K1, 57 ICC K2, Univ. of Illinois (1956); in "Alpha-, Beta-, and Gamma-Ray Spectroscopy" (K. Siegbahn, ed.), p. 1639. North-Holland Publ., Amsterdam, 1965.

For a given nucleus and gamma-ray energy, there are large differences in values with multipolarity, so that a measurement of the conversion coefficient can serve to identify the multipole character of the radiation.

Thus far we have assumed that we are dealing with stable nuclei, which represent only a small fraction of the nuclei known. The decay of radioactive nuclei is primarily by beta decay, while very heavy nuclei also undergo alpha decay. There are also a few nuclei such as ^{252}Cf, which undergo spontaneous fission as well as alpha decay. Since beta and alpha decay may leave the residual nucleus in an excited state, the radioactive decay is usually accompanied by gamma emission as well. In some cases, the decay sequence may not stop after a single nuclear change, but may continue at length as in ^{228}Th, which finally ends at the stable nucleus $^{208}_{82}$Pb after emitting five alpha particles and undergoing two beta decay processes.

Beta decay includes several processes by which the Z of a nucleus is changed by one unit. In β^- decay, $Z \to Z + 1$, but the nucleus may also capture an electron or emit positrons for $Z \to Z - 1$.

A radioactive nucleus generally decays by only one process, but there are exceptions in which a nuclear level will both alpha and beta decay (^{210}Bi, ^{212}Bi), or β^+, electron capture, and β^- decay ($^{152}_{63}$Eu). More rarely, a nuclear state may decay by both particle and gamma emission where the nuclear structure is such that the particle decay is slow enough for the gamma decay to compete. The $\frac{5}{2}^-$ 2.43-MeV level of ^{9}Be decays by neutron emission, as well as by a 0.01% gamma branch, to the ground state. In ^{12}C, the 7.66 MeV 0^+ level undergoes alpha decay and positron–electron pair emission. However, in all these cases, the gamma branch is much less than 1%.

3.1.2. Gamma-Ray Detectors

The detection of gamma rays is usually done by photon counting techniques, in which the individual gamma rays are detected by means of their interaction with matter. Secondary charged products are produced, which are then used to signal the detection of a high-energy photon. The discussion must first start with the principal interactions of gamma rays with matter in which the secondary charged products are created.

3.1.2.1. *Interaction of Gamma Rays with Matter.* As a beam of gamma-ray photons passes through matter, it is absorbed by the matter in the classic Lambert absorption law:

$$I = I_0 e^{-\mu x}, \tag{3.1.7}$$

where I and I_0 are the intensities of the beam at depth x and no depth, respectively, and μ is the attenuation coefficient. Since the product μx must be dimensionless, if we express the depth x in atoms per square centimeter, the attenuation coefficient μ will have units of centimeters squared per atom, and thus is also the atomic cross section for removal of gamma-ray photons from the beam. For gamma rays, the attenuation process is a combination of three dominant interactions and several minor interactions, which do not coherently interfere with each other, so that the attenuation coefficient is the sum of the separate interaction cross sections.

The first strong interaction is the photoelectric effect, in which a photon gives all its energy to a bound electron. The result is a free electron whose kinetic energy equals the photon energy minus the binding energy of the electron. The second strong interaction is the scattering of photons from atomic electrons. In the majority of experimental situations, the gamma-ray photon energy is much greater than the atomic electron binding energy so that it is well approximated by a photon scattering from a free electron at rest, and the scattering is just Compton scattering. The energies of the electron T and scattered photon E' are given by the relations

$$E' = E[1 + (E/m_e c^2)(1 - \cos\theta)]^{-1}, \tag{3.1.8}$$

$$T = E[1 + m_e c^2/E(1 - \cos\theta)]^{-1}, \tag{3.1.9}$$

where E is the incident photon energy and θ is the photon scattering angle which varies from $0°$ to $180°$. The third strong interaction is that of electron–positron pair creation by a photon in the field of a nucleus or an electron.

This process results in an electron–positron pair having a total kinetic energy equal to the photon energy minus the creation energy of the pair.

The cross sections for these three major interactions are complicated functions of both atomic number Z and photon energy E. For a comprehensive discussion of these cross sections as well as other possible interactions, the reader is referred to Davisson[12] and Heitler.[13] For the purposes of discussion of gamma-ray detectors, it is sufficient to realize that the photoelectric effect cross section σ_ϕ is proportional to Z^5 and decreases with E faster than $1/E$. In a similar fashion, the Compton cross section σ_c is proportional to Z and decreases with E slower than $1/E$. Finally, the pair creation cross section σ_π increases with Z faster than Z^2 and increases with energy above the threshold energy, 1.02 MeV. In Fig. 3, the major interaction cross sections[12] are plotted for silicon and sodium iodide. This figure shows that the absorption process may be dominated by one or another of the three major attenuation interactions, depending on the detector material and the gamma-ray energy.

A beam of photons passing through matter may also be attenuated by scatterings where the photon energy is not changed but the direction is. These scattering effects include Rayleigh scattering from bound electrons and Thompson scattering from unbound electrons. Figure 3 shows that these are important for low-energy gamma rays. Nuclear resonance scattering also occurs. The Mössbauer effect is the best-known example of this scattering. Nuclear Delbrück scattering from the nuclear Coulomb potential has also been observed. These last scattering effects are small for high-energy photons and are generally studied only for their own intrinsic interest, rather than as major contributors to gamma-ray detection processes.

3.1.2.2. *Direct Gamma-Ray Detection.* For purposes of the present discussion, we shall define direct gamma-ray detection as detection of gamma rays in such a manner that the energy of the gamma ray may be directly determined from the information obtained from the detector. In this way, we distinguish from another class of gamma ray detectors, which we call indirect gamma-ray detectors, and which have the property that the gamma rays are detected by a specific physical process, and knowledge of that particular process is used to determine the gamma ray energy. Of course, in the sense that all gamma rays are detected by their interaction with matter, producing energetic free electrons whose energy is then absorbed by the detector, all detectors are indirect.

One of the earliest photon detectors was the gas ionization detector. In this detector, the detected gamma ray passes through a chamber of gas,

---

[12] C. M. Davisson, in "Alpha-, Beta-, and Gamma-Ray Spectroscopy" (K. Siegbahn, ed.), pp. 37, 827. North-Holland Publ., Amsterdam, 1965.

[13] W. Heitler, "The Quantum Theory of Radiation." Oxford Univ. Press (Clarendon), London and New York, 1954.

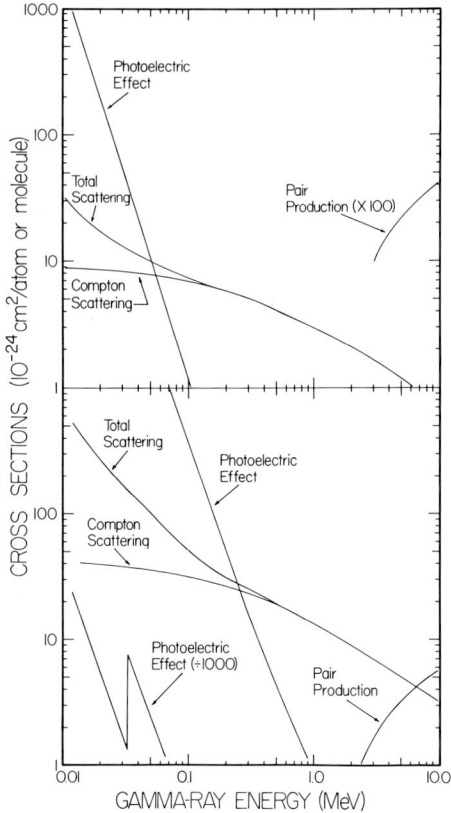

FIG. 3. Interaction cross sections for gamma rays incident on silicon (top) and sodium iodide (bottom). The atomic cross sections for silicon and molecular cross sections for sodium iodide are plotted for scattering, photoelectric effect, and pair production between 10 keV and 10 MeV. The total scattering cross section includes the contribution from scattering processes in which the direction but not the energy of the gamma ray is altered. These figures are based on the calculations of C. M. Davisson, *in* "Alpha-, Beta-, and Gamma-Ray Spectroscopy" (K. Siegbahn, ed.), pp. 37, 827. North-Holland Publ., Amsterdam, 1965.

interacts with the gas atoms, and produces free electrons which in turn produce ionization as they are slowed down and stopped in the gas. An electric field across the gas volume separates the positive ions from the electrons and sweeps them to electrodes, producing an electrical current pulse at the electrodes. This electronic pulse is then amplified and counted. Depending on the strength of the electric field across the gas volume, the counters can operate in one of three general modes of operation. At low electric fields, the ions and electrons are swept out slowly, never accelerating sufficiently between collisions to produce secondary ionization. In this case, the charge collected at the electrode is simply the charge created by the stopping electrons, and the chamber is a gas ionization chamber.

If the electric field is sufficiently high to produce secondary ionization, the ions produce a multiplicative avalanche effect, and the charge collected at the electrode is proportional to the original ionization amplified by a gas multiplication factor. This detector is called a proportional counter. If the electric field is raised yet higher, the charges accelerating toward the electrode produce avalanching ionization and also secondary photons which propagate back and reinitiate the avalanche. In this mode of operation, the charge collected at the electrode contains no information about the original ionization, but simply indicates that an ionizing event has occurred in the gas volume. This detector is the Geiger counter.

All three modes of gas chamber detection have been used in the detection of gamma rays. However, the use of gas chambers for gamma-ray detectors is severely limited by the difficulty of producing a detector with sufficient matter in the sensitive volume so that reasonable detection efficiencies are obtained for high-energy photons. Because of these limitations, the use of gas detectors is generally limited to the detection of low-energy photons (below 0.1 MeV), or to uses where high detector efficiency is not an important consideration.

A second major class of gamma-ray detectors is the scintillation detector. The discovery by Hofstadter[14] of the properties of thallium-activated sodium iodide marks the birth of modern scintillation spectrometry. To this day, NaI(Tl) remains the standard scintillation material for gamma-ray detectors. In the sodium iodide crystal, the energy deposited by the gamma-ray photon, as a result of its interaction with an atom of the crystal, results in a flash of light which is transmitted to the photocathode of a photomultiplier tube which provides electrical amplification to a signal whose total charge is proportional to the energy deposited in the scintillator. The preparation of the scintillator crystal, the optical contact, and the photomultiplier are the subjects of considerable literature. However, for most purposes today, excellent packaged combinations of scintillator, photomultiplier, and electronics are commercially available, and we shall restrict our discussion to the use of a system which has been suitably prepared. These detector combinations generally have their size and their resolution given as the important specifications of the package performance.

In general-purpose packages, the size is specified by the diameter and length of the cylindrical sodium iodide crystal used as a detector in the package. Typical standard sizes of sodium iodide crystals commercially available are 5.1 × 5.1 cm diameter, 7.6 × 7.6 cm diameter, and 12.7 × 12.7 cm

---

[14] R. Hofstadter, *Phys. Rev.* **74**, 100 (1948).

3.1. GAMMA-RAY REGION

diameter cylinders. Of the standard sizes, probably the most widely used and referenced size is the 7.6 × 7.6 cm crystal assembly. Other larger and smaller sizes and noncylindrical geometries are commercially available as special-purpose packages. Probably the most popular geometry other than the standard cylinder is the well counter, in which a small well is located in a large crystal so that low-energy gamma rays from a source may be detected with higher efficiency.

The resolutions available from these detectors vary between 7% and 10% full width at half maximum (FWHM) for 662-keV gamma rays detected from a ^{137}Cs source. These resolutions are mainly dependent upon the crystal package and photomultiplier combinations, although mildly dependent also upon the size of the crystal. The best resolutions are available for the standard 7.6 × 7.6 cm detector, with other sizes typically having slightly poorer resolution.

The pulse height spectrum observed and the efficiency of a detector are much more strongly dependent upon the size of the detector. Two numbers are usually used to represent the range of effects that are encountered here. The first is the total efficiency $T(E)$ of the detector for a given monochromatic gamma-ray energy and source-to-detector distance. In Fig. 4 are displayed

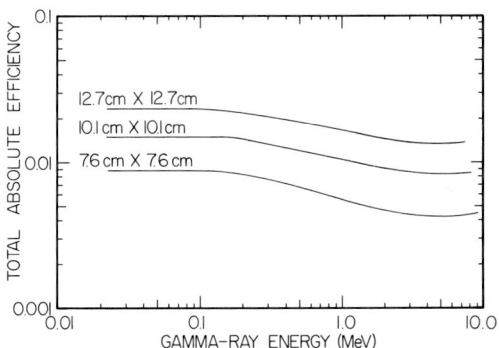

FIG. 4. Total efficients of NaI(Tl) detectors as a function of gamma-ray energy. The total efficiencies of three standard size NaI(Tl) detectors are shown for a source-to-detector distance of 20 cm. This figure is based on the calculations of R. L. Heath, Scintillation Spectrometry: Gamma-Ray Catalogue. USAEC Rep. IDO-16880 (1964).

the efficiencies[15] for various size detectors as a function of gamma-ray energy at a standard source-to-detector separation distance of 20 cm. Of course, the total efficiency is measured for the total number of detector counts independent of pulse height. This is a useful number for monochromatic

[15] R. L. Heath, Scintillation Spectrometry: Gamma-Ray Catalogue. USAEC Rep. IDO-16880 (1964).

gamma rays and is easily calculated from a knowledge of the cross sections for the principal interaction of gamma rays with matter. However, in detecting several gamma rays of different energies, the easiest experimental number to obtain is the number of counts in the full-energy peak, which can be unambiguously attributed to a definite gamma-ray energy.

The ratio p of counts in the full-energy peak to the total number of counts is rather difficult to calculate theoretically in that there are contributions to the full-energy peak from many different processes. For instance, the full-energy peak is the sum of the photoelectric interactions plus Compton scatterings, with the secondary photon undergoing photoelectric absorption, plus pair creation, with the secondary annihilation quanta both undergoing photoelectric absorption, plus other combinations and permutations of these processes. It is obvious that the ratio p should increase with increasing crystal size, but accurate calculations are virtually impossible. For that reason, various workers have experimentally determined, for standard detector configurations, the peak-to-total ratio p. In Fig. 5, the peak-to-total ratio is shown for a 7.6 × 7.6 cm detector as a function of energy.[15] A single curve is shown since Heath[15] has shown that the variation of source-to-detector distance produces less than one percent variations in the peak-to-total ratios for a standard 7.6 × 7.6 cm detector.

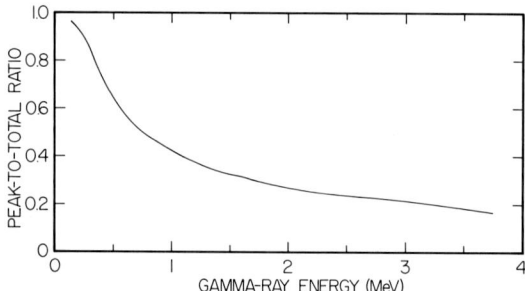

FIG. 5. Peak-to-total ratio p for a 7.6 × 7.6 cm NaI(Tl) detector as a function of gamma-ray energy. This drawing is based on the measurements of R. L. Heath [Scintillation Spectrometry: Gamma-Ray Catalogue. USAEC Rep. IDO-16880 (1964)] who showed that this ratio is relatively independent of source-to-detector distance.

Using the knowledge of the total efficiency $T(E)$ and the peak-to-total ratio p, one can then determine the source strength of various gamma rays using the formula $N = sT(E)p$, where N is the number of counts in a full-energy peak and s is the number of source disintegrations of that energy in a fixed time interval. Although sodium iodide detector pulse heights are very nearly proportional to gamma-ray energy, it is worthy of note that Heath[15] has observed significant deviations from linearity. These results make it imperative that one calibrate the sodium iodide detector with sources

3.1. GAMMA-RAY REGION

of known gamma-ray energy in the region where one is measuring the energies of unknown gamma rays.

In recent years, a new gamma-ray detection method has been introduced into general use. The detection method is that of a solid-state ionization chamber. The virtues of this are that the density, and therefore the photon counting efficiency, are those of a heavy solid rather than a light gas. Further, the energy loss per free electron produced is less than 3 eV, rather than the typical values of 30 eV for a gas ionization chamber and more than 300 eV for a scintillation detector. Thus the solid state detector offers higher efficiency than a gas detector and much better resolution than a scintillation detector.

Both silicon and germanium have been used for solid-state ionization chambers. However, because of the lower atomic number of silicon, its use has been restricted to low-energy photons. In order to achieve the large volumes necessary for practical photon counting efficiencies, the germanium crystals have been subjected to a process called lithium drifting. This process uses lithium impurities to compensate for impurities left in the germanium crystal, and thus achieves high resistivities, approaching that of intrinsic germanium, through a large volume of crystal. The crystal is then processed to form a diode structure, which is then back-biased to provide an electric field through a large volume which accelerates current carriers to the collection electrodes. Very recently, true intrinsic germanium detectors have become available. These detectors have not used the compensation of the lithium drifting process and therefore have improved electrical characteristics as detectors.

The advantages of germanium detectors are illustrated in Fig. 6. In this figure, the gamma rays from a ^{56}Co source have been detected by a 7.6 × 7.6 cm sodium iodide scintillation detector (Fig. 6a) and by a large volume Ge(Li) detector (Fig. 6b). The difference in resolution apparent in this figure is rather dramatic evidence of the virtues of the solid-state detector.

Detectors of this type are commercially available from a number of firms, packaged with their preamplifier, detector, and a variety of mountings in liquid nitrogen cryostats. The Ge(Li) detector must be kept at liquid nitrogen temperature at all times in order to preserve the compensated diode characteristics necessary for good detector operation.

Commercial detectors are available commonly in three types of geometry for the compensated active volume. The first type is the "planar" detector, in which the active volume is a right circular cylinder of large area and limited depth, and it is usually limited to low- and medium-energy gamma-ray detection. A second standard geometry of germanium detectors is the coaxial detector, in which the lithium-compensated volume is a hollow cylinder surrounding an uncompensated small diameter cylinder of germanium. In this way, a large active volume with an approximately uniform electric field is achieved. Finally, there are variations on the coaxial geometry such as a

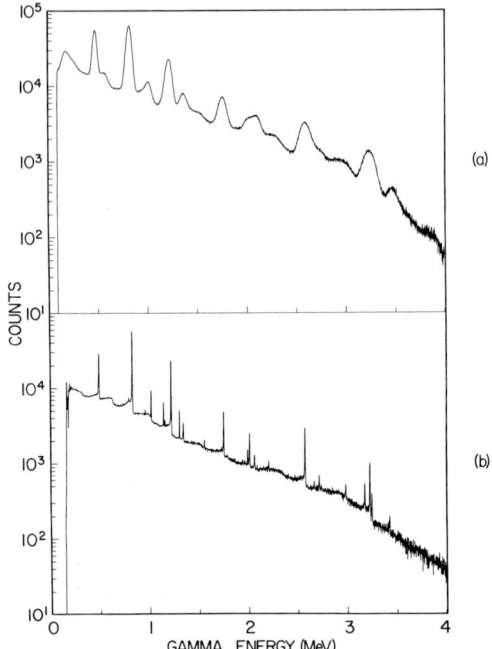

FIG. 6. ^{56}Co source gamma-ray spectrum measured by two detectors. (a), The ^{56}Fe gamma rays produced by the ^{56}Co decay were detected by a 7.6 × 7.6 cm NaI(Tl) detector. (b), The ^{56}Fe gamma rays were detected by a 20% Ge(Li) detector.

cylindrical coaxial geometry with a front end cap, so that the volume becomes a closed-end coaxial geometry, or the trapezoidal coaxial geometry in which a trapezoidal rather than a circular shaped crystal is used and a closed-end coaxial volume is prepared. These last variations on the true coaxial shape produce slightly larger active volume at the expense of nonuniform fields in the active volume, which then result in pulse rise time variations depending upon the location of the gamma-ray detection in the active volume.

The operating characteristics of commercial Ge(Li) detectors are typically specified by the efficiency, the resolution, and the peak-to-Compton height ratio. The efficiency is expressed as a percentage of the efficiency of a standard 7.6 × 7.6 cm NaI(Tl) detector for the full-energy peak of the ^{60}Co 1.33-MeV gamma-ray line. Germanium detectors with efficiencies greater than 20% are commercially available on a regular basis. The resolution of the detector is usually quoted as the full width at half maximum for the ^{60}Co 1.33-MeV gamma ray. Peak widths of less than 2 keV are available commercially on a regular basis. Often the full width at 0.1 maximum peak height is also quoted for the same gamma ray to allow one to determine the deviation from Gaussian shape for the detector. The experimental ratio of full width at tenth maximum to full width at half maximum can be compared with the Gaussian ratio of 1.82. Finally, the ratio of full-energy peak height to Compton peak height is usually measured and quoted for a detector. This ratio is important in the detection of weak gamma rays in the presence of higher-energy gamma rays. The ratio of peak height to Compton height is a complicated function of volume, geometry, and packaging material, and is best determined experimentally. Peak-to-Compton ratios ranging from 15 to over 40 are commercially available.

At high energies, pair production becomes the dominant mechanism for the interaction of gamma rays with matter. Normally, the electron and the positron are stopped in the detector material. However, the positron and an electron then annihilate, producing two 511-keV gamma rays emitted in opposite directions. One or both of these annihilation gamma rays may escape the detector, resulting in three distinct peaks for a single high-energy gamma ray. These peaks are separated by 511 keV, and are designated the full-energy, one-escape, and two-escape peaks. The relative sizes of these peaks are strongly dependent on detector volume and geometry, and on the original gamma ray energy. Figure 7 shows spectra obtained for high-energy

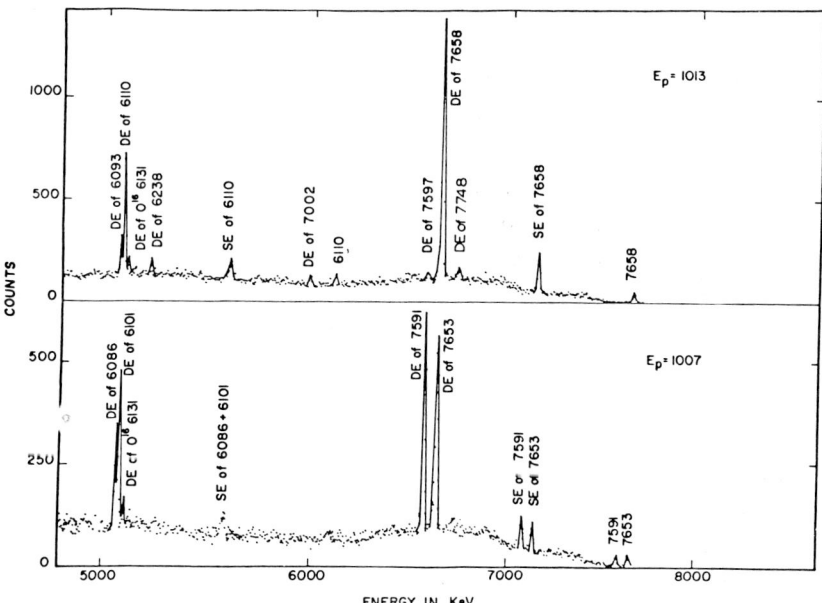

FIG. 7. High energy gamma rays detected by a Ge(Li) detector. High-energy gamma rays resulting from resonant proton capture by ^{48}Ti were detected using a 3% Ge(Li) detector [J. C. Legg, D. G. Megli, D. R. Abraham, L. D. Ellsworth, and S. Hechtl, *Phys. Rev.* **186**, 1138 (1969)]. The two spectra are labeled by proton resonance energy, and the various peaks are labeled by the gamma-ray energy and whether the peak is a single (one)-escape (SE) or double two-escape (DE) peak.

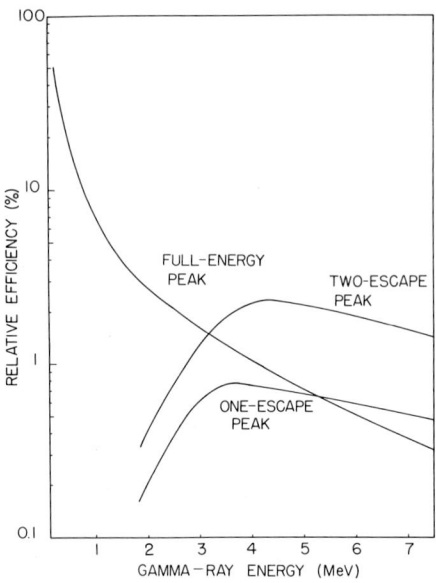

FIG. 8. Relative efficiences for various peaks in a Ge(Li) detector spectrum. The efficiencies of a 5% Ge(Li) detector for producing full-energy, one-escape, and two-escape peaks were experimentally determined over a range of gamma-ray energies. The efficiences are plotted as a percentage of the full-energy peak efficiency of a 7.6 × 7.6 cm NaI(Tl) detector for the ^{60}Co 1.33-MeV gamma ray.

gamma rays using a 3% germanium detector.[16] In Fig. 8 is plotted the experimentally determined full-energy, one-escape, and two-escape peak efficiencies as a function of gamma-ray energy for a 5% germanium detector.

At high counting rates, one may obtain peaks in a pulse height spectrum which do not correspond to any individual gamma ray. If a nucleus emits two or more gamma rays in rapid succession, there is a probability that both gamma rays will enter the detector and be observed. Since nuclear lifetimes are on the order of 10^{-12} sec or less, and the electronic time resolution is on the order of 10^{-6} sec, when both gamma rays are detected, the result is a pulse whose height is the sum of the detector response to the two gamma rays. High counting rates can also produce other effects. Counting-rate dependent variations of gain and base line are very common in the amplification electronics or photomultiplier. These effects can result in broadened peaks and, in extremely severe conditions, multiple peaks. Electronic base-line restorers and gain stabilizers are commercially available.

[16] J. C. Legg, D. G. Megli, D. R. Abraham, L. D. Ellsworth, and S. Hechtl, *Phys. Rev.* **186**, 1138 (1969).

A base-line restorer minimizes the counting rate effects on base-line voltage, while the gain stabilizer maintains the centroid of a preselected pulser peak or source peak at a fixed pulse height. Gain stabilization is also important for measurements at low counting rates over a long period of time in that it compensates for slow drifts in gain over the period of measurement.

Both sodium iodide and germanium detectors are susceptible to neutron-induced degradation of performance. These effects can be observed as spurious gamma rays resulting from the interaction of neutrons with the detector material, and in a permanent long term degradation of resolution due to the production of imperfections in the sodium iodide or germanium crystal. The lithium-drifted germanium detector, in particular, is highly susceptible to permanent damage by neutrons. Measurable broadening of peaks can be observed in germanium detectors after doses of 10^9 neutrons. Consequently, for measurements requiring high-resolution germanium detectors, it is imperative that the germanium detector be shielded from neutron radiation background. A working rule is that a large-volume germanium detector will no longer be useful after exposure to 100 mR of neutrons.

3.1.2.3. Gamma-Ray Spectrometers. One of the standard techniques for measuring the wavelength of any electromagnetic radiation is the diffraction grating. Diffraction grating spectrometers can be used in the measurement of gamma-ray wavelengths. In this use, the diffraction grating will be a crystal, and the measurement is done by using Bragg scattering. Bragg spectrometers will be discussed in great detail in a later section concerning the detection of x rays. We will restrict ourselves here to some general comments about the application of Bragg spectrometers to gamma-ray detection.

The first observation about diffraction spectrometers is that they depend on the Rayleigh scattering of gamma rays from matter. Because this scattering cross section decreases rapidly with energy, the usefulness of diffraction spectrometers in gamma-ray detection is therefore limited to low-energy gamma rays or to high-energy gamma rays which are produced by a very intense source. Further, for most diffracting crystals, the Bragg scattering angle is less than 3° for gamma-ray energies greater than 100 keV. Because of this small Bragg angle for gamma rays, virtually all gamma-ray diffraction spectrometers are of the transmission type.

Most gamma-ray crystal spectrometers are of the bent crystal type and are generally based upon the designs of Cauchois[17] or DuMond.[18] In addition, flat crystal spectrometers have been built[19] and used[20] for gamma-ray

[17] Y. Cauchois, *C. R. Acad. Sci. Paris* **199**, 857 (1934).
[18] J. W. M. Dumond, *Rev. Sci. Instrum.* **18**, 626 (1947).
[19] J. W. Knowles, *Can. J. Phys.* **37**, 203 (1959).
[20] J. W. Knowles, G. Manning, G. A. Bartholomew, and P. J. Campion, *Phys. Rev.* **114**, 1065 (1959).

measurements. The highest detection efficiencies are achieved with the bent crystal spectrometer, while the highest energy resolutions are obtained with the double flat spectrometer. For further information concerning Bragg spectrometers and the applications of crystal spectrometers to gamma-ray measurements, the reader is referred to Chapter 3.2 and to the article by Knowles.[21]

A process competing with the emission of low-energy gamma rays is that of internal conversion. In this process, the electromagnetic transition results in the emission of a bound electron from the atom whose nucleus undergoes electromagnetic decay. One can then measure the properties of electromagnetic decays by using electron spectrometers which measure the kinetic energy of the conversion electron. A knowledge of the kinetic energy of the electron and of the binding energy of the electron in its original orbit then allows one to determine the energy of the electromagnetic decay. Although the measurements associated with internal conversion are outside the subject of this volume, we refer the reader to the article by Rose[10] for further information about the large number of characteristics of electromagnetic decay which may be determined by internal conversion spectrometry. In instances where internal conversion is not a dominant process, one can still use an electron spectrometer to detect gamma rays. This method involves the gamma ray interacting with a thin foil "radiator" and undergoing photoelectric absorption with the photoelectron emerging from the radiator. The electron energy can then be measured by using an electron spectrometer, and, in this way, a gamma-ray energy can be determined by the external conversion spectrometer. In general, this method is restricted to low-energy gamma rays for which the photoelectric cross section is large, and may be used to take advantage of the high precision available with magnetic field electron spectrometers.

Of course, in the case of gamma rays incident on a radiator foil, the photoelectric effect is not the only major interaction of gamma rays with the material in the foil. In addition, Compton electrons are also produced so that one obtains a spectrum of electrons resulting from the Compton effect together with the photoelectron lines. Just as the photoelectron energies yield information about the gamma-ray energy, so also do the Compton electron energies. A careful determination of the highest energy in the Compton spectrum can determine the gamma-ray energy. A refinement of this Compton spectrometer is to define the angle through which the gamma ray is scattered by the Compton process. If one detects the Compton scattered photon and electron in coincidence, one then obtains a single energy "line"

---

[21] J. W. Knowles, in "Alpha-, Beta-, and Gamma-Ray Spectroscopy" (K. Siegbahn, ed.), p. 203. North-Holland Publ., Amsterdam, 1965.

whose energy can be used to determine the gamma-ray energy by use of Eq. (3.1.9). Compton spectrometers using well collimated incident gamma-ray beams and methods to restrict the electron scattering angle have been built and used by Groshev[22] and Motz.[23] These detectors have been used successfully to determine gamma-ray energies resulting from nuclear reactions in a nuclear reactor. Finally, one can use measurements of the angular distribution of the Compton scattering to determine the polarization of the original gamma ray, since this angular distribution depends on the gamma-ray polarization.[24,25]

For high-energy gamma rays, the dominant interaction mechanism is pair creation, and this leads then to yet another type of gamma-ray detector using the magnetic electron spectrometer as the detection agency. The magnetic field in this spectrometer is adjusted so that an electron–positron pair with equal kinetic energies arrive at two detectors in time coincidence. The field is uniform so that the electron and the positron each have half of the original gamma-ray energy minus 1.022 MeV. The use of magnetic field electron spectrometers then allows determination of energies with a resolution of down to 1% in energy, and an absolute energy determination with a precision of better than 1 part in 10^4. Such spectrometers have been used by Alburger[26] and Bent et al.[27] to measure successfully gamma-ray energies with high precision. The efficiency of this type of detection is rather low. Since one is selecting only electron–positron pairs with equal kinetic energy out of the full range of energies possible, the transmission of the spectrometer is related to the resolution used. This transmission is also somewhat dependent on the angular correlation between the electron–positron pair which is emitted. This angular correlation depends upon the multipolarity of the gamma-ray radiation for internal pair production, and has in fact been used as a means of measuring the multipolarity of high-energy gamma rays.[28]

The three-crystal pair spectrometer also depends on the pair-creation mechanism. In this arrangement, an absorption detector, which may be either a scintillator or a solid-state detector, has two other absorption detectors placed on opposite sides. The side detectors are shielded from

[22] L. V. Groshev, A. M. Demidov, V. N. Lutsenko, and A. Malov, *Izv. Akad. Nauk. SSSR Ser. Fiz.* **24**, 791 (1960); *Bull. Acad. Sci. USSR Phys. Ser.* **24**, 794 (1961).
[23] H. T. Motz, *Phys. Rev.* **104**, 1353 (1956).
[24] O. Klein and Y. Nishina, *Z. Phys.* **52**, 853 (1929).
[25] J. W. Olness, A. H. Lumpkin, J. J. Kolata, E. K. Warburton, J. S. Kim, and Y. K. Lee, *Phys. Rev. C* **11**, 110 (1975).
[26] D. E. Alburger, *Phys. Rev.* **111**, 1586 (1958).
[27] R. D. Bent, T. W. Bonner, and R. F. Sippel, *Phys. Rev.* **98**, 1237 (1955).
[28] E. K. Warburton, D. E. Alburger, A. Gallman, P. Wagner, and L. F. Chase, *Phys. Rev.* **133**, B42 (1964).

incident gamma rays, and a triple coincidence is required between the three detectors. This signals that a gamma ray in the central detector has been absorbed by pair creation, with the subsequent emission of the two annihilation quanta in opposite directions which escape to be detected in the side detectors. The energy spectrum from the central detector is analyzed and results in a spectrum consisting solely of two-escape peaks. That is, each gamma-ray peak has a pulse height corresponding to the original gamma-ray energy minus 1.022 MeV. This three-crystal pair spectrometer is particularly useful in analyzing a spectrum which is rich in high-energy gamma rays, since its virtue is that a single gamma ray produces a single peak.

3.1.3. Angular Correlations

Often, the experimenter wishes to determine more than the energies and intensities of gamma rays. It is desirable, in many instances, to gain information about the multipolarity of a gamma ray and the spins and parities of the initial and final states involved in a particular gamma-ray transition. The measurement of angular correlations often allows the experimenter to deduce much of this information.

An angular correlation measurement is simply the measurement of the variation of gamma-ray intensity as a function of angle between the direction of gamma-ray emission and another fixed direction in space. This direction may be fixed in a particular experiment to be the direction of an incident beam which excites the gamma-decaying state, or the direction of emission of a particle or gamma ray which results in the formation of the gamma-decaying state. We sketch an outline of the theory and pertinent results, and then discuss very briefly some of the more important kinds of correlation measurements.

For a more comprehensive discussion of gamma-ray angular correlations, the reader is referred to the articles by Rose and Brink,[29] Frauenfelder and Steffen,[30] and the book by Ferguson.[31]

In general, the gamma-ray intensity as a function of angle, $I(\theta)$, can be expressed as a Legendre polynomial sum:

$$I(\theta) = \sum_{l} a_l P_l(\cos \theta), \qquad (3.1.10)$$

where the coefficients a_l may be determined experimentally and compared to the predictions of reaction models. The coefficient a_0 is related to the

---

[29] H. J. Rose and D. M. Brink, *Rev. Mod. Phys.* **39**, 306 (1967).

[30] H. Frauenfelder and R. M. Steffen, *in* "Alpha-, Beta-, and Gamma-Ray Spectroscopy" (K. Siegbahn, ed.), p. 997. North-Holland Publ., Amsterdam, 1965.

[31] A. J. Ferguson, "Angular Correlation Methods in Gamma-Ray Spectroscopy." North-Holland Publ., Amsterdam, 1965.

3.1. GAMMA-RAY REGION

lifetime of the gamma-decaying state. The experimental values of a_0 for various gamma rays from a single initial state determine the "branching ratios," or relative strengths, of the various transitions. The lifetime of the state is usually determined by separate experiments which will be discussed in the following section. By convention, an angular correlation measurement determines a normalized angular distribution

$$\mathbf{W}(\theta) = 1 + \sum_{k=1} b_k P_k(\cos \theta) \tag{3.1.11}$$

for a particular gamma ray. Throughout the following discussion, we shall assume that no measurement of gamma-ray polarization is made.

Consider a gamma-ray transition between an initial state with spin and parity $J_1^{\pi_1}$ and final state with spin and parity $J_2^{\pi_2}$. The angular distribution will be isotropic, i.e., $b_k = 0$ for all k, if all magnetic substates of the initial state have equal populations. Let us define a population probability parameter $W(M_1)$ such that $\sum_{M_1} W(M_1) = 1$. Then, the case of equal substate populations would be expressed by $W(M_1) = 1/(2J_1 + 1)$ for all M_1. There are departures from this case which can produce anisotropic angular distributions. The first case is that of polarization, for which $W(M_1) \neq W(-M_1)$ for some value of M_1. The second case is that of alignment, for which $W(M_1) = W(-M_1)$ but $W(M_1) \neq 1/(2J_1 + 1)$ for some value of M_1. From these conditions, it is apparent that a spin 0 state cannot be polarized or aligned, a spin $\frac{1}{2}$ state can be polarized but not aligned, and states with spins of 1 or larger can be polarized or aligned. Thus, gamma rays from a spin 0 state must be isotropic, and gamma rays from a spin $\frac{1}{2}$ state can be anisotropic only if the state is polarized.

The normalized angular distribution when the gamma-ray polarization is not observed is given by

$$\mathbf{W}(\theta) = \sum_{\substack{(L\pi)(L'\pi') \\ k_{\text{even}}}} [\{B_k(J_1)R_k(LL'J_1J_2) \delta_L^{(\pi)} \delta_{L'}^{(\pi')} P_k(\cos \theta)\} / \sum_{L\pi} |\delta_L^{(\pi)}|^2], \tag{3.1.12}$$

where

$$B_k(J_1) = \sum_{M_1} W(M_1)(-)^{J_1 - M_1}(2J_1 + 1)^{1/2}(J_1 J_1 M_1 - M_1 | k\ 0), \tag{3.1.13}$$

$$R_k(LL'J_1J_2) = (-)^{1+J_1-J_2+L'-L-k}\{(2J_1 + 1)(2L + 1)(2L' + 1)\}^{1/2}$$
$$\times (LL'1 - 1|k\ 0) W(J_1 J_1 LL'; kJ_2), \tag{3.1.14}$$

and $\delta_L^{(\pi)}$ is the mixing ratio for radiation of multipolarity $L^{(\pi)}$. In this expression, the coefficient of $P_0(\cos \theta)$ is unity, as suggested above. There is a maximum value of k allowed by the properties of the Racah and Clebsch–Gordan coefficients. This maximum value k_{max} is the lesser of $2J_1$ or $2L_{max}$. Tables of R_k are given by Rose and Brink.[29]

In this case, the angular distribution is determined by the values of J_1, J_2, and $W(M_1)$ together with the multipolarities $L^{(\pi)}$ of the gamma-ray transition. In practice, the experimenter usually chooses experimental conditions which restrict possible values of $W(M_1)$, and then uses the experimental correlation coefficients to determine possible values of J_1, J_2, and $\delta_L^{(\pi)}$ for a particular gamma-ray transition. The experimenter, of course, must integrate the theoretical predictions over the finite solid angle subtended by the gamma-ray detector when comparing the predictions to his experimental results. We now shall list briefly various types of experiments and the results appropriate to them.

3.1.3.1. Resonant Capture of an Incident Particle. If a resonant state of spin and parity $J_1^{\pi_1}$ is formed by the capture of a particle with spin i from an unpolarized beam by an unpolarized target nucleus with spin I, then the beam direction is an axis of cylindrical symmetry, with respect to which the orbital angular momentum l has a projection $m_l = 0$. Figure 9 shows the arrangement for such an experiment. The target and projectile spins may be added quantum mechanically to form a channel spin s. Since the nuclear interaction may depend on the channel spin, one must treat the capture probability for each value of s as an independent variable. However, the contribution to the population parameters for different s add incoherently and, for a given s and l,

$$W(M_1) = (sJ_1M_1 - M_1|l\,0)^2. \qquad (3.1.15)$$

If several values of channel spin and orbital angular momentum are involved in forming the resonance, then

$$W(M_1) = \sum_s \left| \sum_l (sJ_1M_1 - M_1|l\,0)A(l,s) \right|^2 T(s), \qquad (3.1.16)$$

where $A(l,s)$ is the complex amplitude with which the partial wave l contributes to the formation of the resonance through channel spin s, and $T(s)$ is the probability of channel spin s contributing to the formation of the resonance. They are normalized by

$$\sum_l |A(l,s)|^2 = 1, \qquad \sum_s T(s) = 1. \qquad (3.1.17)$$

Obviously, these formulas are most useful when the contributing channel spin and orbital angular momentum are restricted to a single value. Then, one, knowing the spin and parity of either the initial or final state, usually

3.1. GAMMA-RAY REGION

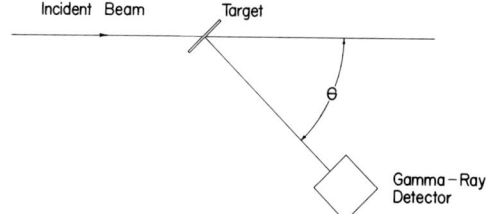

FIG. 9. Experimental arrangement for gamma-ray angular correlation measurements following particle capture. A beam of particls is incident upon a target. The yield of gamma rays resulting from the capture of incident particles is measured at various angles θ with respect to the incident beam direction.

can determine the spin and parity of the other state and the probable mixing ratio of the gamma-ray transition.[16, 32]

3.1.3.2. Nuclear Reactions. The gamma-decaying state may be prepared by a nuclear reaction of the form $A(a, b)B^*$. If the outgoing particle is not observed, cylindrical symmetry around the beam is preserved, but a complete knowledge of the nuclear reaction process is necessary to determine uniquely the population parameters. This method has been useful, however, in using the experimentally determined population parameters of a known gamma decay to determine properties of the reaction, e.g., the spin and parity of a resonance.[33]

In most cases, if the outgoing particle is detected in coincidence with the gamma ray, the cylindrical symmetry is destroyed and the gamma-decaying state may be polarized. These measurements are called triple-correlation measurements. However, in the "Litherland–Ferguson Method II,"[34] cylindrical symmetry and alignment with respect to the beam axis is observed. In this method, the outgoing particle is observed at 0° or 180° with respect to the beam direction. Thus, neither the incoming particle a nor the outgoing particle b can have a nonzero component of orbital angular momentum on the beam axis.

Rather than assuming a specific reaction mechanism, one normally assumes that all possible values of M_1 up to a maximum of the sum of spins of A, a, and b can be populated with parameters to be determined by the experiment. Properties of the gamma-ray transition can often be determined.[35] Obviously this method is most useful when the sum of spins is a small number. In particular, if the three spins add up to 0 or $\frac{1}{2}$, the $W(M_1)$ are uniquely determined.

[32] A. K. Hyder and G. I. Harris, *Phys. Rev. C* **4**, 2046 (1971).
[33] J. P. Schiffer, M. S. Moore, and C. M. Class, *Phys. Rev.* **104**, 1661 (1956).
[34] A. E. Litherland and A. J. Ferguson, *Can. J. Phys.* **39**, 788 (1961).
[35] A. R. Poletti and E. K. Warburton, *Phys. Rev.* **137**, B595 (1965).

3.1.3.3. Beta–Gamma Correlations.

If one has a source which undergoes beta decay followed by a gamma decay, a measurement can be made of the directional correlation between the beta particle and the gamma ray detected in coincidence. Here, the beta decay serves to determine a z-axis and to determine the population parameters $W(M_1)$.

Unfortunately, the beta decay is a three-body process with the neutrino and the daughter nucleus undetected. Further, beta decay is the result of more than one form of interaction with the vector and axial-vector couplings dominating the process. As a result, the correlation can depend upon the observed beta energy as well as on the details of the beta decay.

Consequently, the full theory of beta–gamma correlations is beyond the scope of this article. Readers who are interested in such measurements are referred to Frauenfelder and Steffen[30] for a detailed discussion.

3.1.3.4. Gamma–Gamma Correlation.

Similar measurements to those for beta–gamma correlations can be made for two cascade gamma rays detected in coincidence. In this measurement, the first-emitted gamma ray defines the z-axis and population parameters for the second gamma decay. Since the gamma-ray operator is well known, the situation is much simpler than for beta–gamma correlation measurements.

Poletti and Warburton[35] have used a formalism which is particularly useful for this case in that it easily allows for the possibility of n unobserved gamma transitions between the first- and last-observed gamma rays which are detected in coincidence. If the first-observed gamma-ray transition is between states J_1 and J_2 and the last-observed gamma ray is between states J_m and J_n, the correlation is

$$W(\theta) = \sum_k B_k(J_2) U_k(J_2 J_3) \cdots U_k(J_l J_m) R_k(J_m J_n) P_k(\cos \theta), \quad (3.1.18)$$

where

$$R_k(J_m J_n) = \sum_{(L\pi)(L'\pi')} \left[\{R_k(LL' J_m J_n) \delta_L^{(\pi)} \delta_{L'}^{(\pi')}\} \bigg/ \sum_L |\delta_L^{(\pi)}|^2 \right], \quad (3.1.19)$$

$$U_k(J_l J_m) = \sum_{L_{lm}} (\delta_{L_{lm}})^2 U_k(L_{lm} J_l J_m) \bigg/ \sum_{L_{lm}} (\delta_{L_{lm}})^2, \quad (3.1.20)$$

and

$$U_k(L_{lm} J_l J_m) = (-)^k [W(J_l J_m J_l J_m; L_{lm} k)/W(J_l J_m J_l J_m; L_{lm} 0)]. \quad (3.1.21)$$

If the state J_1 is statistically populated, $W(M_1) = (2J_1 + 1)^{-1}$, then

$$B_k(J_2) = \sum_{(L_{12}\pi)(L'_{12}\pi')} \left[\{(-)^{L_{12}-L'_{12}} R_k(L_{12} L'_{12} J_2 J_1) \delta_{L_{12}}^{(\pi)} \delta_{L'_{12}}^{(\pi')}\} \bigg/ \sum_{L_{12}} \langle \delta_{L_{12}}^\pi \rangle^2 \right] \quad (3.1.22)$$

for k even, and $B_k = 0$ for k odd. If the state J_1 is not statistically populated,

3.1. GAMMA-RAY REGION

then the $B_k(J_2)$ cannot be expressed in such a compact form, and the reader should consult more detailed articles such as Rose and Brink[29] to work out the form of $B_k(J_2)$ in that case.

3.1.3.5. Polarized Targets and Beams. If the gamma-decaying state is prepared by using either polarized targets or beams, the state will usually be polarized. If the gamma-ray polarization is not observed, the correlation will still contain only even-order Legendre polynomials, and the formalism [Eq. (3.1.12) et seq.] based on the population parameters $W(M_1)$ may still be used. However, one no longer can use the alignment condition $W(M_1) = W(-M_1)$ to simplify the sums in this case.

Of course, when the polarization of the emitted radiations are detected, additional information is obtained. In particular, such measurements have been particularly useful in understanding the basic processes of beta decay. Steffen and Frauenfelder[36] give an excellent review of this class of experiments.

3.1.3.6. Triple Correlations. A triple correlation is the determination of coincident counting rate as a function of three directions in space. The directions may be defined by a beam, emitted particles, or emitted gamma rays. As mentioned above, nuclear reactions resulting in an outgoing particle and a gamma ray detected in coincidence are examples of triple-coincidence measurements. The formalism for such measurements has been treated by Kraus et al.,[37] Sheldon,[38] and Satchler and Tobocman[39] by using different assumptions about the nuclear reaction mechanism. Bearse et al.[40] have measured and analyzed the triple correlation for the $^{12}C(p, p'\gamma)^{12}C$ reaction at a strong resonance. Figure 10 shows the apparatus used in this experiment.

The experimental difficulties of detecting three radiations in coincidence have restricted most work on triple correlations of the form $\beta-\gamma-\gamma$ and $\gamma-\gamma-\gamma$ to theoretical studies.[41-44] However, one "triple correlation" involves a gamma–gamma coincidence measurement from a state which is aligned in a third direction. Smith[44] has pointed out that such measurements can yield valuable information.

---

[36] R. M. Steffen and H. Frauenfelder, in "Alpha-, Beta-, and Gamma-Ray Spectroscopy" (K. Siegbahn, ed.), p. 1453. North-Holland Publ., Amsterdam, 1965.

[37] A. A. Kraus, J. P. Schiffer, F. W. Prosser, and L. C. Biedenharn, Phys. Rev. **104**, 1667 (1956).

[38] E. Sheldon, Rev. Mod. Phys. **35**, 795 (1963).

[39] G. R. Satchler and W. Tobocman, Phys. Rev. **118**, 1566 (1960).

[40] R. C. Bearse, J. C. Legg, G. C. Phillips, A. A. Rollefson, and G. Roy, Nucl. Phys. **65**, 545 (1965).

[41] L. C. Biedenharn, G. B. Arfken, and M. E. Rose, Phys. Rev. **83**, 586 (1951).

[42] G. B. Arfken, L. C. Biedenharn, and M. E. Rose, Phys. Rev. **86**, 761 (1952).

[43] V. de Sabbata, Nuovo Cimento **21**, 659, 1058 (1961).

[44] P. B. Smith, "Nuclear Reactions" (P. M. Endt and P. B. Smith, eds.), Vol. 2, p. 248. North-Holland Publ., Amsterdam, 1962.

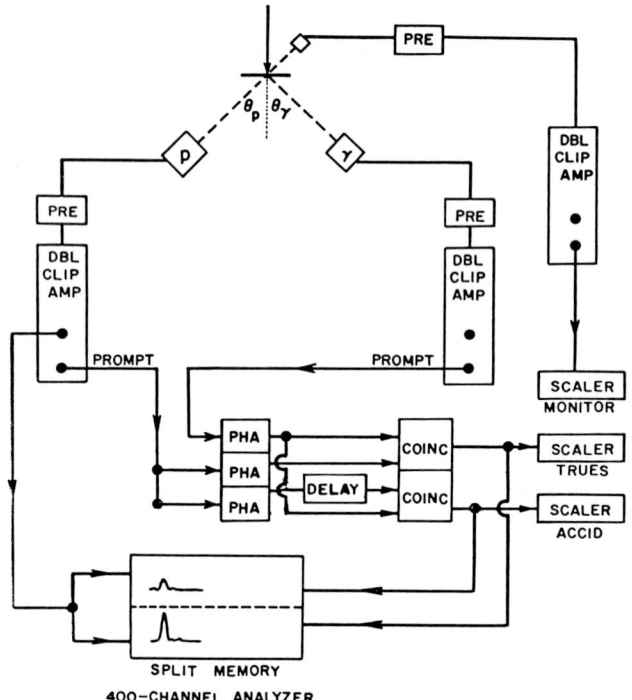

FIG. 10. An experimental arrangement for triple-correlation measurements. The experimental arrangement used by R. C. Bearse, J. C. Legg, G. C. Phillips, A. A. Rollefson, and G. Roy [*Nucl. Phys.* 65, 545 (1965)] in measuring the triple correlation in the ^{12}C(p, p′γ)^{12}C reaction is shown. Protons and gamma rays were detected and time coincidence requirements were imposed. The "true" coincidence spectrum included both true coincidence events and accidental coincidences resulting from the finite time resolution of the electronics. The accidental coincidence had one signal delayed so that it contained only accidental coincidences. The true coincident events are the total minus the accidental events, and were normalized by the number of reaction particles detected in a fixed-angle monitor detector.

3.1.3.7. Perturbed Correlations. All of our discussion concerning correlations assumed that the alignments or polarizations were determined solely by the nuclear interactions. However, if a gamma-decaying state is subjected to external magnetic or electric fields, these fields can affect the population parameters, resulting in a perturbed angular correlation.

A number of review articles [45–47] give complete discussions of the experimental and theoretical details. We shall simply note that measurements of

[45] R. M. Steffen, *Advan. Phys.* 4, 293 (1955).
[46] L. Grodzins, *Annu. Rev. Nucl. Sci.* 18, 291 (1968).
[47] E. Karllson, E. Matthias, and K. Siegbahn (eds.), "Perturbed Angular Correlations." North-Holland Publ., Amsterdam, 1964.

the perturbed correlations can result in the determination of the multipole moments of nuclear excited states, or in the determination of crystalline fields which produce the perturbation.

3.1.4. Transition Rate and Lifetime Measurements

3.1.4.1. General Considerations. Measurements of transition probabilities of nuclear states may be made from the lifetime of the state, or from the width of the state since $\Gamma\tau = \hbar$. However, the widths of the states are very narrow, so that width measurements are generally more difficult. The widths in electron volts are given by $\Gamma = 0.658 \times 10^{-15}$ eV/τ(sec) so that for lifetimes of 10^{-6} sec and longer, one has widths of 10^{-9} eV and less. Lifetime measurements have been very extensive, and have been extended down to values as low as 10^{-11} sec in direct measurements and down to 10^{-16} sec by using slowing-down processes as a clock.

The probability of decay of a group of identical nuclei is proportional to the number of nuclei present, so that a measurement of the number that decay as a function of time provides a value for the lifetime. The activity $-dN/dt = \lambda N$, where λ is the decay constant, is integrated to obtain the number as a function of time, $N = N_0 e^{-\lambda t} = N_0 e^{-t/\tau}$, where N_0 is the number at time $t = 0$ and τ is the mean lifetime. The exponential decay law is then

$$dN/dt = dN/dt|_{t=0}\, e^{-t/\tau}, \tag{3.1.23}$$

so that a graph of $\ln(dN/dt)$ versus time will yield the mean lifetime of the decaying nuclei.

The half-life, i.e., the time for $N = \tfrac{1}{2}N_0$, is equal to $\tau/\ln 2$, and is customarily used for lifetimes of nuclei in the range of seconds and longer. The activity of a quantity of radioactive nuclei is measured in curies, where 1 Ci = 3.7×10^{10} distingrations per second. In order to determine the number of gamma rays from a particular source, the decay scheme must be known. Although there are long-lived radioactive sources that emit gamma rays of a single energy, many are more complex, and the number of gamma rays per distingration needs to be taken from the decay scheme.[6, 48, 49]

Considerable complications can occur in measured activities as a function of time if the decay of a parent nucleus to a daughter nucleus is followed by decay to a granddaughter. The activity of the daughter can even increase with time, depending upon the relative lifetimes of parent and daughter. For the case of decay A to B to C, with decay constants λ_A and λ_B, the activities are

$$dN_A/dt = -\lambda_A N_A, \tag{3.1.24}$$

$$dN_B/dt = +\lambda_A N_A - \lambda_B N_B. \tag{3.1.25}$$

---

[48] P. M. Endt and C. von der Leun, *Nucl. Phys.* **A214**, 1 (1973).
[49] *Nucl. Data Sheets* **1–10** (1966–1973).

For $N_A = N_0$ and $N_B = 0$ at $t = 0$, the solution is

$$N_B = [\lambda_A/(\lambda_B - \lambda_A)]N_0(e^{-\lambda_A t} - e^{-\lambda_B t}), \tag{3.1.26}$$

which can change sign, depending upon λ_A, λ_B, and t. More complicated activities can occur if nucleus C also decays. These considerations apply to individual levels in the same nucleus as well as to separate species of nuclei.

3.1.4.2. *Direct Timing.* Measurement of the lifetime of a nuclear state can be made by determining the number of decays in a time range smaller than the lifetime but comparable to it. The slope of a graph of the logarithm of number of decays versus time then gives the decay constant or reciprocal lifetime. For half-lives in the range of minutes and longer, a stop-watch is a suitable timing mechanism. For lifetimes down to about 10^{-3} sec, mechanical means can be employed, and for shorter values, electronic systems can be used.

Considerable variation in technique is involved in the choice of time $t = 0$ for the measurement. Clearly, if one has a number of nuclei that decay by the transition of interest, and no additional nuclei are added, then the choice of $t = 0$ is arbitrary. For a transition that is the secondary one, such as the 1.52-MeV ^{39}Ar (see Fig. 11) gamma ray fed by beta decay of ^{39}Cl with a half-life of 55.5 min, the decay of the state is governed by the parent nucleus. In this case, one could use the occurrence of the β^- particle as a signal of the population of the state, and measure the time difference between the β^- particle and the gamma ray to obtain the decay rate of the ^{39}Ar level. In a similar fashion, outgoing particles from nuclear reactions are commonly used to indicate the population of individual states in the residual nuclei, such as the outgoing protons inelastically scattering from ^{40}Ca and leaving ^{40}Ca in an excited state. Time difference measurements are conveniently made with a time-to-amplitude converter. These are commercially available, and can be employed for measurements down to mean lives of about 10^{-10} sec.

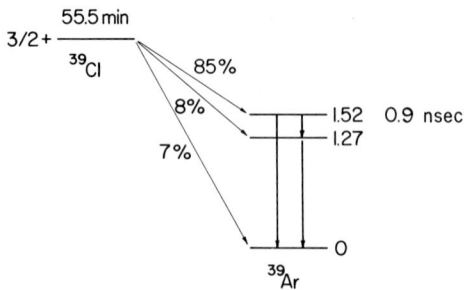

FIG. 11. Decay scheme for ^{39}Cl. The lifetime of the 1.52-MeV level in ^{39}Ar may be obtained by determining the number of gamma rays as a function of time after the β^- particle is emitted in the decay of ^{39}Cl.

3.1.4.3. Recoil Distance Method.

In a nuclear reaction in which a projectile is captured by a target nucleus or in which a compound system is formed that breaks up by emitting a particle to leave a residual nucleus, the final nucleus recoils with a velocity that can be calculated. If the recoiling nucleus excited in the reaction recoils in a vacuum, then the number of gamma rays emitted as a function of distance from the original target position depends upon the lifetime of the state.

Since gamma rays are not easily collimated, the number decaying in flight can be determined by measuring the Doppler-shifted gamma-ray energy with a high-resolution Ge(Li) detector. The Doppler-shifted energy is given by $E_\gamma = E_0(1 + (v/c) \cos \theta)$, where E_0 is the energy at rest, v the velocity of the source, and θ the angle between source and detector. If the recoils are stopped in some material after a certain flight path, they decay with the unshifted gamma-ray energy (see Fig. 12). Thus, the number decaying in flight and the number decaying after reaching the stopper are obtained from the number at the shifted and unshifted gamma-ray energies. By varying the distance from target to stopper, a decay curve can be obtained.

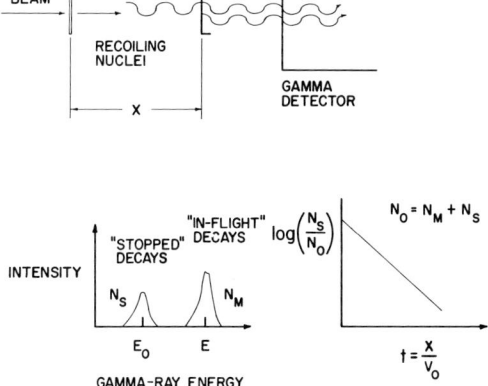

FIG. 12. Recoil-distance measurement of lifetime of gamma-decaying nuclear state. The number of gamma rays emitted in flight and the number of gamma rays emitted in the stopping material are determined from the number that are Doppler-shifted in energy and the number that are not shifted. The decay curve is obtained as a function of distance from the target to the stopping material.

The range of lifetimes that can be studied by this method depends upon the uncertainty in position of the stopper, nonuniformities in target and stopping material surfaces, velocity of the recoiling nucleus, and resolution of the gamma-ray detector. Of course, the lifetime of the nuclear state must be much greater than the stopping time of the recoiling nucleus, which is of the order of 10^{-13} sec. For a nuclear velocity of 10^9 cm/sec, uncertainties of 0.001 mm in position, and 10 measurements in a 0.01-mm flight path, lifetimes of the order of 10^{-12} sec can be determined. Since Ge(Li) detectors can be obtained with 2-keV resolution, the gamma-ray energy shift should be greater than about 6 keV, which would occur for gamma rays of about 200 keV and higher at velocities of 10^9 cm/sec.

3.1.4.4. Doppler Shift Attenuation. The energy loss per unit distance, or "stopping power," of moving ions in materials has been extensively studied for many years and is well-known for a large number of ions and materials. A recent article[50] uses all available data and extrapolations based on the theory to present the stopping power of all elements in a wide range of materials. At velocities below $0.01c$ there are few measurements, but these show that the stopping power is proportional to the velocity of the ion:

$$dE/dx = -kv. \tag{3.1.27}$$

The theory of Lindhard et al.[51] gives a general formula for k, which is accurate to better than 30%.[52]

The linear dependence upon velocity is known as the electronic stopping power, and, according to this process, the ion moves in a straight line. However, at very low velocities, the ions lose energy by scattering in "billiard-ball" collisions and no longer move in a straight line. This process is known as nuclear stopping, in which scattering occurs between nuclear centers, and is also calculated by Lindhard et al.[51]

A gamma ray emitted by a nucleus that is slowing down in matter will have an energy that depends upon the velocity at the time of emission. For gamma rays emitted before slowing down, the maximum Doppler shift is obtained, and for gamma rays emitted after the nucleus has come to a stop, the "rest energy" is obtained. The average attenuation of the Doppler shift, $\langle S \rangle$, is found by integrating the Doppler shift over the velocity and decay dependence upon time:

$$\langle S \rangle = \int_0^\infty S \, dn/n_0 = (\cos \theta)/c \int v(t) \, dn/n_0, \tag{3.1.28}$$

where $n = n_0 e^{-t/\tau}$ so that

$$\langle S \rangle = (\cos \theta)/c\tau \int_0^\infty v(t) e^{-t/\tau} \, dt. \tag{3.1.29}$$

The average shift is usually expressed in terms of a fraction of the full shift,

$$\langle S \rangle = F(\tau)(v_0/c) \cos \theta, \tag{3.1.30}$$

so that $F(\tau)$ is measured experimentally and compared with calculations for the stopping process involved as a function of the nuclear lifetime τ.

---

[50] L. C. Northcliffe and R. F. Schilling, *Nucl. Data* **A7**, 233 (1970).

[51] J. Lindhard, M. Scharff, and H. E. Schiott, *Kgl. Dan. Vidensk. Selsk. Mat. Fys. Medd.* **33**, No. 14 (1963).

[52] J. H. Ormrod, J. R. Macdonald, and H. E. Duckworth, *Z. Naturforsch.* **21a**, 130 (1966).

For lifetimes much shorter than the stopping time, $F(\tau) \sim 1$, while for lifetimes much longer, $F(\tau) \sim 0$. As noted above, the nuclear stopping process changes the direction of the ion so that $\cos \theta$ is not known at very low velocities. Procedures for correcting for the nuclear stopping process have been carried out for both the additional slowing down and the change in direction.[53, 54] The range of lifetimes studied in this process is 10^{-13}–10^{-15} sec.

3.1.4.5. Coulomb Excitation. Nuclei can be excited to the first few excited states by the Coulomb field of an incident nuclear projectile. If the energy of the projectile is sufficiently low that the nuclei do not come within the range of the nuclear force, the orbit can be calculated from just the Coulomb repulsion to a high degree of accuracy. The excitation probability can then be separated into a kinematic portion and a nuclear matrix element for absorption of energy to the final nuclear state.

Excitations directly from the ground state to a final state have been calculated by quantum mechanical and semiclassical methods.[55] Multistep processes, such as from a 0^+ to 2^+ to 4^+ state, rather than the $L = 4$ absorption directly from 0^+ to 4^+ state, have been considered both in particular models of the nucleus and in the general case, and have been treated in several papers.[55-57] Many measurements have shown the existence of transitions which have rates that are as high as 50 to 100 times the single-particle transition probability, thus demonstrating the collective nature of many electromagnetic transitions.

3.1.4.6. Fluorescence Measurements

3.1.4.6.1. RESONANCE FLUORESCENCE. Absorption measurements can be made with a source of gamma rays that is continuous with energy, such as from Bremsstrahlung radiation or the Compton distribution of scattered monoenergetic gamma rays. The nuclear level width can be obtained from the probability for absorption from the incident gamma-ray flux and emission and detection in a direction away from the incident flux direction. However, the nuclear level widths are extremely small, so that the number of gamma rays in the appropriate energy range is small. The gamma-ray background from the necessarily strong incident flux makes these kinds of measurements

---

[53] A. E. Blaugrund, *Nucl. Phys.* **88**, 501 (1966).

[54] W. M. Currie, *Nucl. Instrum. Methods* **73**, 173 (1969).

[55] K. Alder, A. Bohr, T. Huus, B. Mottelson, and A. Winther, *Rev. Mod. Phys.* **28**, 432 (1956).

[56] K. Alder, A. Winther, *Kgl. Dan. Vidensk. Selsk. Mat. Fys. Medd.* **32**, No. 8 (1960).

[57] A. Winther and J. de Boer, in "Coulomb Excitation" (K. Alder and A. Winther, eds.), p. 303. Academic Press, New York, 1960.

difficult. Improvements have been made[58] by obtaining variable energy monochromatic gamma rays from a bent crystal spectrometer.

Another procedure for producing nuclear fluorescence is to use gamma rays from a radioactive source to study the same gamma-ray transition. The difficulty is that the energy of the emitted gamma ray is shifted as the nucleus recoils. From momentum and energy conservation,

$$(M - (E_\gamma/c^2))v_R = E_\gamma/c, \quad v_R \sim E_0/M_c; \quad (3.1.31)$$

$$E_\gamma = E_0 - E_R = E_0 - \tfrac{1}{2}Mv_R^2, \quad (3.1.32)$$

where E_0 is the excitation energy of the level, E_γ the gamma-ray energy, and v_R the recoil velocity. For a 100-keV transition in a mass 100 nucleus, the recoil velocity is 8.6×10^4 cm/sec, which gives an energy shift of 0.053 eV. This is considerably greater than level widths, and so no overlap occurs, even when broadening due to the thermal motion of individual atoms is included.

An interesting variation is the use of gamma rays produced in a nuclear reaction in which the emitting source is a recoiling nucleus. By varying the angle between the recoiling nucleus and the absorbing nucleus, the Doppler-shifted energy can be scanned through the resonance. This procedure has been used to study a transition in ^{28}Si by producing gamma rays in ^{28}Si by the ^{27}Al(p, γ) reaction.[59]

Mechanical motion of the source or absorbing material may be employed to shift the gamma-ray energy. By equating total energies before and after emission of a photon, it can be shown[60] that recoil-free emission occurs at relative velocity of approximately $\tfrac{1}{2}v_R$. For the example given above, this corresponds to 3.4×10^4 cm/sec, which could be achieved for the tangential velocity of a 10-cm radius rotor moving at 5400 rps. Such methods have been employed to measure nuclear resonance fluorescence, and are discussed in several review articles.[61,62]

3.1.4.6.2. MÖSSBAUER EFFECT. Most nuclear resonance measurements now rely on the Mössbauer effect, in which an emitting nucleus located in a crystal lattice does not recoil as much due to the large effective mass of the nucleus in its environment. If the emitting and absorbing nuclei are strongly bound to a crystal lattice, they do not individually recoil, and so

---

[58] E. J. Soppi and F. Boehm, *Bull. Amer. Phys. Soc.* **116**, 503 (1961).

[59] P. B. Smith and P. M. Endt, *Phys. Rev.* **110**, 397 (1958).

[60] P. Marmier and E. Sheldon, "Physics of Nuclei and Particles," Vol. I, p. 448. Academic Press, New York, 1969.

[61] S. Devons, in "Nuclear Spectroscopy" (F. Ajzenberg-Selove, ed.), Part A, p. 512. Academic Press, New York, 1960.

[62] K. G. Malmfors, in "Alpha-, Beta-, and Gamma-Ray Spectroscopy" (K. Siegbahn, ed.), p. 1281. North-Holland Publ., Amsterdam, 1965.

the energy and width of the natural gamma-ray line are undisturbed. The work of Mössbauer with ^{191}Ir showed that the number of recoil-free nuclei increased as the temperature of the crystal was lowered,[63] as would be expected since the lattice vibration would decrease. However, the vibrations of the crystal produce a continuous background that cannot be fully eliminated with cooling because of the zero-point vibration energy. For the gamma rays with the natural line width, a small relative motion between source and absorber, of the order of a few centimeters per second, can be employed to determine the width of the resonance.

The Mössbauer effect has been studied in many nuclei and has been found to be appreciable at room temperature in some cases, notably that of ^{57}Fe. In this nucleus, the first-excited state at 14.4 keV, populated in the decay of ^{57}Co, has a relatively long half-life of 1.4×10^{-7} sec, so that the natural line width is very narrow, at 5×10^{-9} eV. The large internal magnetic field in Fe produces a nuclear Zeeman effect, splitting the $(2J + 1)$-fold degeneracy of the initial and final nuclear states in the gamma decay. Thus, one can determine the spin g-factor and magnetic moment of the 14.4-keV level. External fields can be applied to make magnetic dipole measurements in other nuclei, and even electric quadrupole moments.[64, 65]

The range of lifetimes that can be studied by the Mössbauer technique is limited to less than about 10^{-10} sec because extranuclear fields cause broadening of narrower lines, thus leading to erroneously low values of the lifetime. However, the very narrow lines from long lifetimes have been useful for other purposes such as measuring moments, as mentioned above, measurements of changes in nuclear radii for excited states,[66] and minute effects of chemical binding on nuclear level structure when source and absorber have different environments.[64–67] Many review articles[68–70] discuss these effects in greater detail.

[63] R. L. Mössbauer, *Z. Phys.* **151**, 124 (1958).
[64] O. C. Kistner and A. W. Sunyar, *Phys. Rev. Lett.* **4**, 412 (1960).
[65] S. DeBenedetti, G. Lang, and R. Ingalls, *Phys. Rev. Lett.* **4**, 60 (1961).
[66] L. R. Walker, G. K. Wertheim, and V. Jaccarino, *Phys. Rev. Lett.* **6**, 98 (1961).
[67] T. A. Carlson, P. Erman, and K. Fransson, *Nucl. Phys.* **A111**, 371 (1968).
[68] H. Frauenfelder, "The Mössbauer Effect." Benjamin, New York, 1962.
[69] A. J. F. Boyle and H. E. Hall, *Rep. Progr. Phys.* **25**, 441 (1962).
[70] R. L. Mössbauer, *in* "Alpha-, Beta-, and Gamma-Ray Spectroscopy" (K. Siegbahn, ed.), p. 1293. North-Holland Publ., Amsterdam, 1965.

3.2. X-Ray Region*

3.2.1. Introduction

A vast amount of information about x rays has been accumulated since Roentgen first discovered the x ray in 1895.[1] Because its wavelength is of the same order as atomic dimensions, it has been a powerful tool in probing the atom. Today, its primary importance is still the investigation of atomic and molecular processes. The x-ray region is defined here as radiation in the wavelength range between approximately 62 and 0.1 Å, or in the corresponding energy range between approximately 200 eV and 124 keV. This region extends from the carbon K x rays and argon L x rays to the uranium K x rays. The lower limit is suggested by the limitation of techniques which are normally used in x-ray studies, whereas the upper limit is set by the atomic radiation produced by naturally occurring materials. Atomic radiation from man-made transuranium elements extends the region of interest for x-ray studies. Atomiclike transitions in mu-mesonic atoms leading to muonic x rays have energies far above the x-ray region but, because of their nature, are included in this section.

Because of space limitations, all of the information concerning x rays cannot hope to be included in this section. The traditional fields of x-ray spectroscopy are reviewed in several excellent and comprehensive studies.[2-5] In this treatment, we discuss the production and identification of x rays emitted from various systems. X-ray spectra produced by ion–atom collisions, hot plasmas, laser-induced plasmas, and high-voltage sparks are discussed, together with the more conventional electron- and photon-induced spectra. Some of the newly observed x-ray features discussed include quasi-molecular x rays, radiative electron capture x rays, radiative Auger electron

[1] W. K. Roentgen, *Sitzungsber. Phys. Med. Ges.*, Würzburg 1895; *Ann. Phys. Chem.* **64**, 1 (1898).

[2] L. V. Azároff, "X-Ray Spectroscopy." McGraw-Hill, New York, 1974.

[3] A. E. Sandström, "Handbuch der Physik" (S. Flügge, ed.), Vol. 30, pp. 78–245. Springer-Verlag, Berlin and New York, 1957.

[4] A. H. Compton and S. K. Allison, "X Rays in Theory and Experiment." Van Nostrand-Reinhold, Princeton, New Jersey, 1935.

[5] M. Siegbahn, "The Spectroscopy of X Rays." Oxford Univ. Press, London, and New York, 1925.

---

* Chapter 3.2 is by Robert L. Kauffman and Patrick Richard.

x rays, and hypersatellite x rays. Basic techniques of x-ray detection and x-ray spectroscopy are summarized. The field of absorption spectroscopy is not discussed here, although these studies have provided much information about atoms and especially solid effects.[2] With the development of x-ray synchroton radiation, the interest in this field should greatly expand. X-ray scattering phenomena such as x-ray diffraction and Compton scattering are not discussed here.[4]

The units used for x-ray studies have varied over the years. Some of the difficulties in units exist because there is no easy connection between the x-ray region and the optical region which uses a multiple of the meter to measure wavelengths. Two units were commonly used, the x-ray unit xu and the angstrom Å. The angstrom was defined as 10^{-10} m, but, in practice, no test could be made of its accuracy. The x-ray unit was defined in terms of the crystal spacing in calcite. This original definition was believed to be such that the xu was equal to 10^{-3} Å. The accepted conversion factor is[6]

$$1 \text{ kxu} = 1.002056 \pm 0.000005 \text{ Å}. \tag{3.2.1}$$

A new unit, designated by Å*, has been defined in terms of a standard x-ray line. The standard x-ray line selected is the $K\alpha_1$ transition of tungsten. Its wavelength is defined as

$$\lambda(W_{K\alpha_1}) = 0.2090100 \text{ Å*}. \tag{3.2.2}$$

This unit may differ from the angstrom by ± 5 ppm. The convenience of this standard is that it can easily be obtained by the spectroscopist for calibration purposes. A convenient formula for converting wavelength to energy is

$$E\lambda = 12398.10 \pm 0.13 \text{ eV-Å*}, \tag{3.2.3}$$

where E is the photon energy in electron volts and λ is the photon wavelength in (angstrom units)*. Recent precision studies of the x ray to visible wavelength ratios by Deslattes and Henins,[7] using an x-ray/optical interferometer, have shown that the conversion factor, using Cu $K\alpha_1$ and Mo $K\alpha_1$ reference x-ray lines, varies in parts per hundred thousand, whereas the experimental accuracies are in parts per million.

3.2.2. Detectors and Spectrometers

Because the energy of a single x ray is large compared to a typical ionization energy, individual photon counting is possible in the x-ray region. Energy information can be obtained from the detector signal itself if high

---

[6] For a discussion concerning wavelength units, see J. A. Bearden, *Rev. Mod. Phys.* **39**, 78 (1967).

[7] R. D. Deslattes and A. Henins, *Phys. Rev. Lett.* **31**, 972 (1973).

resolution is not required. For high resolution work, some type of dispersion device must be used. Normally, these are combinations of crystals or gratings. A few of the more common types of crystal spectrometers will be discussed. The gratings are more applicable in the vacuum ultraviolet region and will not be discussed here. A photoelectron spectrometer has recently been developed. This instrument and its applications are described. Finally, a Doppler-tuned spectrometer is described that has been used to measure x rays in beam-foil experiments.

3.2.2.1. Gas Counters. Three types of gas-filled detectors (ionization chamber, proportional counter, and Geiger–Müller counter) have been used for x-ray detection.[8,9] The three detectors are similar in their design and operation. Each consists of a gas-filled chamber with a thin, centrally located anode wire held at a constant positive voltage. When a quantum of radiation enters the detector it is absorbed by the gas, usually by the photoelectric process. The electrons created during the process are collected on the anode wire. For the ionization chamber, the number of electrons collected is equal to the energy of the quantum of radiation divided by the average energy required to ionize each gas atom. The total charge collected per photon in an ionization chamber is small, on the order of 10^{-17} C; therefore, they cannot be used for individual photon counting. They are usually operated in a dc mode, detecting the total current for large x-ray doses. In the mode of operation for the proportional counter and the Geiger–Müller counter, the anode voltage is sufficient to cause electron avalanche, amplifying the electronic pulse from the x ray. The primary electrons created by the x ray can acquire sufficient energy in their mean free path to cause secondary ionization. This avalanche effect is great enough to cause signals 10^4 times as great as the primary ionization. In both counters, pulses from individual photons are readily detected. The signal from a Geiger–Müller counter is a function of the detector parameters only and does not contain any energy information. The proportional counter,[10,11] which is the most commonly used gas detector, produces a signal which is dependent on the x-ray energy.

The geometry of the proportional counter usually approximates a cylinder with the anode placed along the axis. Many designs have been tested to improve resolution and efficiency. These include the addition of shields and

---

[8] S. A. Korff, "Electron and Nuclear Counters," pp. 6–194. Van Nostrand-Reinhold, Princeton, New Jersey, 1955.

[9] M. A. Blokhin, "Methods of X-Ray Spectroscopic Research," pp. 85–119. Pergamon, Oxford, 1965.

[10] D. West, *Prog. Nucl. Phys.* **3**, 18–62 (1953).

[11] S. C. Curran, "Handbuch der Physik" (S. Flügge, ed.), Vol. 45, pp. 174–221. Springer-Verlag, Berlin and New York, 1958.

grids and the use of multiple wires. The anode wire is on the order of 50 μm in diameter. A thin foil window is placed on one side or at one end of the detector. The windows usually contain only light elements which are transparent to the x rays of interest. The inside surface of the window is coated with a metallic material to prevent surface charge buildup. The counter is filled with a gas which does not readily form negative ions. Inert gases such as argon and xenon are commonly used for proportional counter gases, and are mixed with small quantities of an organic quenching gas such as methane or carbon dioxide.

The increase in the pulse height from a proportional counter is usually given in terms of the gas amplification factor. Several efforts have been made, both theoretically and experimentally, to determine this factor.[12-14] The gas amplification factor is strongly dependent on the electric field in the vicinity of the wire. It is in this region where most of the avalanche occurs. For a proportional counter, the avalanche is only created locally in the vicinity of the radiation, giving a signal proportional to the primary ionization, whereas, for a Geiger–Müller counter, the avalanche continues everywhere along the anode, causing a long saturated pulse. The detector voltage, the gas pressure and purity, the wire diameter, and the detector geometry all affect the gas amplification factor.

The energy resolution† $\Delta E/E$ for a proportional counter can be given by

$$\Delta E/E = CE^{-1/2}, \qquad (3.2.4)$$

where C is a proportionality constant and is proportional to the Fano factor[15] F and to the square root of the average energy required to produce an electron–ion pair, $w^{1/2}$. For an argon filled counter, C ranges from a lower limit of 0.14 keV$^{1/2}$ for a good counter. Factors which affect the resolution of the counter are variations in gas pressure, gas impurities, electronic noise, variations in wire thickness, end effects of the wire, and counting rate. Most of the effects can be minimized by careful experimental technique. The end effect is due to a decrease of the electric field surrounding the wire at its union to the counter, and can be minimized by the use of shielding and grids. Gain shifts which can affect the resolution have been observed for high count rates. When a pulse occurs in the detector, the voltage in the region of the pulse is temporarily lowered. If the count rate

---

[12] A. J. Burek and R. L. Blake, *Advan. X-Ray Anal.* **16**, 37–52 (1973).
[13] G. D. Alkhazov, *Nucl. Instrum. Methods* **89**, 155 (1970).
[14] A. Zastawny, *J. Sci. Instrum.* **43**, 179 (1966).
[15] U. Fano, *Phys. Rev.* **72**, 26 (1947).

---

† The energy resolution is the reciprocal of the resolving power R commonly stated for other regions.

is sufficiently high such that the pulses occur before the detector can recover, the second pulse will have a lower pulse height than the first pulse since the gas amplification at the lowered detection voltage will be smaller. Even if all of the resolution effects are minimized, detector resolution of between 50 and 6% can be realized for the normal operating region of the proportional counters (x-ray energies between 0.5 and 15 keV). This resolution is not as good as that which can normally be obtained from a Si(Li) detector at these energies.

The efficiency of the proportional counter is determined by the number of x rays absorbed in the window and by the number of photons stopped in the detector gas.[16] Calculated efficiency curves for some commercially produced proportional counters and Geiger–Müller counters are shown in Fig. 1. The low energy region of the curves is dominated by the window

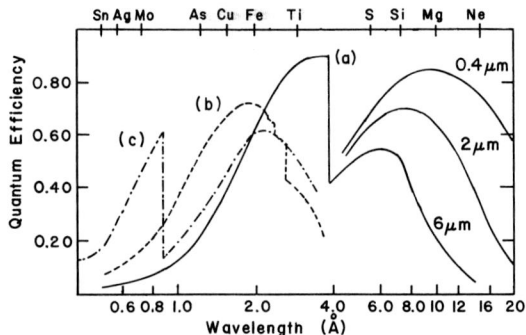

FIG. 1. Calculated quantum efficiency curves for several gas counters. These are (a) a proportional counter filled with 90% argon and 10% methane and having entrance windows of 6-, 2-, and 0.4-μm thickness, (b) a proportional counter with a 15-μm mica window and filled with xenon at 0.27 bars, and (c) a Geiger–Müller counter with a 7-μm mica window filled with krypton at 0.2 bars (Courtesy of Siemens Corp.).

absorption. The fraction of the x rays transmitted, I/I_0, through a window of a given material having a thickness t, is given by

$$I/I_0 = e^{-(\mu/\rho)\rho t}, \tag{3.2.5}$$

where μ/ρ is the mass absorption coefficient of the material[17] and ρ is the mass density of the material. The high-energy portion of the efficiency curves is dominated by the absorption of the x ray in the active volume of the gas. Absorption varies approximately as Z^4 for different ionizing target gases, which is the reason that the Kr and Xe filled counters are more efficient than Ar for the higher-energy x rays. If the detector has a dead region of

---

[16] J. Taylor and W. Parrish, *Rev. Sci. Instrum.* **26**, 367 (1955).
[17] See, for example, B. L. Henke and R. L. Elgin, *Advan. X-ray Analy.* **13**, 639–65 (1970).

gas, this must also be included in the calculations of efficiency. In the dead region, the x rays can be absorbed by the gas but the electrons created in the process are not collected at the anode. The absorption in the dead region can be calculated in the same manner as absorption in the window if the effective thickness of the dead layer is known. The abrupt changes in efficiency at about 4 Å for the argon filled counter, and at 0.9 Å for the krypton filled counter, are due to the K absorption edge of the gases. Efficiencies in the region immediately surrounding the absorption edge are difficult to determine.

3.2.2.2. Solid-State Detectors. The solid-state detector is the most widely used x-ray detector in experiments not employing dispersive devices such as crystal spectrometers.[18–21] The two most common solid-state detectors for x rays are the lithium-drifted silicon detector, Si(Li), used in the lower-energy x-ray region, and the lithium-drifted germanium detector, Ge(Li), used in the high-energy x-ray region. The solid-state detectors have better resolution and are more efficient than proportional counters for most x-ray energies. Solid-state devices have not been used in conjunction with dispersive devices. Both Si(Li) and Ge(Li) detectors need to be kept at liquid nitrogen temperatures (77°K), necessitating that the detectors be accompanied by a large Dewar, which makes them too cumbersome to be used with dispersive devices.

The principle of photon detection in a solid-state detector is similar to that of a gas detector. The solid-state detectors are composed of single crystals of silicon or germanium which have been doped with lithium. The doping is to minimize the effect of the impurities in the solid. When a photon is absorbed in the crystal, the photon energy is converted to a number of electron–hole pairs, similar to the creation of a number of electron–ion pairs in a gas counter. A pair of electrodes are attached to opposite sides of the crystal, and a high voltage is applied across the crystal to collect the electronic pulse created by the photon.

The resolution of solid-state detectors is superior to that obtained in a proportional counter. The primary reason is that the average energy needed to create an electron–hole pair in a semiconductor is 3–4 eV, compared with about 30 eV required to create an electron–ion pair in argon.[22, 23]

[18] G. Dearnaley and D. C. Northrop, "Semiconductor Counters for Nuclear Radiations." Wiley, New York, 1966.

[19] J. M. Taylor, "Semiconductor Particle Detectors." Butterworth, London and Washington, D.C., 1963.

[20] S. Deme, "Semiconductor Detectors for Nuclear Radiation Measurement." Wiley, New York, 1971.

[21] *Proc. Symp. Semicond. Detectors Nucl. Radiat.*, 2nd, Munich, 6–9 September 1971 (*Nucl. Instrum Methods* **101**, No. 1 (1972)).

[22] E. M. Gunnerson, *Rep. Progr. Phys.* **30**, 27 (1967).

[23] A. H. G. Muggleton, *J. Phys. E* **5**, 390 (1972).

This smaller energy per electron–hole pair increases the average number of charges created per photon and decreases the statistical fluctuation in the number of charges, thereby improving the resolution. The resolution has the same $E^{-1/2}$ dependence as for a gas counter predicted in Eq. (3.2.4). The resolution of a 5.9 keV x ray (Mn Kα) for a Si(Li) detector is about 3%, compared with 6% for a very good proportional counter.[24] Some of the factors which limit the resolution in a solid-state detector are incomplete charge collection, variations in the absorbing layer thicknesses, and electrical noise. The collection of charge in the solid depends primarily on the electron–hole lifetime and impurity concentrations. The electron–hole lifetime is long in both silicon and germanium because the transitions from the bottom of the conduction band to the top of the valence band are forbidden by selection rules. The impurity concentration affects charge collection by providing trapping centers for the electron–hole pairs and by inducing recombination. These effects are reduced by lithium drifting. If the absorption layer at the surface of the crystal due to one electrode is not constant across the surface, variations in the pulse height occur, resulting in loss of resolution. The electronic noise is more important in solid-state detectors than in proportional counters because in solid-state detectors, there is no signal amplification at time of detection so that the signal-to-background ratio is smaller. Variations in background would show up more readily in the solid-state detector.

The efficiency of a solid-state detector depends upon the number of x rays absorbed by the window and detector dead layer, and upon the number of photons stopped in the active region of the solid. Most detectors are guarded by a beryllium window. Absorption by the window and by the electrode and dead layer on the front of the crystal affect the efficiency of the detector for the lower-energy x rays. These effects can be calculated using Eq. (3.2.5). The efficiency in detecting high-energy x rays is a function of the thickness of the crystal. Because of its higher Z, germanium is better for stopping high-energy x rays. The size of the detector is limited by the ability to grow and dope large, pure, single crystals. Also, for larger crystals, the resolution is worse because of incomplete charge collection since the charges must travel further to reach the electrodes. The proper crystal size to be used in an experiment depends upon whether good resolution or high efficiency is needed. Efficiencies for commercially produced Ge(Li) and Si(Li) detectors having the same thickness are shown in Fig. 2. The low-energy end is dominated by the absorption in the Be window. The difference in efficiency for the high-energy x rays is due to the difference in mass of germanium and silicon.

[24] Ortec, Inc., Oak Ridge, Tennessee.

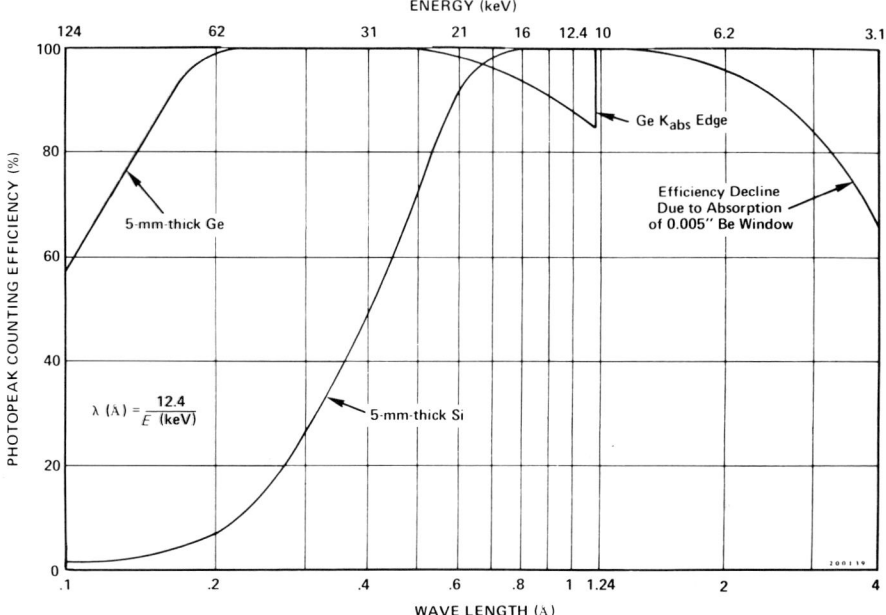

FIG. 2. Calculated efficiencies for Si(Li) and Ge(Li) detectors having 5-mm-thick crystals. The effects of the window absorption are seen for the low-energy region. The high-energy region of the curves depends on the stopping power of the crystals. (Reproduced by permission of Ortec, Inc., Midland Rd., Oak Ridge, Tennessee.)

An important feature in Ge(Li) and Si(Li) detectors is the escape peak. An inner-shell vacancy in Ge or Si can be created by the incident x ray. If the atom decays by x-ray emission and the x ray is not reabsorbed in the crystal, the electronic pulse detected at the electrodes will have less total charge than the pulse detected if all of the incident x-ray energy remained in the detector. This causes spurious peaks called escape peaks which are present in the pulse height spectra at energies equal to the normal peak minus the energy of the emitted x ray. The escape peaks can have as much as 10% of the intensity of the normal peak.[25] Such features could cause incorrect interpretation of data.

3.2.2.3. Photographic Film. Despite the developments made in electronic detection, photographic film is still important in x-ray detection.[26, 27]

[25] J. M. Palms, P. V. Rao, and R. E. Wood, *Nucl. Instrum Methods* **64**, 310 (1968).

[26] J. A. R. Samson, "Techniques of Vacuum Ultraviolet Spectroscopy," pp. 209–211. Wiley, New York, 1967.

[27] D. H. Tomboulian, 'Handbuch der Physik," (S. Flügge, ed.), Vol. 30, pp. 269–70. Springer-Verlag, Berlin and New York, 1957.

Its primary importance is in use with a diffraction device to detect x rays from a transient event such as a laser pulse. The largest advantage that photographic film has is that it can record all diffraction angles from a device like a Bragg spectrometer at once. The disadvantages are that it is usually less efficient than an electronic detector and the efficiency versus x-ray intensity is highly nonlinear.

3.2.2.4. Single Crystal Spectrometers. The basic principle which governs the diffraction of x rays from a crystal is Bragg's law,

$$n\lambda = 2d \sin \theta, \tag{3.2.6}$$

which relates the wavelength λ of the incident x ray to the angle θ at which a coherent diffraction pattern is formed; d is the crystal lattice spacing constant or grating constant of the crystal, and n is an integer which gives the order of reflection. One of the first spectrometers built using a crystal as a dispersive element is the Bragg spectrometer. A schematic of its operation is shown in Fig. 3. The x rays from the source F pass through the slit S. The divergent beam strikes the crystal in position A. If the source is monochromatic, only one ray will strike the crystal at the Bragg angle θ. Assume

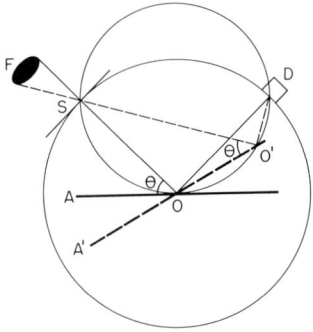

FIG. 3. A schematic of a Bragg spectrometer. Monochromatic x rays pass through the slit S, are Bragg diffracted at point O from the crystal in position A, and are detected at point D. If the crystal is rotated to position A', the monochromatic x ray will be Bragg diffracted at O', which lies on the auxiliary circle SOD, and the x rays are detected at D.

that this ray hits at O and is diffracted onto the detecting device at D. If the crystal is rotated by an arbitrary amount to position A', the monochromatic x ray will strike the crystal at the Bragg angle at O'. If a circle is drawn through points S, O, and D, the point O' will lie on this circle. It can be shown from the geometry of the circle that the x ray will be diffracted from O' to the position D, so that a monochromatic x ray will always be diffracted to the same position D, regardless of the crystal angle. The auxiliary circle is analogous to the Rowland circle in concave gratings. If a photographic plate is placed along the primary circle, an entire spectrum can be generated by rotating the crystal. In this way, the errors due to the crystal imperfections and source variations tend to be averaged out.

The angular dispersion D_θ for a Bragg spectrometer can be obtained directly from Eq. (3.2.6) by

$$D_\theta = d\theta/d\lambda = (\tan \theta)/\lambda = (n/2d) \sec \theta. \tag{3.2.7}$$

From the above equation it can be seen that high dispersion is obtained for larger θ, or at backward angles. For a given crystal, better resolution can be obtained, in general, by measuring the x rays in a higher order.

If an extended source is used, the slit S can be replaced by a Soller collimator so that the entire crystal can be illuminated by a source of parallel x rays. The Soller collimator consists of a set of parallel plates that absorbs any x rays not propagating parallel to the plates. If x is the distance between two adjacent plates and l is the length of each plate, the angular divergence ϕ of the x rays through the collimator is

$$\phi \simeq 2x/l. \tag{3.2.8}$$

To generate a spectrum, the crystal is rotated through an angle θ and the detector is rotated by 2θ.

Another type of spectrometer, first developed by Siegbahn and Larsson, in which the slit is placed after the crystal, is called a tube spectrometer.[28] A principle ray diagram is shown in Fig. 4. The x ray from the source F strikes the crystal at O, and is diffracted through the slit S and strikes the photographic plate P. The crystal is then rotated through an angle $180° - 2\phi$, where ϕ is approximately θ. The source is rotated through approximately 4θ to F'. The x ray strikes the crystal at O' and is again diffracted onto the photographic plate. The distance between the two lines on the plate is given by a and the distance from the slit to the photographic plate is R. Angle θ can be determined by measuring the crystal rotation angle $180° - 2\phi$, and using the formula

$$180° - 2\phi = 180° - 2(\theta \pm \Delta\theta), \tag{3.2.9}$$

where $\Delta\theta$ is determined by

$$\Delta\theta = a/2R. \tag{3.2.10}$$

The most critical measurement is the angle of rotation of the crystal. The correction $\Delta\theta$ is only about 1% of θ, and thus the errors in this measurement are usually small. Normally it is difficult to change the position of the source, so that usually the photographic plate and the crystal are moved.

---

[28] M. Siegbahn, *Ark. Mat. Astron. Fys.* **A21**, No. 21 (1929); A. Larsson, *Phil. Mag.* **3**, 1136 (1927).

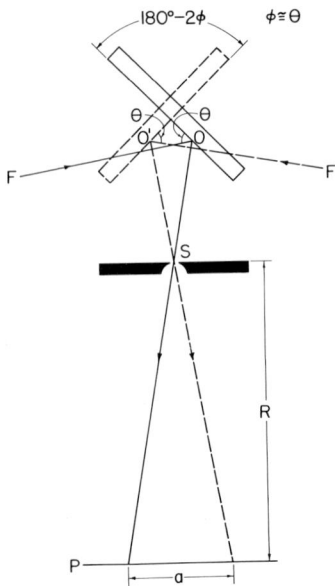

Fig. 4. A schematic showing the operation of a tube spectrometer. X rays from a source F strike a crystal at O and are Bragg diffracted through the slit S. The source is rotated, through an angle approximately equal to 4θ, to F', and the crystal is rotated through an angle approximately equal to $180°-2\phi$. The angle through which the crystal is rotated and the angle of separation of the lines on the photographic plate determine the measured wavelength.

3.2.2.5. Focusing Methods. To improve upon the intensity obtained from a plane crystal spectrometer, the crystals can be bent, thus providing some focusing. The bending process can destroy the crystalline structure and in general results in a loss of resolution. The advantage in using a curved crystal is that the intensity is increased without increasing the intensity of the x-ray source.

One type of focusing called vertical focusing is obtained if the crystal is bent perpendicular to the direction of dispersion. The design employed by Hámos is shown in Fig. 5a.[29] The x rays from a source F pass through the slit S and strike the crystal at A. They are diffracted from the crystal and detected at P. If the crystal is bent with a radius of curvature r equal to the line segment NA, all x rays impinging on the crystal along the arc passing through A will focus to the point P. The linear dispersion, $D_L = |\Delta L|/|\Delta \lambda|$, along the detection region is given by

$$D_L = (4dr/n\lambda^2)[1 - (n\lambda/2d)^2]^{-1}. \qquad (3.2.11)$$

[29] L. Von Hámos, *Naturwissenschaften* **20**, 705 (1932); *Ann. Phys. (Leipzig)* **17**, 716 (1933).

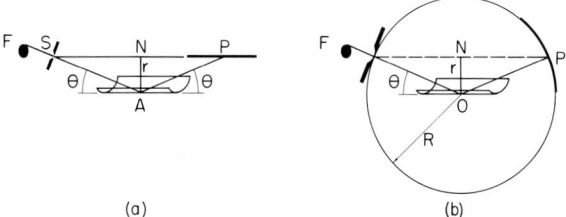

FIG. 5. Schematics showing two types of vertical focusing crystal spectrometers.

Another type of vertical focus spectrometer, which was designed by Kunzl,[30] is similar to the Bragg spectrometer. A schematic is shown in Fig. 5b. The flat crystal of a Bragg spectrometer has been replaced by a curved crystal whose radius of curvature r is equal to ON as shown in Fig. 5b. In terms of the wavelength, r is given by

$$r = (n\lambda/2d)R, \qquad (3.2.12)$$

where R is the radius of the primary circle of the spectrometer. This provides exact vertical focusing at point P. Approximate focus is obtained in the neighborhood of P. To provide focusing at every wavelength, Dolejsek and Tayerle[31] designed a spectrometer which changes the radius of curvature to satisfy the focusing criterion at each angle. Van der Berg and Brinkman[32] have built a spectrometer which adjusts the distance between the source and the crystal and between the crystal and the detector to satisfy the focusing conditions for each wavelength.

Focusing along the plane of dispersion of the crystal, called horizontal focusing, was first discussed by DuMond and Kirkpatrick.[33] To focus a divergent source to a point, both the source and the focusing point should be on the auxiliary circle, as seen from the Bragg spectrometer in Fig. 3. If the radius of the auxiliary circle is r, the crystal should lie along the auxiliary circle, but the diffracting planes should be bent to a radius $r = 2R$. A spectrometer constructed by Johansson,[34] which used these principles, is shown in Fig. 6. The crystal is bent to a radius r and is further ground to a radius of R. Earlier, a spectrometer providing approximate focusing was constructed by Johan.[35] The crystal is bent to a radius r but is not ground. Since the crystal face does not lie on the auxiliary circle, the focusing is only approximate. Sometimes the reflection properties of a crystal are destroyed by the

---

[30] V. Kunzl, *C. R. Acad. Sci.* (*Paris*) **201**, 656 (1935).
[31] V. Dolejsek and M. Tayerle, *C. R. Acad. Sci.* (*Paris*) **205**, 605, 1143 (1937).
[32] C. B. Van der Berg and H. Brinkman, *Physica* **21**, 85 (1955).
[33] J. W. M. DuMond and H. A. Kirkpatrick, *Rev. Sci. Instrum.* **1**, 88 (1930).
[34] T. Johansson, *Z. Phys.* **82**, 507 (1933).
[35] H. H. Johann, *Z. Phys.* **69**, 185 (1931).

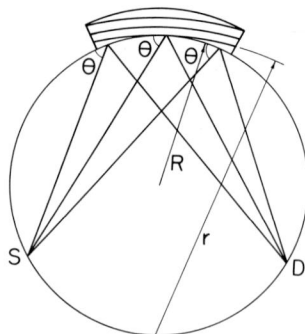

FIG. 6. A schematic of the Johansson type of curved crystal spectrometer. The crystal is bent to a radius r and ground to a radius R.

grinding process. All crystals cannot therefore be used in a Johansson-type spectrometer.

3.2.2.6. *Double Crystal Spectrometer.* The double crystal spectrometer employs two crystals to obtain a higher dispersion than can be attained for single crystal spectrometers. A principal ray diagram for the two different modes of operation of the double crystal spectrometer is shown in Fig. 7. X rays from a broad x-ray source are diffracted in mth order from the first crystal at an angle θ_m. The diffracted x rays strike a second crystal and are Bragg diffracted in nth order at an angle θ_n. If the x rays are diffracted in the same sense as shown in Fig. 7a, the spectrometer is said to be in the plus

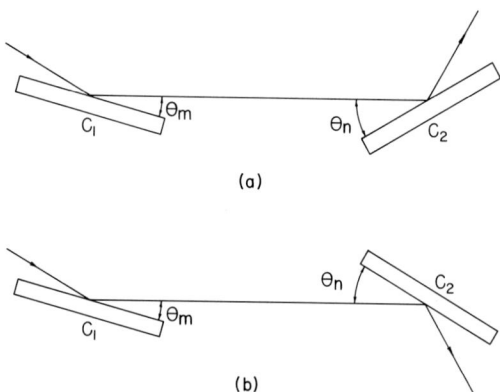

FIG. 7. A schematic showing the two possible modes of operation for a double crystal spectrometer: (a) the plus position (m, n) and (b) the minus position $(m, -n)$. The x rays are diffracted in mth order from the first crystal C_1 at an angle θ_m, and in nth order from the second crystal C_2 at an angle θ_n.

position and is denoted by (m, n). If the x rays are diffracted in the opposite sense from each crystal, as shown in Fig. 7b, the spectrometer is in the minus position which is denoted by $(m, -n)$. The $(n, -n)$ position is called the parallel position.

The spectrometer is ordinarily operated by fixing the first crystal at a given angle and varying the angle of the second crystal. The first crystal is often operated in the first order to obtain maximum possible intensity. The second crystal many times is used in higher orders to obtain greater resolution. The first crystal acts as a wavelength filter which reduces the noise and background radiation since the second crystal sees only the narrow range of Bragg diffracted x rays from the first crystal. The first crystal also acts as a collimator because the diffracted x rays emerge parallel at an angle θ_m from the first crystal. When the double crystal spectrometer is operated in the plus mode, the dispersion is the sum of the dispersion of the single crystals.

3.2.2.7. Crystal Properties. The grating constant d of a crystal depends upon the order of reflection in which the crystal is used. The dependence of the grating constant upon the order of reflection is due to the refraction of the x ray as it penetrates the crystal. The index of refraction for x rays is almost, but not quite, 1. If it is assumed that Bragg's law is exactly correct in the interior of the crystal and that the optical principles of refraction are valid, the nth-order grating constant d_n can be given in terms of the true grating constant d_∞ by

$$d_n = d_\infty(1 - (4d_\infty/n^2)[(1 - \mu)/\lambda^2], \qquad (3.2.13)$$

where μ is the index of refraction. The quantity $(1 - \mu)/\lambda^2$ is on the order of 10^{-6} so that the corrections are typically one part per thousand. A correction due to the thermal expansion of the crystal may also be necessary. For very high precision work the crystal atomic spacing should be measured for each crystal. The spacing may differ among different crystal specimens and even crystals cut from the same specimen, depending on their growth conditions, the impurity concentration, and their previous history. The most commonly used crystals for high precision work are quartz, calcite, and silicon.

The intrinsic resolution can be defined in terms of its "rocking curve." The rocking curve is normally measured by mounting the crystal of interest as the second crystal in a double crystal spectrometer. The first crystal should be a nearly perfect specimen of the same species. The spectrometer is operated in the $(1, -1)$ mode, or in the parallel position. In this mode there is no dispersion. A line profile is obtained by rotating the second crystal about the parallel position. The line width observed is the width of the diffraction pattern from the two crystals. This width, called the rocking curve width, is usually measured in seconds or minutes of arc of rotation. The diffraction width has also been determined using a single crystal spectrometer and a narrowly collimated source of a single line with a narrow width.[36, 37]

The reflection properties of the crystal dictate the intensity. Darwin has derived the reflectivity for a perfect crystal assuming no absorption,[38] and Prins has modified the derivation to include absorption.[39] The theories are used to calculate the integrated reflection coefficient for potassium acid phthalate (KAP) in four orders, as shown in Fig. 8 along with several experimental points.[40] It can be seen that the reflectivity decreases rapidly with the order of reflection.

3.2.2.8. Transmission Spectrometers. Bragg diffraction can also occur for x rays transmitted through the crystal. Measurements using transmission spectrometers are usually limited to the shorter wavelengths because the longer wavelengths are greatly attenuated in the crystal. The properties of the single- and double-crystal spectrometers are essentially the same in transmission as in reflection, with one or both crystals being used in the transmission mode. Cauchois has constructed a bent crystal transmission spectrometer analogous to the Johan reflecting type.[41] Another method for focusing in a transmission spectrometer which requires a point source has been designed by DuMond.[42]

3.2.2.9. Photoelectrons for Analysis of X Rays (PAX). The photoelectron spectrometer employs the method of ESCA (Electron Spectroscopy Analysis) to obtain information concerning x rays.[43] In the ESCA technique, a known x-ray source is used to photoionize the material of interest. The photoelectrons are measured by an electron analyzer to obtain information concerning the material. The PAX system uses a converter atom having known photoionization properties to study the incident x rays.[44] The converter atoms are usually inert gases such as helium, neon, and argon, chosen because they have distinct, narrow energy levels and because there are no chemical effects or surface potentials. The energy of the photo-

---

[36] G. A. Sawyer, A. J. Bearden, I. Henins, F. C. Jahoda, and F. L. Ribe, *Phys. Rev.* **131**, 1891 (1963).

[37] R. L. Blake and T. Passin, *Advan. X-Ray Anal.* **14**, 293–310 (1971).

[38] C. G. Darwin, *Phil. Mag.* **27**, 315, 675 (1914).

[39] J. A. Prins, *Z. Phys.* **63**, 477 (1930).

[40] A. J. Burek, D. M. Barrus, and R. L. Blake, *Astrophys. J.* **191**, 533 (1974).

[41] Y. Cauchois, *J. Phys. Et. Radiat.* **3**, 320 (1932); **4**, 61 (1933).

[42] J. W. M. DuMond, *Rev. Sci. Instrum* **18**, 626 (1947).

[43] For a discussion of ESCA, see K. Siegbahn *et al.*, ESCA, *Nova Acta Regiae Soc. Sci. Upsal. Ser. IV* **20** (1967) and North-Holland Publ., Amsterdam, 1972; D. A. Shirley "Electron Spectroscopy." North-Holland Publ., Amsterdam, 1972.

[44] M. O. Krause, *Phys. Fenn. Suppl. S1* **9**, 281 (1974); *Advan. X-Ray Anal.* **16**, 74–89 (1973).

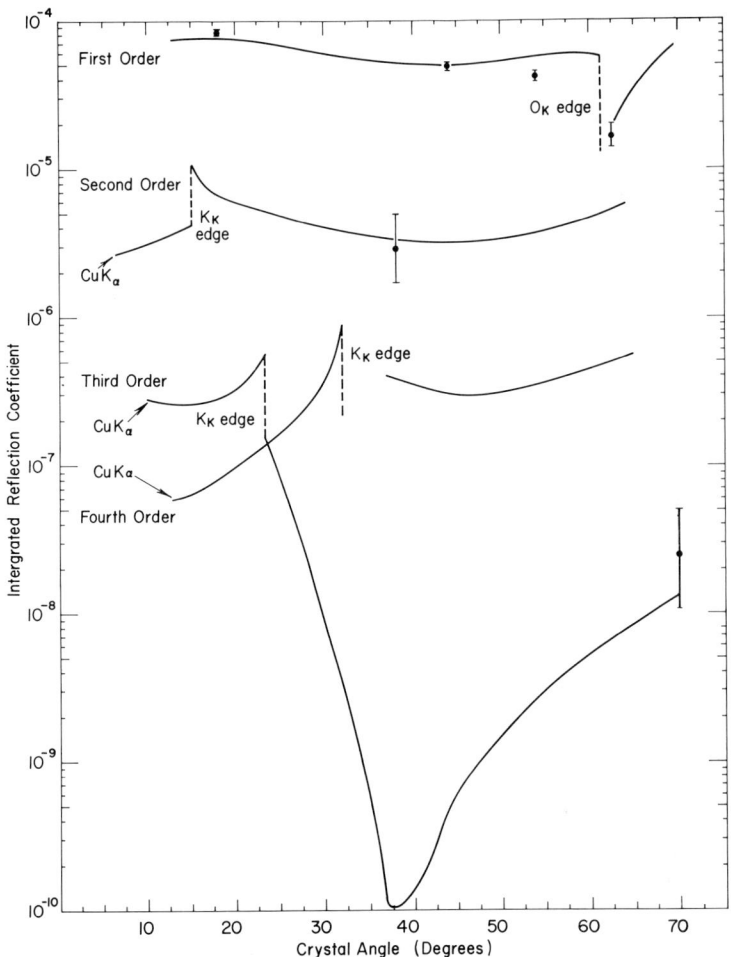

FIG. 8. Integrated crystal reflectivity for four orders of reflection from a potassium acid phthalate (KAP) crystal [taken from A. J. Burek, D. M. Barrus, and R. L. Blake, *Astrophys. J.* **191**, 533 (1974), by permission of The University of Chicago Press, Chicago, Illinois].

electron, E_{kin}, is given, from conservation of energy, by

$$E_{kin} = h\nu - E(n, l), \quad (3.2.14)$$

where $h\nu$ is the energy of the photon of interest and $E(n, l)$ the energy of the converter level. The spectrometer can be calibrated by using a photon of known energy or a known Auger electron line.

The photoelectron spectrometer can be used to measure x rays from 20 eV to 20 keV. This spans the region between gratings and crystals. The range and resolution are determined in large part by the converter atoms. Typical resolution obtained using neon and helium as converter gases is shown in Fig. 9. The different curves correspond to different resolutions of the electron analyzer, R_e. The other important factors which determine the energy resolution are the width of the energy level of the converter atom and the broadening due to thermal motion of the gas. The solid and open squares in Fig. 9 are typical resolutions obtained with a crystal spectrometer

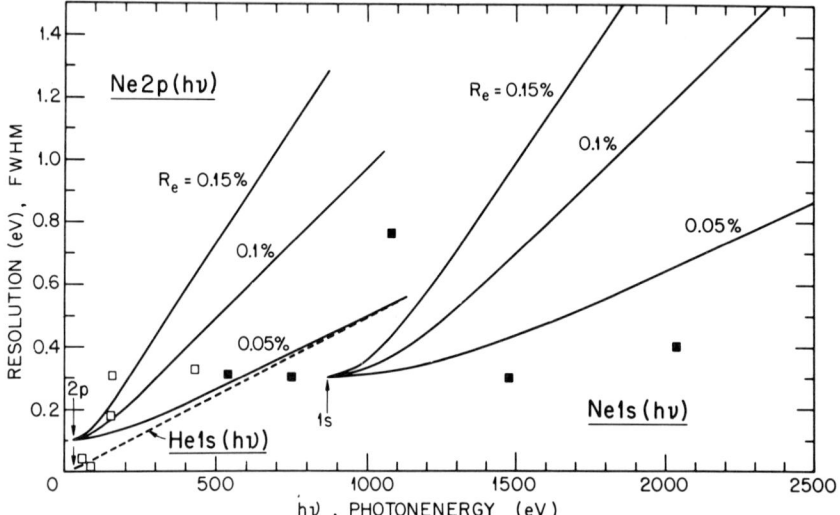

FIG. 9. Resolution of a photoelectron spectrometer as a function of x-ray energy for different electron analyzer resolution R_e. The solid lines are using the 2p and 1s levels of neon as converter levels, and the dashed line is that obtained using the He 1s level. Resolution of some grating spectrometers (open squares) and crystal spectrometers (solid squares) are shown [taken from M. O. Krause, *Advan. X-Ray Anal.* **16**, 74 (1973)].

and grating spectrometer, respectively. The efficiency of the spectrometer depends upon the efficiency of the electron analyzer to detect the photoelectrons and upon the photoionization cross sections of the converter atom. Efficiency on the order of 10^{-5} has been obtained, which is comparable to crystal spectrometers.

3.2.2.10. *Doppler-Tuned Spectrometer.* In beam-foil studies, the x rays of energy E_0 emitted from a fast moving beam of ions of velocity v has a different energy E in the laboratory frame of reference due to the Doppler shift. The relation between the two energies is

$$E_0 = E[1 - (v/c)\cos\theta][1 - (v^2/c^2)]^{-1/2}, \qquad (3.2.15)$$

where c is the speed of light and θ is the angle that the emitted x ray makes with the beam, as shown in Fig. 10. The Doppler shift is used as the dispersive element in the spectrometer diagrammed in Fig. 10.[45] A beam of ions from an accelerator is passed through a foil, exciting the ions. These ions decay by emitting an x ray at the angle θ in the laboratory frame of reference. A filter is inserted between the beam and detector which has a well-defined absorption edge having an energy E_{abs} in the energy region of interest. The angle of detection is varied so that the energy varies from $E < E_{abs}$ to $E > E_{abs}$. For $E < E_{abs}$, the x ray will pass through the filter into the detector.

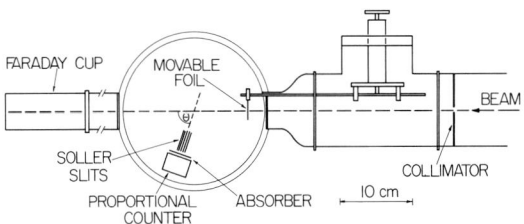

FIG. 10. Diagram of a Doppler-tuned spectrometer used in beam-foil studies. The x rays emitted by the beam are detected at an angle θ by a proportional counter. An absorber having an absorption edge with an energy in the range of interest is inserted in front of the proportional counter. Spectra are taken by rotating the proportional counter about the angle θ [taken from C. L. Cocke, B. Curnutte, J. R. Macdonald, and R. Randall, *Phys. Rev.* **9**, 57 (1974)].

For $E > E_{abs}$, the x ray is attenuated in the filter before reaching the detector. By comparing the count rate above and below E_{abs} and by using knowledge of the filter, the intensity of the x-ray energy E_0 can be determined. By scanning over the region of interest, the angle θ_0 at which the x ray is absorbed can be measured and the energy of the x ray determined by

$$E_0 = E_{abs}[1 - (v/c)\cos\theta_0][1 - (v^2/c^2)]^{-1/2}. \qquad (3.2.16)$$

A typical spectrum obtained from such a technique is shown in Fig. 11.

The resolution for such a spectrometer is normally around 10^{-3}. This depends upon the knowledge of the absorption edge and the angular collimation of the x ray. The efficiency is high compared to dispersive devices, and is determined by the attenuation in the filter and the detector efficiency. A correction must be made to the data to obtain absolute intensities, due to the cosecant brightening which occurs because, as the detector is rotated to more acute angles, the spectrometer views a larger cross section to the beam. This varies approximately as $\csc\theta$. Beam focusing and misalignment of the spectrometer, as well as anisotropies in the angular distribution of the x rays, may also affect the measurements.

[45] R. W. Schmieder and R. Marrus, *Nucl. Instrum. Methods* **110**, 459 (1973).

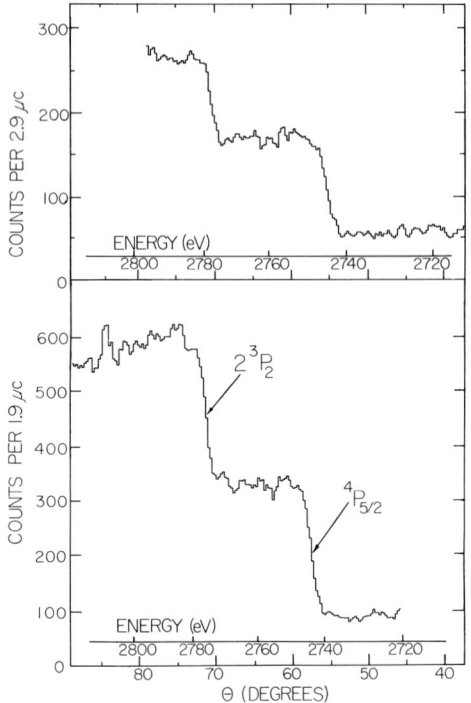

FIG. 11. Spectra of 50.5-MeV (lower) and 44.0-MeV (upper) Cl taken with a Doppler-tuned spectrometer using a 50-μm vinylidene chlorine absorber. The exciting carbon foil is 1.5 cm upstream. The $(1s2p)^3P_2$–$(1s^2)^1S_0$ and $(1s2s2p)^4P_{5/2}$–$(1s^22s)^2S_{1/2}$ metastable decays are observed. The intensity of the x rays is proportional to the height of the observed step [taken from C. L. Cocke, B. Curnutte, J. R. Macdonald, and R. Randall, *Phys. Rev.* **9**, 57 (1974)].

3.2.3. X-Ray Spectra

3.2.3.1. Diagram Lines. The dominant features in photon induced spectra and electron induced spectra are the diagram lines. These x rays are emitted from atoms having, initially, one inner shell vacancy and no other excitation. The spectra of diagram lines are one of the simplest x-ray systems and can be interpreted using the one-electron atomic model. The x rays are emitted during transitions of electrons in an upper energy state to a vacant state of lower energy. The transitions also can be thought of as a vacancy changing from one energy level to another.

The diagram lines were measured and catalogued before the x-ray spectra were completely explained using the atomic theory. This has led to a notation which is often cumbersome and confusing. A schematic of the energy level diagram is shown in Fig. 12. The one-electron quantum numbers are given on the right. The transitions are governed by the electric dipole selection

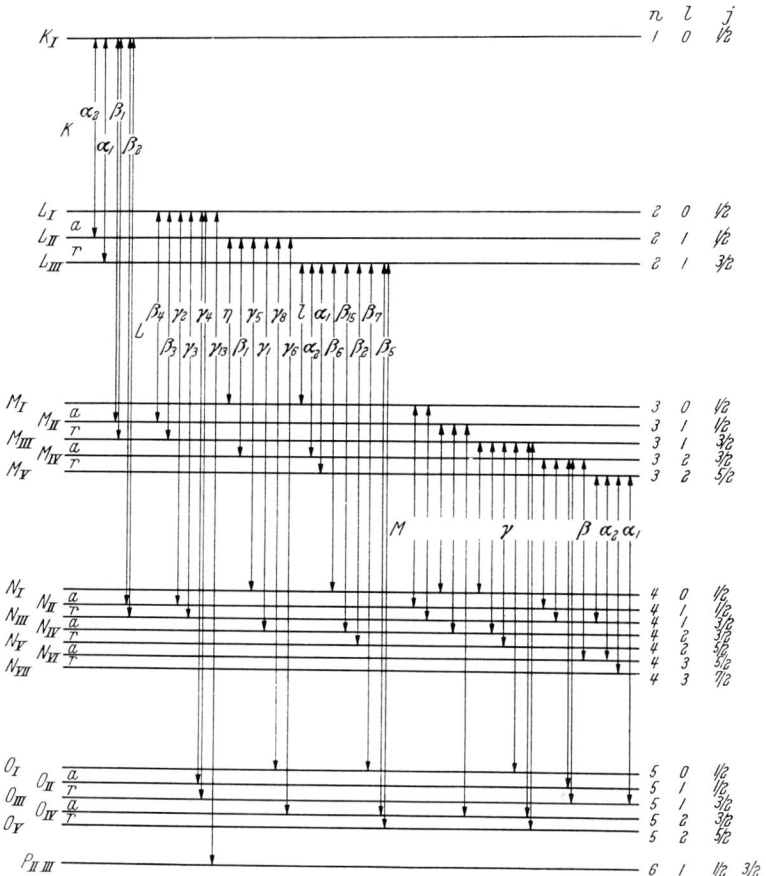

FIG. 12. A schematic energy level diagram showing the observed x-ray transitions of the K, L, and M series. The one-electron quantum numbers are listed on the right. Labels of the energy levels are given on the left. The Siegbahn notation of the x-ray transitions is also shown [taken from A. E. Sandström, "Handbuch der Physik" (S. Flügge, ed.), Vol. 30, p. 78–245. Springer-Verlag, Berlin and New York, 1957].

rules and are grouped into different series depending upon the energy level which contains the initial hole. Figure 12 shows the transitions for the K, L, and M series, which correspond to transitions having initial holes in the $n = 1$, 2, and 3 shells, respectively. The N and O series x rays have been observed also. Each series is composed of a number of lines labeled by a Greek letter and a subscript. For example, the electron transitions $2p_{3/2}$–$1s_{1/2}$ and $2p_{1/2}$–$1s_{1/2}$ are labeled $K\alpha_1$ and $K\alpha_2$, respectively. The numerical subscript denotes the relative intensity of the line within a given series (e.g., the $K\alpha_1$ x ray is more intense than the $K\alpha_2$ x ray); therefore, no simple correlations exist between the notation for the various lines and the quantum numbers of the energy levels involved in the transition. A more lucid system is to label the energy levels involved in the transition. The labels of the energy levels are given on the left in Fig. 12. In this notation the $K\alpha_1$ transition is labeled $K_I L_{III}$. The energies of the diagram lines are well known for most elements, and have recently been reviewed and tabulated.[46]

An active area of x-ray spectroscopy is the study of the effect that the chemical bonding and the solid medium have on the x-ray spectrum. These effects usually appear as a small energy shift and some broadening of the x-ray line. If a valence energy band is involved in the transition, then x-ray bands are observed. The measurement and interpretation of these phenomena have had much scrutiny,[47,48] and will not be discussed here.

Intensity measurements of the diagram line can be used to measure the number of atoms of a given element present, or to measure the number of collisions which produce an inner-shell vacancy. In the field of trace element analysis, the diagram x ray serves as a signature of the element and is therefore used to detect its presence. If the cross sections for x-ray production of the different elements are known, quantitative information can be obtained from the intensity measurements. If the number of target atoms present is known, as in a scattering experiment, the intensity measurements of the diagram lines can be used to determine the cross section for producing atoms having inner-shell vacancies. These measurements are discussed in greater length in Section 3.2.4.

3.2.3.2. Satellite Lines. Many other less intense lines in addition to the diagram lines are observed in photon and electron induced spectra. These lines appear on the short wavelength side of the diagram lines and are called satellite lines. They were also called "nondiagram" lines because at the time they were observed, they did not fit conveniently into an energy level diagram such as the one in Fig. 12. The energies of the lines can be predicted by using an energy level diagram of a multiply ionized atom, so that the lines are usually referred to as satellite lines instead of "nondiagram" lines. Satellite lines have received increased attention since these lines are the most intense lines observed in spectra from hot plasmas and heavy-ion–atom collisions.

The origin of the satellite lines has been investigated since the lines were first observed by Siegbahn and Stenstrom[49] in 1921. In Fig. 13 the photon-induced spectrum of Ca is shown. The $K\alpha_1$ and $K\alpha_2$ diagram lines are shown at 3.3584 Å* (3351.5 xu) and 3.3617 Å* (3354.8 xu). The satellites are

[46] J. A. Bearden, *Rev. Mod. Phys.* **39**, 78 (1967).

[47] L. G. Parratt, *Rev. Mod. Phys.* **31**, 616 (1959).

[48] D. J. Nagel and W. L. Baun, "X-Ray Spectroscopy" (L. V. Azároff, ed.), pp. 445–532. McGraw-Hill, New York, 1974.

[49] M. Seigbahn and W. Stenstrom, *Phys. Z.* **17**, 48, 318 (1916).

3.2. X-RAY REGION

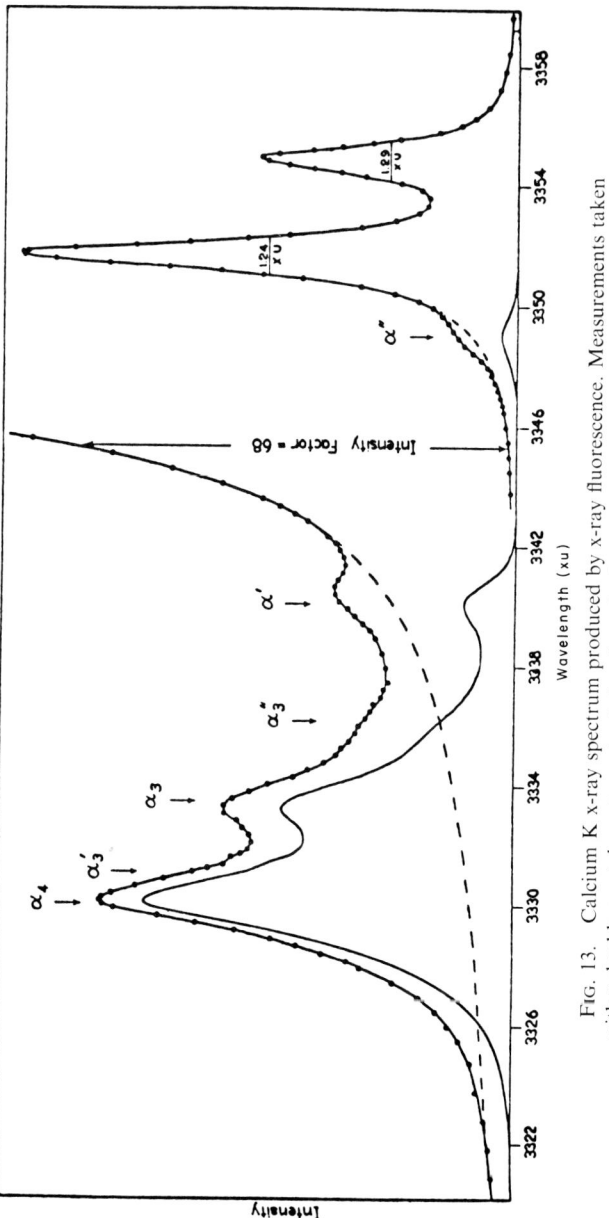

FIG. 13. Calcium K x-ray spectrum produced by x-ray fluorescence. Measurements taken with a double crystal spectrometer [taken from L. G. Parratt, *Phys. Rev.* **49**, 502 (1936)].

denoted by α'', α', α_3'', α_3', α_3 and α_4. The same notation that is used for the diagram lines is used for the satellite lines. The satellites are numbered consecutively in order of decreasing intensity, with primes denoting less intense lines detected in the spectra. Not all of the expected satellite lines are observed in each spectrum, but rather the number observed depends upon the atomic number of the elements being investigated. Other $K\alpha$ satellite lines are present at wavelengths shorter than that shown in the spectrum. Satellite lines are also observed accompanying $K\beta$ x rays and L and M series x rays. An experimental and theoretical summary of the early work on satellites is given in a review by Hirsh.[50]

The satellite lines arise from the deexcitation of multiply ionized atoms. The $K\alpha$ satellite lines have received the most comprehensive theoretical investigations. By using Hartree–Fock calculations, Kennard and Ramberg[51] correctly identified the $K\alpha_3$ and $K\alpha_4$ satellites as originating from an initial configuration having one K-shell hole and one L-shell hole. They also showed that the other less intense satellite lines at shorter wavelengths are transitions from atoms having one K-shell hole and multiple L-shell holes. Candlin[52] has analyzed the double hole states for $Z = 19$ to $Z = 42$ in jj-coupling. Horák[53] has used LS-coupling to confirm the identifications of Kennard and Ramberg.

It has recently been shown that the spectra produced by high-energy heavy-ion–atom collisions are dominated by the satellite transitions. This was first pointed out as an energy shift in Si(Li) detector spectra of the $K\beta$ x rays.[54] Si(Li) detector spectra of Ca and V, produced by 15 MeV oxygen ion bombardment, are compared with proton-induced spectra in Fig. 14. By using high resolution crystal spectrometers, it was quickly verified that these are caused by satellite lines which dominate the spectra.[55–57] Spectra of Ca produced by bombardment of protons, alpha particles, and oxygen are shown in Fig. 15. The resolution in this spectrum is not as good as in Fig. 13. In Fig. 15 the $K\alpha_1$ and $K\alpha_2$ peaks are not resolved. The notation used in Fig. 15 gives the number of 2p electrons present in the atom at the time of the x-ray emission. The $K\alpha(2p)^6$ peak in the oxygen-induced spectrum is shifted to slightly higher energy with respect to the

[50] F. R. Hirsh, Jr., *Rev. Mod. Phys.* **14**, 45 (1942).
[51] E. H. Kennard and E. G. Ramberg, *Phys. Rev.* **46**, 1040 (1934).
[52] D. J. Candlin, *Proc. Phys. Soc.* **A68**, 322 (1955).
[53] Z. Horák, *Proc. Phys. Soc.* **77**, 980 (1961).
[54] P. Richard, I. L. Morgan, T. Furuta, and D. Burch, *Phys. Rev. Lett.* **23**, 1009 (1969).
[55] M. E. Cunningham *et al.*, *Phys. Rev. Lett.* **24**, 931 (1970).
[56] D. Burch, P. Richard, and R. L. Blake, *Phys. Rev. Lett.* **26**, 1355 (1971).
[57] A. R. Knudson, D. J. Nagel, P. G. Burkhalter, and K. L. Dunning, *Phys. Rev. Lett.* **26**, 1149 (1971).

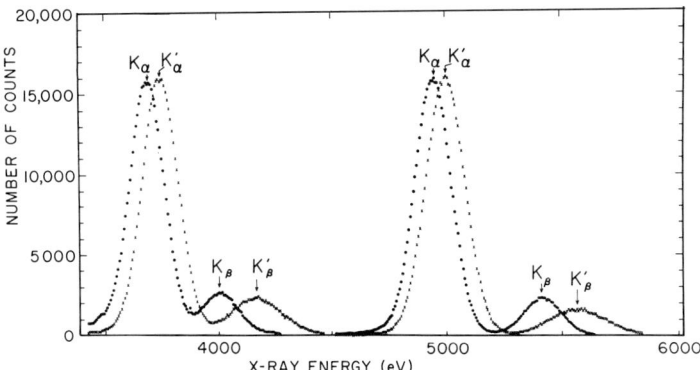

FIG. 14. K x-ray spectra of Ca and V produced by proton and oxygen bombardment. [left: (·) Ca + p (6 MeV), (×) Ca + O^{4+} (15 Mev); right: (·) V + p (6 MeV), (×) V + O^{4+} (15 MeV)]. Spectra taken with a Si(Li) detector having resolution of 200 eV FWHM. Note the large energy shifts in the oxygen-induced spectra and the increased width of the Kβ spectra. (D. Burch and P. Richard, *Phys. Rev. Lett.* **25**, 983 (1970).)

K$\alpha_{1,2}$ energy. This shift is caused by multiple M-shell ionization accompanying the K-shell ionization. Similarly, the Kα(2p)^5 peak is shifted slightly from the K$\alpha_{3,4}$ energy. For the proton- and alpha-induced spectra, no energy shifts are detected.

The bottom spectrum in Fig. 15 is typical of most spectra observed from heavy-ion–atom collisions. The satellite peaks are usually resolved only well enough to identify the peaks belonging to a given multiple ionization peak. The individual lines such as those observed in Fig. 13 are washed out by multiple M-shell ionization. The target Z dependence of the energy shift of the multiple ionization peaks relative to the K$\alpha_{1,2}$ diagram line is plotted in Fig. 16. The notation KL^n designates one K-shell and n L-shell vacancies in the state. This notation is more desirable than that used in Fig. 15 because one cannot tell from the spectra whether the holes are initially in the 2s or 2p subshell. Also plotted in Fig. 16 are the energy shifts predicted by a Hartree–Fock–Slater calculation,[58] and a Hartree–Fock calculation.[59] Both calculations predict satisfactorily the energy shifts for the smaller multiple ionization states, but neither predicts accurately the shifts for the larger multiple ionization states.

Similar x-ray satellite structure is observed in other collisional environments such as plasmas and laser-induced plasmas. As in the ion–atom

---

[58] F. Herman and S. Skillman, "Atomic Structure Calculations." Prentice-Hall, Englewood Cliffs, New Jersey, 1963.

[59] C. Froese Fischer, *Comp. Phys. Comm.* **4**, 107 (1972).

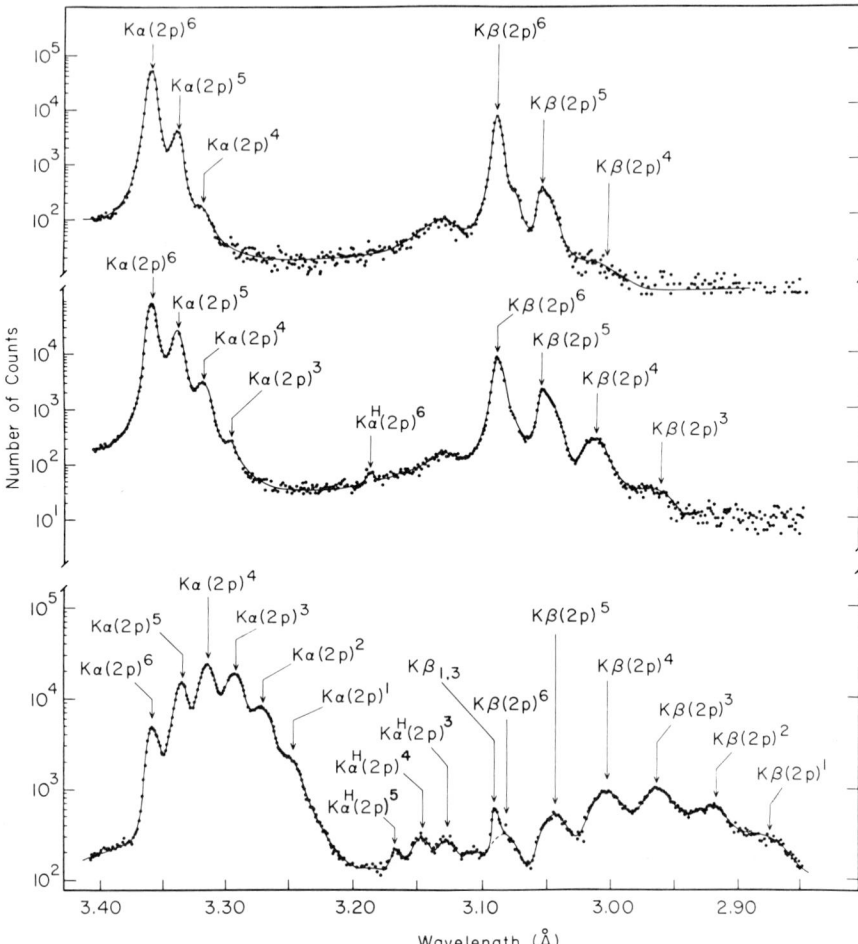

FIG. 15. High resolution spectra of Ca produced by bombardment of 0.8-MeV protons (top), 3.2-MeV alpha particles (center), and 30-MeV oxygen (bottom). The Kα and Kβ satellites are resolved, and indicated in the spectra by arrows; Kα hypersatellites (designated $K\alpha^H$) are also observed [taken from J. McWherter J. Bolger, C.F. Moore, and P. Richard, Z Phys. **263**, 283 (1973)].

collisions, the satellite structure are the most intense lines in the spectra. Comparison of plasma-induced spectra and heavy-ion-induced spectra is discussed in Section 3.2.4.

3.2.3.3. Hypersatellite Lines. Hypersatellite transitions have only recently been observed. The term "hypersatellite" was first used to designate x rays from atoms having two holes in the inner shell. Most observations

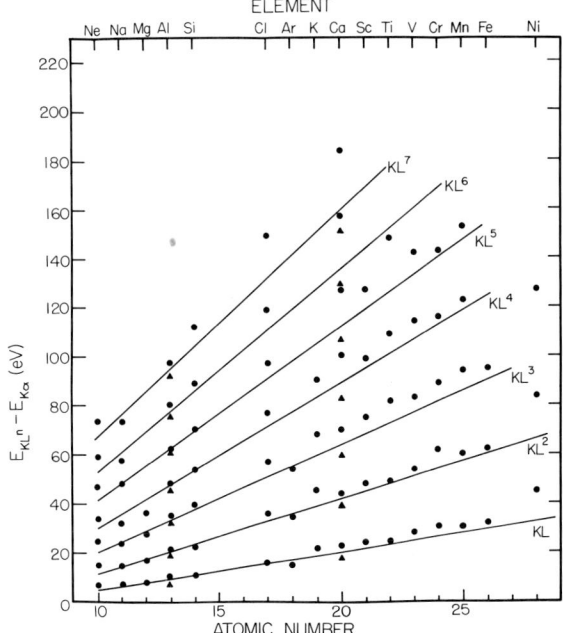

FIG. 16. Calculated and observed energy shifts of the Kα satellite lines produced by heavy-ion bombardment. The shifts are measured with respect to the $K\alpha_{1,2}$ line. The Hartree–Fock (▲) and Hartree–Fock–Slater(—) calculations predict energy shifts of the same magnitude for the various satellite lines [●: experimentally observed shifts].

have been of K hypersatellite lines where the atom initially has two holes in the K shell. Hypersatellites were not observed in early crystal spectroscopy measurements because the probability for creating such states in fluorescence-induced or electron-induced spectra is negligible. K hypersatellites were first detected from a radioactive source of ^{71}Ge, which decays primarily by K-electron capture to ^{71}Ga.[60] The two K x rays were observed in coincidence from the decay of the double K-vacancy state. The coincidence spectrum obtained with two Si(Li) detectors is shown in Fig. 17. The hypersatellite coincidences are shown in the shaded area. These are denoted by a superscript H, i.e., $K\alpha_{1,2}^{H}$ is the $K\alpha_{1,2}$ hypersatellite line. The $K\alpha_{1,2}$ and Kβ transitions observed in the spectrum result from random coincidences between the two detectors. The large energy shift between the normal lines and the hypersatellite lines is due to the reduced screening from the extra K-shell vacancy.

[60] J. P. Briand, P. Chevallier, M. Tavernier, and J. P. Rozet, *Phys. Rev. Lett.* **27**, 777 (1971).

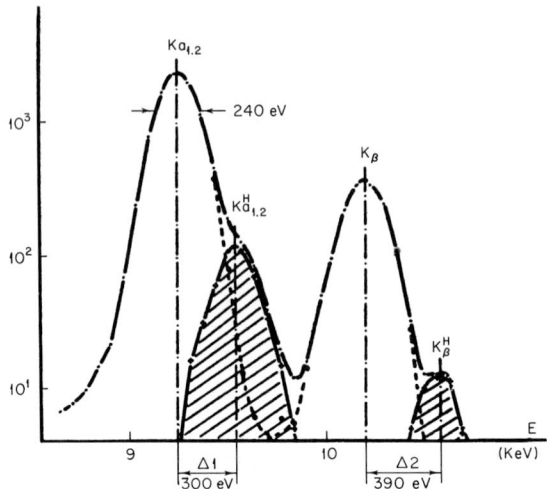

FIG. 17. K x-ray spectrum of Ga from a radioactive source of ^{71}Ge. The spectrum is obtained by observing in coincidence the x rays from two Si(Li) detectors. The shaded portion is the true coincidences of the Kα and Kβ hypersatellite x rays. The presence of the K$\alpha_{1,2}$ and Kβ lines is due to random coincidences [taken from J. P. Briand, P. Chevallier, M. Tavernier, and J. P. Rozet, *Phys. Rev. Lett.* **27**, 777 (1971)].

The hypersatellite transitions have been readily observed in heavy-ion-induced spectra.[61,62] The Kα hypersatellite lines are indicated in Fig. 15. The hypersatellites observed in these spectra contain multiple L-shell vacancies in addition to the two K-shell vacancies. The energy shifts of the hypersatellite relative to the diagram K$\alpha_{1,2}$ line observed in heavy-ion–atom collisions are shown in Fig. 18. These shifts are compared with the shifts predicted by using Hartree–Fock and Hartree–Fock–Slater calculations. It is seen that the Hartree–Fock calculations predict approximately the correct energy shifts, whereas the Hartree–Fock–Slater calculations predict consistently smaller energy shifts than observed and could lead to misidentification of the transitions. Identification of the Kα hypersatellites of Ca by McWherter *et al.*,[63] who used the Hartree–Fock–Slater calculations as a guide, are in error. These data are given in Fig. 15 with the correct Kα^H assignments.

3.2.3.4. *X Rays from the Few-Electron System.* X rays from the few-electron system have been observed from beam-foil excitations, some plasma sources, some heavy-ion collisions, and in solar flares. The transitions from

[61] P. Richard, W. Hodge, and C. F. Moore, *Phys. Rev. Lett.* **29**, 393 (1972).
[62] P. Richard, D. K. Olsen, R. Kauffman, and C. F. Moore, *Phys. Rev. A* **7**, 1437 (1973).
[63] J. McWherter, J. Bolger, C. F. Moore, and P. Richard, *Z. Phys.* **263**, 283 (1973).

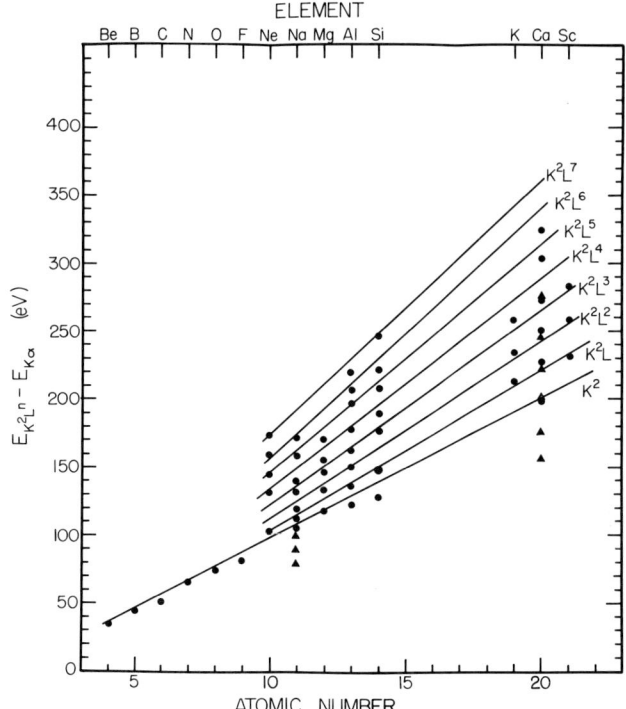

FIG. 18. The energy shifts of the $K\alpha$ hypersatellite lines with respect to the $K\alpha_{1,2}$ line observed in heavy-ion–atom collisions are plotted as a function of the atomic number. The Hartree–Fock–Slater predictions (▲) do not agree with the experimentally observed shifts (●). The Hartree–Fock calculations (—) are in better agreement.

few-electron systems are interesting because the x-ray spectra are relatively simple and easy to interpret. The measured energies also provide a good test for various atomic approximations. Several of the states are forbidden to decay by electric dipole selection rules. These metastable states decay by configuration mixing with other states or by higher-order multipole decay. These few-electron systems are one of the few cases in atomic physics in which the higher-order multipole decays can be observed. For further discussion of the metastable states and their measurement by using the beam-foil technique, see Chapter 5.1 and other references.[64,65]

Spectra of few-electron systems are shown in Figs. 19 and 20. Figure 19 gives the spectra of fluorine-beam x rays from 12- and 22.2-MeV fluorine

[64] B. C. Fawcett, *Advan. At. Mol. Phys.* **10**, 223–293 (1974).
[65] I. Martinson and A. Gaupp, *Phys. Rep.* **15C**, 113 (1974).

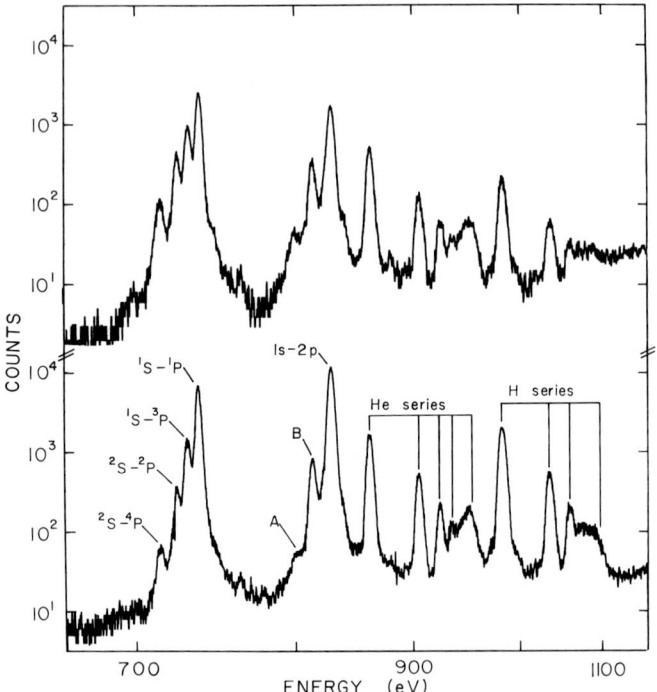

FIG. 19. Fluorine x rays from 12-MeV (upper) and 22.2-MeV (lower) fluorine beams excited by a thin carbon foil. The spectra contain mostly transitions from the heliumlike and hydrogenlike systems. The peaks labeled ^2S–^2P and ^2S–^4P are from the lithiumlike system [taken from R. L. Kauffman, C. W. Woods, F. F. Hopkins, D. O. Elliott, K. A. Jamison, and P. Richard, J. Phys. B **6**, 2197 (1973)].

beams excited by a thin carbon foil. Figure 20 is a spectrum of the few-electron system of aluminum from a laser-induced plasma. The dominant features of both spectra are the hydrogenlike and heliumlike series. The hydrogenlike series is composed of the transitions np–$1s$, where $n \geq 2$. Each peak in the hydrogenlike spectrum is actually an unresolved doublet from the decay of the $np_{3/2}$ and $np_{1/2}$ states. As n approaches infinity, the transition of an electron near the continuum fills the $1s$ vacancy. This x-ray energy is called the series limit. The hydrogenlike series limit is the highest-energy x ray which can be emitted by a given element. The hydrogenlike transitions have been calculated for H through Ca by Garcia and Mack,[66] who used relativistic Dirac energy levels. The known quantum electrodynamic corrections are included.

[66] J. D. Garcia and J. E. Mack, J. Opt. Soc. Amer. **55**, 654 (1965).

3.2. X-RAY REGION

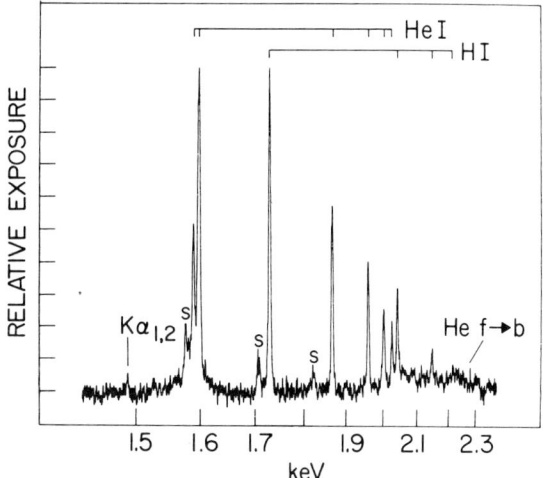

FIG. 20. Aluminum x rays produced in a laser-induced plasma. The helium and hydrogen-like series similar to those in Fig. 19 are observed. The feature labeled He $f \to b$ should be labeled H $f \to b$ and corresponds to the hydrogen series limit [taken from D. J. Nagel et al., Phys. Rev. Lett. **33**, 743 (1974)].

The heliumlike series is also observed in Figs. 19 and 20. The series is composed of transitions of the form $(np1s)^1P_1-(1s^2)^1S_0$. The end-point energy of the heliumlike series is equal to the ionization potential of a 1s electron in the $(1s^2)^1S_0$ ground state. The $(2p1s)^3P_1-(1s^2)^1S_0$ intercombination transition is also observed at a slightly lower energy than the $(2p1s)^1P_1-(1s^2)^1S_0$ transition. The $^3P_1-^1S_0$ transition is spin forbidden; however, the $^3P_1$ state decays by electric dipole radiation through mixing with the $^1P_1$ states. The lifetimes of these states are on the order of a nanosecond for elements near fluorine.[67,68]

Most of the other lines observed in Figs. 19 and 20 are from the lithiumlike systems or from the doubly excited heliumlike systems. These lines are sometimes called satellite or spectator transitions in few-electron spectra. The lines on the low energy side of the 2p 1s hydrogenlike transition are probably due to doubly excited heliumlike configurations. These transitions are labeled B (at 813.2 eV) in Fig. 19 and S (at 1.71 keV) in Fig. 20. The transition labeled A in Fig. 19 is probably due to a highly excited lithiumlike configuration, as is the line labeled S at 1.82 keV in Fig. 20. These highly

---

[67] G. W. F. Drake and A. Dalgarno, *Astrophys. J.* **157**, 459 (1969).
[68] P. Richard, R. L. Kauffman, F. F. Hopkins, C. W. Woods, and K. A. Jamison, *Phys. Rev. Lett.* **30**, 888 (1973); *Phys. Rev. A* **8**, 2187 (1973).

excited states have been investigated[69, 70]; however, their assignments are not as definite as other transitions in the spectra. The other lithiumlike transitions observed in the spectra are decays of the 1s2s2p and 1s2p² configurations. The resolution in either spectrum is not sufficient to observe the lines from each multiplet state. The peak labeled ²S–²P in Fig. 19 and that labeled S at 1.58 keV in Fig. 20 are from the decay of the doublet states in the lithiumlike system. These include the decays (1s2s2p)²P–(1s²2s)²S, (1s2p²)²S–(1s²2p)²P, (1s2p²)²P–(1s²2p)²P, and (1s2p²)²D–(1s²2p)²P. In Fig. 19, another line is observed which is labeled ²S–⁴P. This peak is a composite of the two spin forbidden transitions (1s2s2p)⁴P–(1s²2s)²S and (1s2p²)⁴P–(1s²2p)²P. These transitions, which occur as a result of configuration mixing, have been studied, and their lifetimes have been measured.[68] The (1s2s2p)⁴P level can decay also by autoionization. The ejected electron spectrum from the three-electron system has also been investigated.[71]

3.2.3.5. Radiative Auger Transitions. For many years, it was believed that x-ray and Auger-electron decay processes were the only modes of decay of atomic states. Recently, it has been demonstrated through experiment and calculation that a third process, called "radiative Auger," competes with the other two.[72] The radiative Auger decay is similar to the normal Auger decay in which an electron of higher energy fills the inner-shell vacancy and another electron, carrying away the excess energy, is ejected. For the radiative Auger process, the electron, instead of carrying away all of the excess energy, receives only part of the energy, and the remainder is carried away by an x ray. The energy of the x ray, $h\nu$, is given by the relation

$$h\nu + E_{\text{kin}} = E_{\text{Auger}}, \qquad (3.2.17)$$

where E_{kin} is the kinetic energy of the ejected electron and E_{Auger} is the energy of the normal Auger transition involving the same three levels. The x-ray spectrum for the Kα radiative Auger transitions in Al is shown in Fig. 21. The radiative Auger decay leads to a continuous spectrum of x-ray energies whose distribution depends upon the energy distribution of the ejected electron. The high-energy cutoff occurs at approximately the energy of the normal Auger transition, or when the emitted electron is ejected at zero energy. The end points and their corresponding Auger transition are indicated in Fig. 21. The actual end-point edge energy is influenced by solid

---

[69] D. L. Matthews, W. J. Braithwaite, H. H. Wolter, and C. F. Moore, *Phys. Rev. A* **8**, 1397 (1973).

[70] R. L. Kauffman, C. W. Woods, F. F. Hopkins, D. O. Elliott, K. A. Jamison, and P. Richard, *J. Phys. B* **6**, 2197 (1973).

[71] H. H. Haselton, R. S. Thoe, J. R. Mowat, P. M. Griffin, D. J. Pegg, and I. A. Sellin, *Phys. Rev. A* **11**, 468 (1975); D. J. Pegg *et al. Phys. Rev. A* **8**, 1350 (1973).

[72] T. Åberg and J. Utriainen, *Phys. Rev. Lett.* **22**, 1346 (1969).

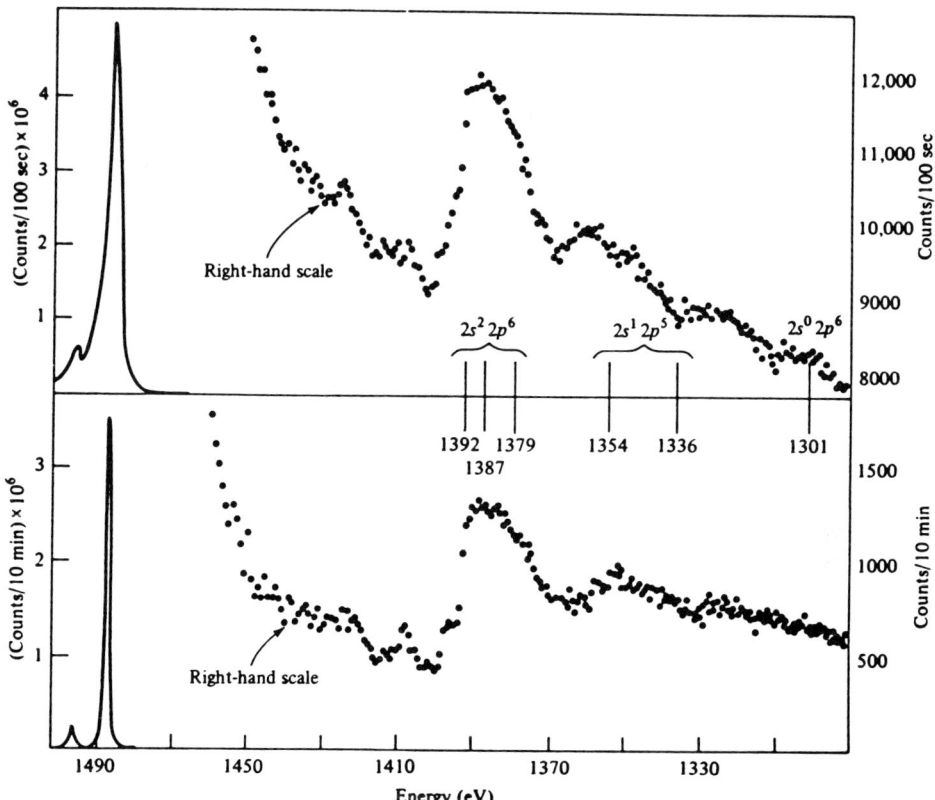

Fig. 21. Radiative Auger spectrum of Al. The spectra show the low-energy range below the Al Kα peaks. The upper figure is produced by primary excitation and the lower figure is produced by secondary excitation. The various KLL Auger transitions are indicated [taken from J. Siivola, J. Utriainen, M. Linkoaho, G. Graeffe, and T. Åberg, *Phys. Lett.* **32A**, 438 (1970)].

effects and chemical effects so that the energy may differ slightly from the Auger energy.[73] Discrete peaks are observed in addition to the continuous spectrum. The peaks correspond to an electron being transferred to an unoccupied bound state instead of being ejected into the continuum. The intensity of the radiative Auger lines is approximately 0.05% of the x-ray intensity for the Kα diagram line[72] and about 10% of the diagram line for L x rays.[74] The probability for production, calculated on the basis of a

---

[73] J. Utriainen and T. Åberg, *Phys. Lett.* **50A**, 263 (1974); *J. Phys. (Paris) Suppl. C4* **32**, 295 (1971).

[74] J. W. Cooper and R. E. LaVilla, *Phys. Rev. Lett.* **25**, 1745 (1970).

shake-off theory, have predicted intensities of the correct order of magnitude.[75] Configuration interaction should also be important in these multielectron processes.

3.2.3.6. Plasmon Satellites. Structure on the low-energy side of some x-ray emission bands from metals has been observed. These features can be explained by assuming that a plasmon is excited during x-ray emission. The x ray is thus emitted at an energy $\hbar\omega_p$ below the parent radiation, where ω_p is the plasmon frequency. These plasmon satellites were first observed in the L x-ray emission spectra of Na, Mg, and Al.[76] They have also been observed in the Kα emission of beryllium,[77] graphite,[78] and the Kβ emission of magnesium, aluminum, and silicon.[79] These plasmon satellites are approximately 1% of the parent line. Because the plasmons are a many-body effect, the plasmon satellites are interesting theoretically.[80, 81] The plasmons are created during the x-ray emission process as a result of the sudden disturbance of the electron distribution. Plasmon satellites have been observed only in transitions involving electrons originally in the conduction band of the solid.

If a plasmon exists in the solid at the time of x-ray emission, a high-energy plasmon satellite may be observed. These satellites would have an energy equal to the normal x-ray transition energy plus the plasmon energy. The intensity of these satellites would be small for the L-emission because the plasmon lifetime is much shorter than the L-shell vacancy lifetime. The K-shell vacancy lifetimes are comparable to the plasmon lifetime, and the satellites could possibly be observed. The Mg Kβ^V and Al Kβ^{VI} satellites observed by Kunzl[82] may possibly be plasmon satellites.[83]

3.2.3.7. Low-Energy Satellite Lines Observed in Heavy-Ion–Atom Collisions. Low-energy satellite lines have been observed in spectra produced by heavy-ion beams of oxygen[84, 85] and chlorine.[86] A spectrum of Si K x rays produced by a 30-MeV oxygen projectile is shown in Fig. 22.

[75] T. Åberg, *Phys. Rev. A* **4**, 1735 (1971).

[76] G. A. Rooke, *Phys. Lett.* **3**, 234 (1963).

[77] L. M. Watson, R. K. Dimond, and D. J. Fabian, "Soft X-Ray Band Spectra" (D. J. Fabian, ed.), pp. 45–58. Academic Press, New York, 1968.

[78] O. Aita, I. Nagakura, and T. Sagawa, *J. Phys. Soc. Japan* **30**, 516 (1971).

[79] C. Senemaud and M. T. Costa Lima, *Phys. Lett.* **47A**, 395 (1974); *Phys. Fenn. Suppl. S1* **9**, 373 (1974).

[80] F. Brouers, *Phys. Lett.* **11**, 297 (1964); *Phys. Status Solidi* **22**, 213 (1967).

[81] P. Longe and A. J. Glick, *Phys. Rev.* **177**, 526 (1969).

[82] V. Kunzl, *Z. Phys.* **99**, 481 (1936).

[83] K. S. Srivastava, S. P. Singh, and R. L. Srivastava, *Phys. Lett.* **47A**, 305 (1974).

[84] C. F. Moore, D. K. Olsen, B. Hodge, and P. Richard, *Z. Phys.* **257**, 288 (1972).

[85] P. Richard, C. F. Moore, and D. K. Olsen, *Phys. Lett.* **43A**, 519 (1973).

[86] P. Richard, C. F. Moore, D. L. Matthews, and F. Hopkins, *Bull. Amer. Phys. Soc.* **19**, 570 (1974).

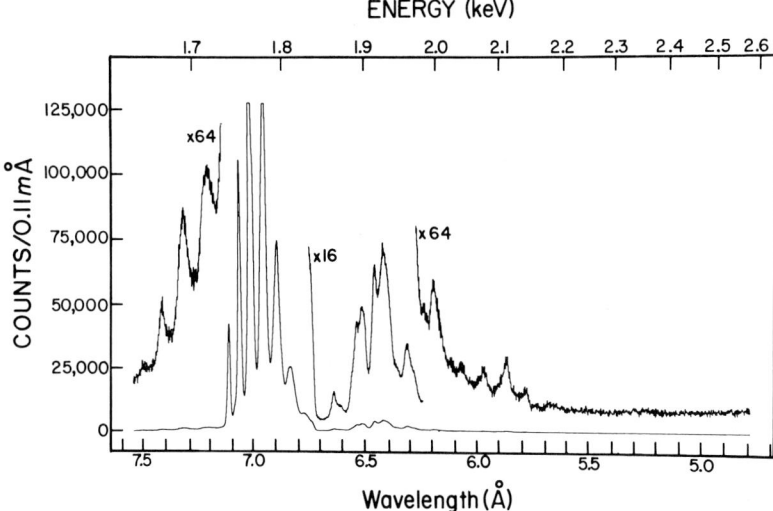

FIG. 22. Silicon K x rays produced by 30-MeV oxygen bombardment. The peaks below the normal $K\alpha_{1,2}$ peak at 1.486 keV have not been observed in spectra produced by any production mechanism other than heavy-ion collisions [taken from P. Richard, C. F. Moore, and D. K. Olsen, *Phys. Lett.* **43A**, 519 (1973)].

The low-energy satellite lines, which appear as peaks below the $K\alpha_1$ diagram lines, have a width larger than that observed in the high-energy satellite peaks, and are approximately evenly spaced. The energy shift of the low-energy satellite lines with respect to the diagram $K\alpha$ transition for elements Na to Ti is shown in Fig. 23. The energy shift is approximately linear with atomic number. Recent investigation by Richard et al.[87] have clarified the origin of these peaks. The heavy-ion induced spectra were compared to H- and He-induced spectra of Si and Al where the radiative Auger transitions discussed in Section 3.2.3.5 were observed as well as the lowest energy of the four low-energy satellites. By comparing the relative intensities of the low-energy satellites to the normal satellites, it was found that the ratio of the intensities of the lowest satellite to the first normal $K\alpha$ satellite was independent of the mode of excitation. It was concluded that these two lines are transitions from the same initial state. The low-energy satellites were then assigned as $(1s)^{-1}(2p)^{-n}$ to $(2s)^{-2}(2p)^{-n+1}$ electron rearrangement transitions for $n = 1$ to 4. The observed energies are in agreement with Hartree–Fock calculations of the electron rearrangement transition energies.

[87] P. Richard, J. Oltjen, K. A. Jamison, R. L. Kauffman, C. W. Woods, and J. M. Hall, *Phys. Lett.* **54A**, 169 (1975); K. A. Jamison, J. M. Hall, and P. Richard *J. Phys. B: Atom. Molec. Phys.* **8** (1975).

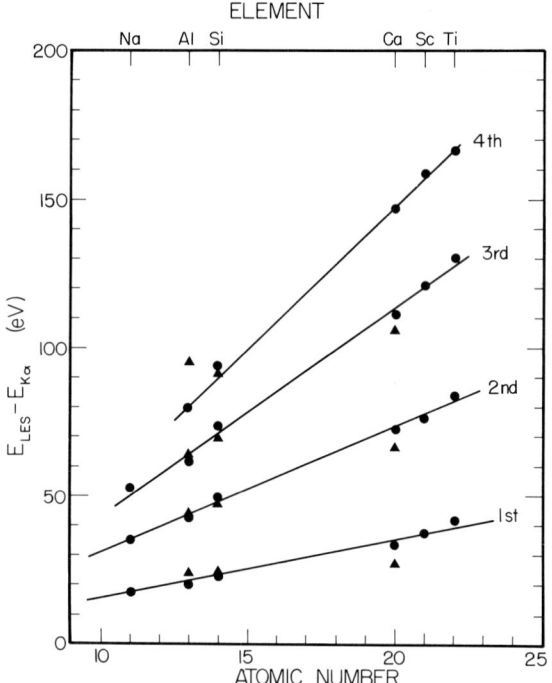

FIG. 23. The energy shift of the low energy satellite lines as a function of the atomic number. The shifts are measured with respect to the $K\alpha_{1,2}$ line observed in heavy-ion–atom collisions [● : oxygen; ▲ : chlorine]. The shifts vary monotonically as a function of the atomic number.

3.2.3.8. Quasi-Molecular X Rays. The x-ray spectra discussed in the previous sections generally refer to transitions in an excited atom or ion for which the energies of the initial and final states are time-independent. A new type of x-ray spectroscopy, for which the initial and final orbitals of the electron transition are undergoing large time-dependent variations, has received much attention recently, both experimentally and theoretically. Saris et al.[88] first reported the observation of a broad x-ray band in slow collisions between an incident ion and a stationary atom. This band, which is not characteristic of either the target atom or the projectile alone, is due to the time-dependent perturbed orbitals of the colliding system. The model incorporated to explain the x-ray bands is the molecular orbital (MO) model as discussed by Fano and Lichten[89] and Lichten.[90] The MO model

[88] F. W. Saris, W. F. van der Weg, H. Tawara, and R. Laubert, *Phys. Rev. Lett.* **28**, 717 (1972).
[89] U. Fano and W. Lichten, *Phys. Rev. Lett.* **14**, 627 (1965).
[90] W. Lichten, *Phys. Rev.* **164**, 131 (1967).

3.2. X-RAY REGION

was originally suggested as a mechanism for promotion of one electron from a lower-energy orbital to a vacant higher-energy orbital during the collision. After separation of the two colliding partners, the atom which carries away the vacancy can emit characteristic radiation. This model explained very well the large ionization cross sections for low-energy ion–atom collisions, as discussed by Saris,[91] Saris and Onderlinden,[92] Brandt and Laubert,[93] and Briggs and Macek.[94] The observed cross sections for these low-energy heavy-ion–atom collisions are typically orders of magnitude larger than predicted by direct Coulomb ionization.[95]

The energy level diagram of the molecular orbitals for the case of symmetric Ar–Ar collisions is shown in Fig. 24. At large distances, the molecular levels are given by the energy levels of the separated atoms, as shown on the right of the figure. If the collisional velocity is small compared to the average orbital electron velocities, it is then possible for the electrons to adjust to the two-centered molecular orbitals. The united atom limit is approached as the internuclear separation r goes to zero. The energy levels of the united atom Kr are shown on the left of the figure. The broad x-ray (quasi-molecular x ray) band observed by Saris et al.[88] can be explained as an L x-ray transition in which a vacancy in the $2p\pi$ orbital is filled during the collision. An energy band is observed because of the rapid increase in the binding energy of the $2p\pi$ orbitals as the internuclear separation decreases. The $2p\pi$ vacancy is brought into the collision by a vacancy in the 2p orbital of the projectile formed in a previous collision.

The search for such x-ray bands has been conducted by several groups: Mokler et al.,[96] Macdonald et al.,[97, 98] Meyerhoff et al.,[99] Davis and Greenberg,[100] Bissinger and Feldman,[101] and Cairns et al.[102] Figure 25a gives recent results of Wölfli et al.[103] for Ni–Ni and Ni–Fe collisions. The large x-ray continuum above the $K\beta$ x ray of Ni is in part due to the MO x-ray band. The united atom $K\alpha$ x-ray energies of 33 and 29 KeV are the expected

[91] F. W. Saris, Physica 52, 290 (1971).
[92] F. W. Saris and D. Onderdelinden, Physica 49, 441 (1970).
[93] W. Brandt and R. Laubert, Phys. Rev. Lett. 24, 1037 (1970).
[94] J. S. Briggs and J. Macek, J. Phys. B 5, 579 (1972).
[95] E. Merzbacher and H. W. Lewis, "Handbuch der Physik" (S. Flügge, ed.), Vol. 34, pp. 166–192. Springer Verlag, Berlin and New York, 1958.
[96] P. H. Mokler, H. J. Stein, and P. Armbruster, Phys. Rev. Lett. 29, 827 (1972).
[97] J. R. Macdonald and M. D. Brown, Phys. Rev. Lett. 29, 4 (1972).
[98] J. R. Macdonald, M. D. Brown, and T. Chiao, Phys. Rev. Lett. 30, 471 (1973).
[99] W. E. Meyerhof, T. K. Saylor, S. M. Lazarus, W. A. Little, B. B. Triplett, and L. F. Chase, Jr., Phys. Rev. Lett. 30, 1279 (1973).
[100] C. K. Davis and J. S. Greenberg, Phys. Rev. Lett. 32, 1215 (1974).
[101] G. Bissinger and L. C. Feldman, Phys. Rev. A 8, 1624 (1973).
[102] J. A. Cairns, A. D. Marwick, J. Macek, and J. S. Briggs, Phys. Rev. Lett. 32, 509 (1974).
[103] W. Wölfli, Ch. Stoller, G. Bonani, M. Suter, and M. Stöckli, Phys. Rev. Lett. 35, 656 (1975).

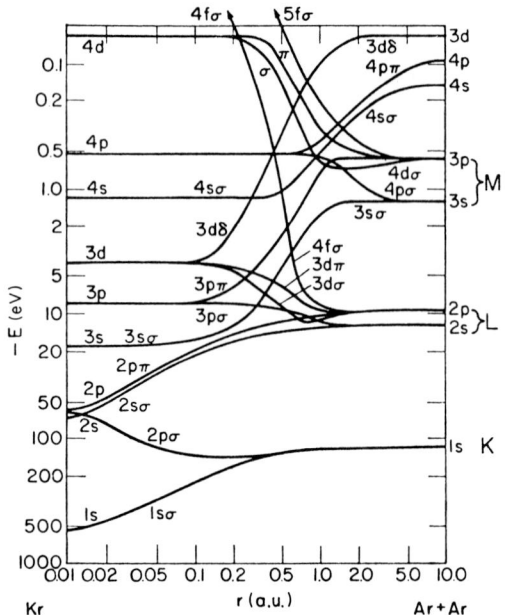

FIG. 24. Molecular-orbital diagram for the argon–argon system. The quasi-molecular L x-ray band observed by F. W. Saris, W. F. van der Weg, H. Tawara, and R. Laubert [*Phys. Rev. Lett.* **28**, 717 (1972)] is attributed to a 3d to 2pπ transition, which varies from a few hundred electron volts up to 1.6 keV as the internuclear distance r decreases [taken from W. Lichten, *Phys. Rev.* **164**, 131 (1967)].

end-point energies for the quasi-molecular 2p-1s transition in the single MO model for Ni–Ni and Ni–Fe transitions, respectively. Two other sources of continuous radiation from the collision can contribute significantly in this energy range and extend the continuum above the united atom limit. These sources are nucleus–nucleus bremsstrahlung[100] and inner-shell electron bremsstrahlung.[104] The latter is discussed in the next section.

A motivation for the further study of quasi-molecular x rays has come from the theoretical investigations of atomic levels in super-heavy elements by Pieper and Greiner,[105] Popov,[106] and Rafelski et al.,[107] who predict supercritical fields near atomic number $Z_{cr} = 169$. Müller et al.[108] propose that high-energy heavy-ion collisions could be used to create quasi-molecules

[104] P. Kienle et al., *Phys. Rev. Lett.* **31**, 1099 (1973).
[105] W. Pieper and W. Greiner, *Z. Phys.* **218**, 327 (1969).
[106] V. S. Popov, *Zh. Eksp. Theor. Fiz.* **59**, 965 (1970) [English transl.: *Sov. Phys.-JETP* **32**, 526 (1971)].
[107] J. Rafelski, L. P. Fulcher, and W. Greiner, *Phys. Rev. Lett.* **27**, 958 (1971).
[108] B. Müller, H. Peitz, J. Rafelski, and W. Greiner, *Phys. Rev. Lett.* **28**, 1235 (1972).

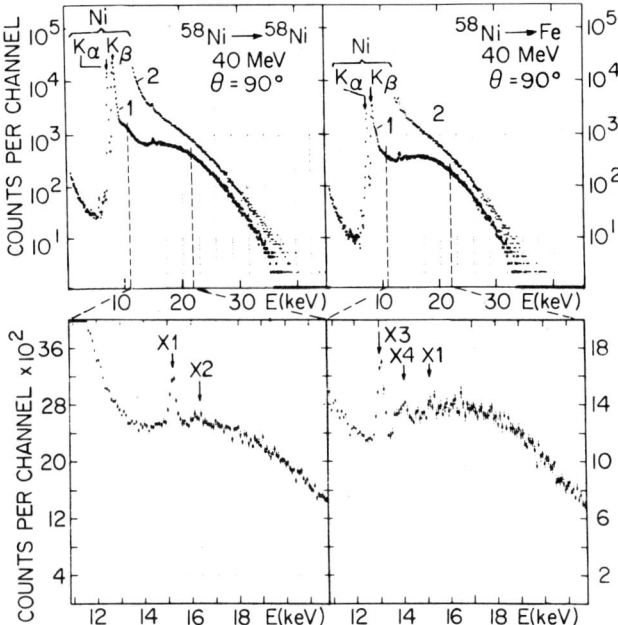

FIG. 25 (a) Quasi-molecular x-ray spectrum taken with a Ge(Li) detector for Ni–Ni and Ni–Fe collisions at 40 MeV bombarding energy. (b) Expanded region shows small peaks believed to be due to double electron filling of two K-shell holes [taken from W. Wölfli, Ch. Stoller, G. Bonani, M. Suter, and M. Stöckli, *Phys. Rev. Lett.* **35**, 656 (1975)].

of super-heavy elements, and thereby be a means of investigating the quantum electrodynamics of these super-heavy elements. One prediction of the theory is that, for $Z \geq Z_{cr}$, the K level dives into the negative-energy continuum so that a vacancy in such a shell can be filled by an electron from the negative continuum under simultaneous emission of a positron. This new process competes with normal x-ray emission. No experimental evidence of positron emission has been reported.

Müller *et al.*[109] also have predicted a directional anisotropy in the quasi-molecular K x-ray emission which is peaked at 90° to the beam direction and is greatest near the united atom limit. Several investigators[110–112] have recently presented experimental evidence for this anisotropy. The results of Thoe *et al.*[112] are given in Fig. 26 for symmetric Al–Al collisions at two

[109] B. Müller, R. Kent-Smith, and W. Greiner, *Phys. Lett.* **49B**, 219 (1974); B. Müller and W. Greiner, *Phys. Rev. Lett.* **33**, 469 (1974).
[110] J. S. Greenberg, C. K. Davis, and P. Vincent, *Phys. Rev. Lett.* **33**, 473 (1974).
[111] G. Kraft, P. H. Mokler, and H. J. Stein, *Phys. Rev. Lett.* **33**, 476 (1974).
[112] R. S. Thoe *et al.*, *Phys. Rev. Lett.* **34**, 64 (1975).

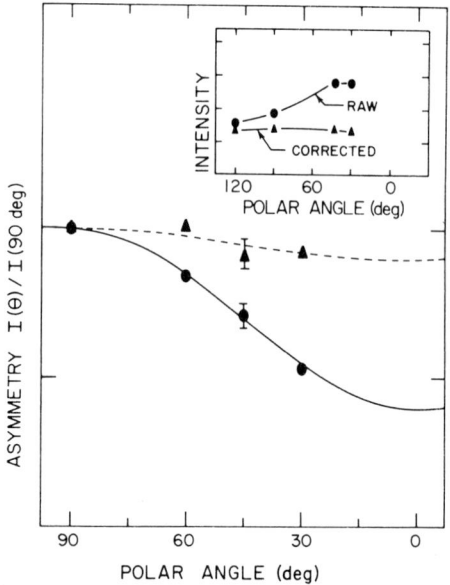

FIG. 26. Angular distribution of quasi-molecular x rays from Al–Al collisions in the x-ray energy interval 5–6 keV [● : 30 MeV,—: $0.4 + 0.6 \sin^2 \theta$; ▲ : 10 MeV,—: $0.9 + 0.1 \sin^2 \theta$]. The united atom Kα x-ray limit is 6.4 keV. The inset depicts the distribution for the 1–3 keV radiation from 30-MeV Al–Al collisions. After corrections for Doppler-shift-window absorption corrections, the characteristic radiation exhibits approximate isotropy [taken from R. S. Thoe et al., Phys. Rev. Lett. **34**, 64 (1975)].

incident energies. The asymmetry b in the form $1 + b \sin^2 \theta$ is $+1.5$ for 30-MeV collisions and $+0.11$ for 10-MeV collisions in the x-ray interval 5–6 keV. It has also been demonstrated that the anisotropy increases as the x-ray interval approaches the united atom limit.

Figure 25b shows an expanded view of the region between 12 and 18 keV. The peaks denoted X1 and X2 are believed to be transitions due to double electron filling of two K-shell hole states in the Ni. X1 results from both electrons originating from the L shell. This transition should have an energy approximately equal to the sum of the energy of the Kα and the Kα hypersatellite transition. X2 results from one electron originating from the M shell and one electron from the L shell. X3 and X4 are similar transitions from the Fe target. This is the first report of the observation of these transitions. More systematics are needed to verify these identifications.

3.2.3.9. Radiative Electron Capture and Bremsstrahlung. The results of Saris et al.[88] in 1972, describing quasi-molecular x rays in ion–atom collisions, were closely followed by the observation of other new features in the x-ray spectra of charged particle collisions. In 1972, Schnopper et al.[113] reported the observation of a broad x-ray band whose peak energy shifted proportionally with the incident beam energy. They attributed this feature to radiative electron capture (REC) of target electrons into bound orbitals of the fast projectile. The emitted photon derives its energy from the kinetic energy of the captured electron, relative to the projectile, plus the difference in its binding energies in the initial and final states.

The basic properties of the REC energy distribution can be obtained from a simple model which assumes a velocity distribution of loosely bound target electrons which are captured into a shell of the projectile with binding energy $-E_f$. The target electron velocity $\bar{V}$ in the rest frame of the projectile is

$$\bar{V} = \bar{v}_e + \bar{v}_i, \quad (3.2.18)$$

where $\bar{v}_e$ is the velocity of the electron in the target rest frame and $\bar{v}_i$ is the laboratory velocity of the projectile. The REC photon energy distribution is

$$h\nu = \tfrac{1}{2}m_e v^2 + E_f = T' + E_f, \quad (3.2.19)$$

where

$$T' = \tfrac{1}{2}m_e(v_i^2 + v_e^2 + 2\bar{v}_i \cdot \bar{v}_e) = T_r + T + 2(TT_r)^{1/2} \cos\theta \quad (3.2.20)$$

and

$$T_r = \tfrac{1}{2}m_e v_i^2. \quad (3.2.21)$$

For ion velocities such that $v_i \gg v_e$, the peak of the REC spectrum occurs near

$$(h\nu)_p \simeq T_r + E_f = (m_e/M_i)E_i + E_f, \quad (3.2.22)$$

where E_i is the incident projectile energy. The term $\bar{v}_i \cdot \bar{v}_e$ determines the shape of the REC peak so that it is related to the electron momentum distribution of the target electrons.

This is demonstrated in Fig. 27 for 65-, 55-, 45-, and 30-MeV oxygen on various gaseous targets. For the 65-MeV case, the available kinetic energy is 2.2 keV, and the binding energy for the K- and L-shell capture supplies an additional 0.742 and 0.175 eV, respectively. The REC peaks centered at 2.94 and 2.38 keV are thus predicted by Eq. (3.2.22). Similarly, the 55-, 45-, and 30-MeV spectra should show REC K-capture peaks centered at 2.61, 2.27, and 1.76 keV, respectively. The data bear out these systematics. The two low-energy peaks in Fig. 27 are from oxygen projectile radiation and from Si Kα x rays.

[113] H. W. Schnopper et al., Phys. Rev. Lett. **29**, 898 (1972).

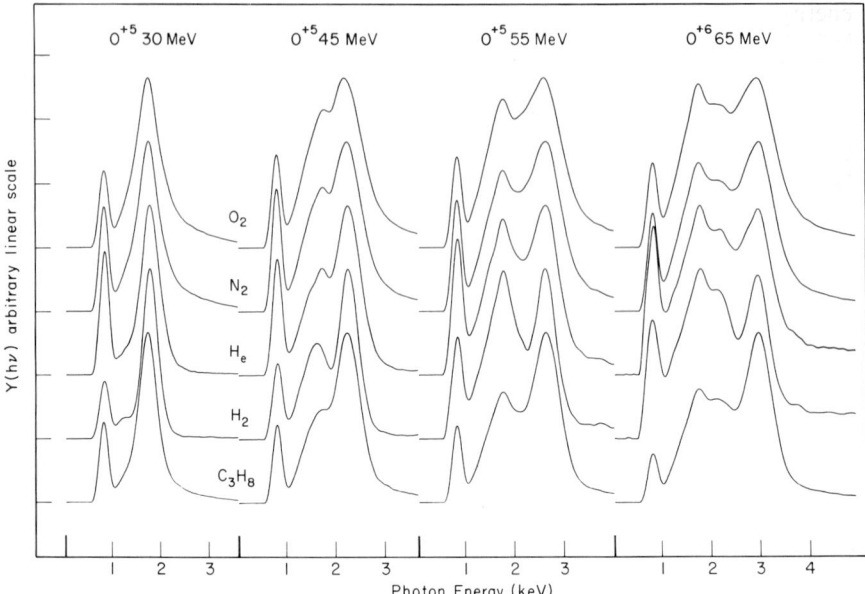

FIG. 27. X-ray yields for oxygen projectiles incident on various gaseous targets. The peak energies for radiative electron capture in the K shell of O^{+7} are expected at 2.94, 2.61, 2.27, and 1.76 keV for the 65-, 55-, 45-, and 30-MeV projectiles, respectively [taken from H. W. Schnopper and J. P. Delvaille, *Proc. Int. Conf. At. Collisions Solids, 5th*, Gatlinburg, (1973)].

Two contributions to the x-ray continuum due to electron bremsstrahlung are discussed by Schnopper and Delvaille[114] and by Kienle et al.[104] Schnopper and Delvaille discuss the bremsstrahlung resulting from almost free target electrons undergoing transitions to a continuum state, as opposed to the REC photons which arise when the electrons are captured into a bound state of the projectile. A continuous spectrum will result with endpoint energy T_r for electrons in the final state at rest in the projectile frame. This is demonstrated in Fig. 28 for 4- and 10-MeV proton beams with endpoint energies of 2.18 and 5.45 keV, respectively, and 65-MeV O^{8+} with end-point energy of 2.2 keV. The bremsstrahlung end point corresponds to the lowest-energy REC peak position.

Kienle et al.[104] discuss the bremsstrahlung from tightly bound electrons in the target. Since it is possible to deliver more momentum to these electrons, the bremsstrahlung continuum extends to energies above the REC. Kienle et al.[104] compare photon spectra from 140-MeV Ne on He and Ne targets,

---

[114] H. W. Schnopper and J. P. Delvaille, in *Proc. Int. Conf. At. Collisions Solids, 5th*, Gatlinburg (1973).

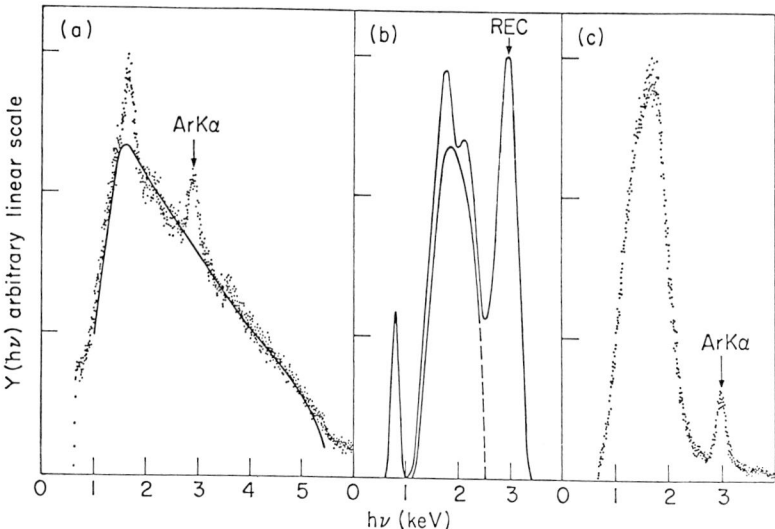

FIG. 28. (a) Experimental and theoretical curve for bremsstrahlung from 10-MeV p on H_2, (b) experimental yield for 65-MeV O^{8+} on H_2, showing REC and bremsstrahlung (centered near 1.8 keV), and (c) experimental yield for 4-MeV p on H_2, showing bremsstrahlung. Curves (b) and (c) were measured for matched velocities that should yield bremsstrahlung spectra of the same shape [taken from H. W. Schnopper and J. P. Delvaille, *Proc. Int. Conf. At. Collisions Solids, 5th, Gatlinburg* (1973)].

and show that the ratio of bremsstrahlung to REC in the peak is 0.2 for the Ne target and that there is a high-energy tail due mainly to bremsstrahlung. The continuum–continuum bremsstrahlung discussed by Schnopper and Delvaille has an end-point energy of 3.8 keV, far below the main intensity of the bound-continuum bremsstrahlung discussed by Kienle *et al.*, which occurs at ~4.5 keV.

The noncharacteristic radiation in ion–atom collisions has contributions from bremsstrahlung and quasi-molecular x rays. One is therefore cautioned to consider carefully all of these x-ray bands in analyzing x-ray spectra from charged particle collisions.

3.2.3.10. Muonic X-Rays. When a negatively charged muon μ^- is stopped in matter, it can be captured by a nucleus in an atomic orbit forming a muonic atom.[115–118] X rays emitted during a cascade of the muon were

[115] S. Devons and I. Duerdoth, *Advan. Nucl. Phys.* **2**, 295–423 (1969).

[116] E. H. S. Burhop, "High Energy Physics," Vol. 3, pp. 109–281. Academic Press, New York, 1969.

[117] C. S. Wu and L. Wilets, *Annu. Rev. Nucl. Sci.* **19**, 527–606 (1969).

[118] Y. N. Kim, "Mesic Atoms and Nuclear Structure." North-Holland Publ., Amsterdam, 1971.

first observed by Chang[119] using naturally occurring muons, and by Fitch and Rainwater[120] using accelerator-produced muons. The study of muonic x rays is interesting because information concerning both atomic and nuclear structure can be obtained.

A muon is a spin $\frac{1}{2}$ particle having a negative or positive charge equal in magnitude to that of an electron. The mass of a muon is approximately 207 times the mass of an electron. Because of this large mass ratio, the energy of the x rays from a muonic transition is much larger than from electronic transitions. The muonic transition energies can be as high as a few MeV. The lifetime of the muon is 2.2 nsec, long compared with the time estimated for the capture and subsequent cascade to the ground state of a muon.[121] These cascade transitions are readily observed.

The Schrödinger equation for the muonic atom, treating the nucleus as a point charge, is the same as the Schrödinger equation for a one-electron atom except that the muon reduced mass is substituted for the electron reduced mass. For muon states having principal quantum number n less than 14, the radius of the orbitals is less than the K-shell electron radius. In this region, the energy levels can be obtained by substituting the muon reduced mass for the electron reduced mass in the H-atom formula. For outer orbitals, the electron screening of the nucleus must be included in the calculations. For better estimates of the energy levels, solutions to the Dirac equation should be used.

To obtain accurate calculations of the energy levels, the finite size of the nucleus must be taken into account. The finite nuclear size decreases the muon binding energy. This is due to the fact that part of the muon orbit is located inside the nucleus. The energy level shifts have been used extensively to investigate the size of the nucleus.[120, 122, 123] Other corrections to the energy level calculations are due to vacuum polarization,[124] nuclear static quadrupole moment,[122] and nuclear polarization.[125]

Other negatively charged elementary particles may also form "mesic" atoms. Two systems which have received much attention are the pionic[126] and kaonic atoms.[127] Their treatment is similar to muonic atoms except that the pion and kaon are bosons. Their behavior is governed by the Klein–

---

[119] W. Y. Chang, *Rev. Mod. Phys.* **21**, 166 (1949); *Phys. Rev.* **95**, 1288 (1954).
[120] V. L. Fitch and J. Rainwater, *Phys. Rev.* **92**, 789 (1953).
[121] E. Fermi and E. Teller, *Phys. Rev.* **72**, 399 (1947).
[122] J. A. Wheeler, *Rev. Mod. Phys.* **21**, 133 (1949); *Phys. Rev.* **92**, 812 (1953).
[123] D. L. Hill and K. W. Ford, *Phys. Rev.* **94**, 1617 (1954).
[124] A. B. Mickelwait and H. C. Corben, *Phys. Rev.* **96**, 1145 (1954).
[125] L. N. Cooper and E. M. Henley, *Phys. Rev.* **92**, 801 (1953).
[126] G. Backenstoss, *Annu. Rev. Nucl. Sci.* **20**, 467–508 (1970).
[127] C. E. Wiegand and G. L. Godfrey, *Phys. Rev. A* **9**, 2282 (1974).

Gordon equation instead of the Dirac equation, and they can have strong interactions in addition to the electromagnetic interactions with the nucleus. The study of the x-ray energies from these "mesic" atoms is used as a tool for investigating the strong interactions.

3.2.4. Selected Topics

The space limitation in this chapter does not allow us to discuss, or even mention, all the regions of interest in x-ray spectroscopy; therefore, we have chosen two selected topics of recent interest to conclude our discussion of the x-ray region. One section is the comparison of spectra from five different x-ray sources. Each type of x-ray source seems to have created an x-ray field within a field and has established its own techniques and expertise. The other section reviews some of the important aspects of x-ray cross-section measurements from ion–atom collisions which have attracted a large group of researchers in the last five years.

3.2.4.1. Comparison of X-Ray Spectra from Solar Plasmas, Laser-Induced Plasmas, Spark Plasmas, Tokamaks, and Accelerators. The investigation of the properties of an x-ray emitting system requires correct identification of the spectral lines and accurate measurement of the relative intensities. The observation or nonobservation of various transitions can be a clue to the reaction mechanism leading to the formation of ionic states, whereas the relative intensities can be used to measure such quantities as the electron temperature, electron number density, or relative formation rates.

Astrophysical plasmas such as that found in the sun are at sufficiently high temperature and are of sufficient densities as to form an appreciable flux of K x-ray radiation from few-electron atoms. A summary of the interpretation and use of the heliumlike and lithiumlike spectra is discussed. A more thorough review of this subject is given by Gabriel and Jordan.[128] Figure 29 is a solar spectrum of Mg taken in 1971 on a Sun-stabilized Skylark rocket, SL1101, from Woomera, South Australia.[129] This is an example of a solar x-ray spectrum, showing the low-lying few-electron transitions. The line identification is given in Table I. The peaks labeled R_1 and R_2 were not identified in Parkinson,[129] but are likely due to the resonance transition (R) with spectator electrons in higher orbits. The model used to describe this spectrum assumes that the dominant excitation is by electron impact, while ion–ion collisions, photoexcitation, and direct ionization are assumed to be unimportant.

---

[128] A. H. Gabriel and C. Jordan, "Case Studies in Atomic Collision Physics," Vol II, pp. 209–291. North-Holland Publ., Amsterdam, 1972.

[129] J. H. Parkinson, *Nature (London) Phys. Sci.* **236**, 68 (1972).

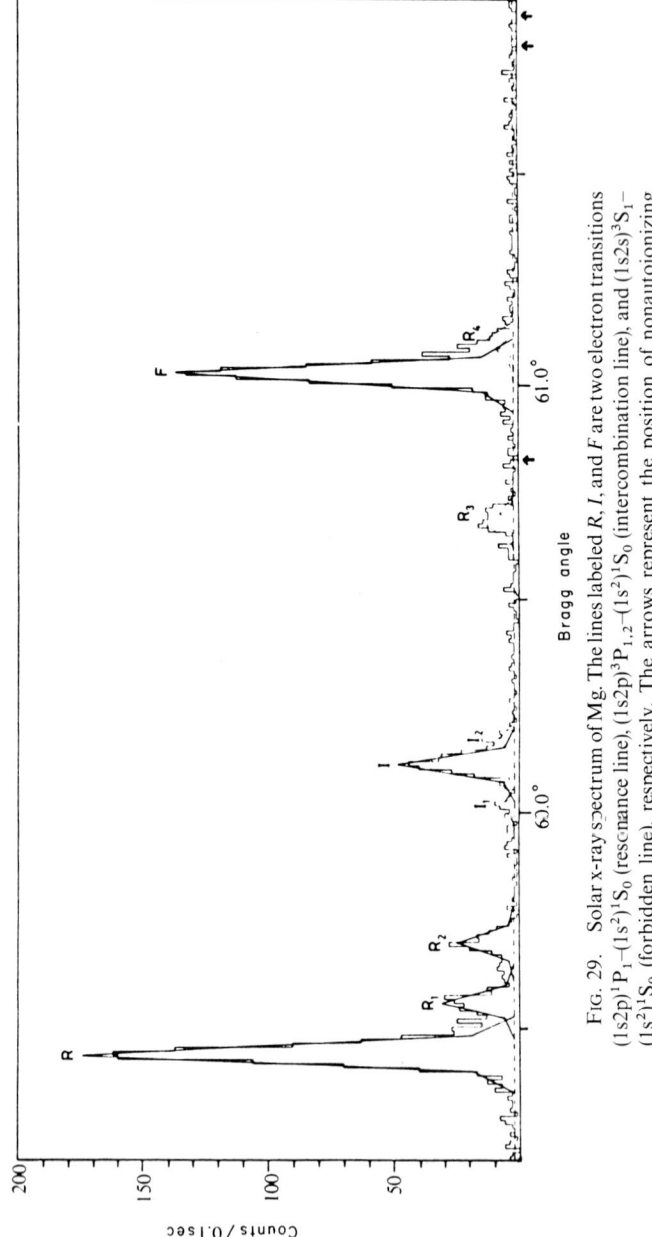

FIG. 29. Solar x-ray spectrum of Mg. The lines labeled R, I, and F are two electron transitions $(1s2p)^1P_1-(1s^2)^1S_0$ (resonance line), $(1s2p)^3P_{1,2}-(1s^2)^1S_0$ (intercombination line), and $(1s2s)^3S_1-(1s^2)^1S_0$ (forbidden line), respectively. The arrows represent the position of nonautoionizing three-electron transitions, and the lines labeled R_4, R_3, I_1, and I_2 refer to autoionizing three-electron transitions (see text) [taken from J. H. Parkinson, Nature (London) Phys. Sci. **236** 68 (1972)].

TABLE I. Solar X-Ray Spectrum

Ion species	Transition	Characterization	Label
Heliumlike	$(1s2p)^1P_1–(1s^2)^1S_0$	Resonance line	R
	$(1s2p)^3P_{1,2}–(1s^2)^1S_0$	Intercombination line	I
	$(1s2s)^3S_1–(1s^2)^1S_0$	Forbidden line	F
Lithiumlike	$(1s2p^2)^2D–(1s^22p)^2P$	Upper level autoionizing	R_4
	$(1s2p^2)^2S–(1s^22p)^2P$	Upper level autoionizing	I_1
	$(1s2p^2)^2P–(1s^22p)^2P$	Upper level nonautoionizing	$\uparrow^a$
	$(1s2p^2)^4P–(1s^22p)^2P$	Upper level nonautoionizing	$\uparrow$
	$(1s2p^1P)2s^2P–(1s^22s)^2S$	Upper level autoionizing	R_3
	$(1s2p^3P)2s^2P–(1s^22s)^2S$	Upper level autoionizing	I_2
	$(1s2s2p)^4P–(1s^22s)^2S$	Upper level nonautoionizing	$\uparrow^b$

[a] Highest energy.
[b] Lowest energy.

First, we consider the information obtained from the heliumlike spectrum. The level scheme for the two-electron system for the case of O VII is given in Fig. 30. The six $n = 2$ levels are $(1s2p)^1P_1$, $(1s2p)^3P_{0,1,2}$, $(1s2s)^1S_0$ and $(1s2s)^3S_1$. The $^1P_1$ state is excited by direct electron impact and has only one decay mode, the prominent, so-called "resonance line" labeled R in Fig. 29. The $^1S_0$ state is formed by electron impact and decays to the ground state via double $E1$ emission, which leads to a continuum radiation pattern

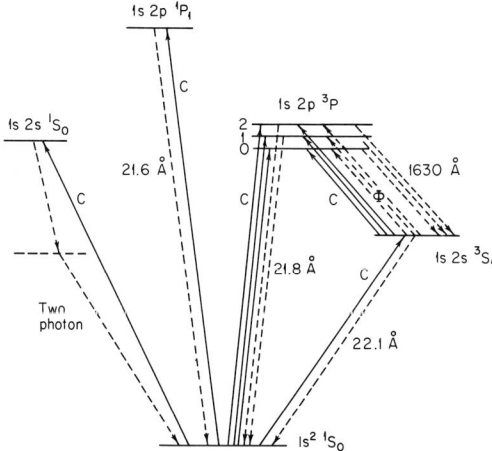

FIG. 30. The two-electron levels of O VII ions for $n = 2$ and $n = 1$. The excitation modes having rates C and Φ, and the decay modes denoted by the wavelength for the transition, are schematically represented [taken from A. H. Gabriel and C. Jordan, "Case Studies in Atomic Collision Physics," Vol. II, pp. 209–291 (1971)].

not observable in Fig. 29. The three states $^3P_{0,1,2}$ are formed by electron impact and are assumed to be statistically populated (i.e., 1:3:5 population for $J = 0, 1$, and 2, respectively). The $^3P_1$ and $^3P_2$ states decay to the ground state via $E1$ (mixing with $L = 1, J = 1$ states) and $M2$ radiation, respectively, and yield the peak labeled I in Fig. 29. The $^3P_{0,1,2}$ states each have an $E1$ radiative branch to the $^3S_1$ state. The $^3S_1$ state is formed by direct electron impact and by radiative decay of the $^3P_{0,1,2}$ states. The state decays by relativistic $M1$ radiation to the ground state, the so-called forbidden line labeled F in Fig. 29.

If the density is sufficiently low, only the above excitation and decay processes are important. The x-ray intensity ratio r_0 of the forbidden line F to the intercombination line I under these conditions is

$$r_0 = \frac{C(^3S) + C(^3P)B(^3P-^3S)}{C(^3P)[1 - B(^3P-^3S)]}, \qquad (3.2.23)$$

where the C's refer to the electron impact formation cross sections, and B refers to the ^3P branching ratio to the ^3S state which is

$$B(^3P-^3S) = \frac{1}{3}\frac{A(^3P_1-^1S_0)}{[A(^3P_1-^1S_0) + A(^3P_1-^3S_1)]} + \frac{5}{9}\frac{A(^3P_2-^1S_0)}{[A(^3P_2-^1S_0) + A(^3P_2-^3S_1)]}. \qquad (3.2.24)$$

This ratio r_0 is generally referred to as the *low density limit*, and depends on the radiative rates A and cross sections C only. This points to the need for accurate measurements of these x-ray rates and electron impact cross sections. In this low density limit, the ratio r_0 is independent of electron temperatures T_e and electron number density N_e, assuming that the C's are, at most, weakly dependent on T_e. For O VII, r_0 is approximately 3.9.

If the density is increased, eventually the $^3S_1$ level will be quenched through the ^3P level by electron impact and photoexcitation. For the case of O VII, this depletion of the $^3S_1$ population and corresponding increase in the $^3P_{0,1,2}$ population occurs at electron densities N_e near 10^{10} cm^{-3}. The ratio r of the forbidden line to the intercombination line in this region, called the *low density region*, is

$$r = \frac{B'\{C(^3S) + C(^3P)B(^3P-^3S)\}}{\{C(^3P) + C(^3S)(1 - B')\}(1 - B(^3P-^3S))}. \qquad (3.2.25)$$

The quantity B' is the branching ratio of the $^3S_1$ level to the $^1S_0$ level. This branching ratio is

$$B' = \frac{A(^3S_1-^1S_0)}{A(^3S_1-^1S_0) + N_eC(^3S-^3P) + \Phi}. \qquad (3.2.26)$$

The last two terms in the denominator are the $^3S_1$ quenching terms. When these terms are negligible, then $B' = 1$ and Eq. (3.2.25) reduces to Eq. (3.2.23).

3.2. X-RAY REGION

If all the decay rates and cross sections are known, the ratio r can then be used to determine the electron density N_e. In the *low density region* and the *low density limit*, there is no depopulation of the ^3P level by collisional excitation, so that the triplet to singlet intensity ratio G is given by

$$G = \frac{I + F}{R} = \frac{C(^3P) + C(^3S)}{C(^1P)}. \quad (3.2.27)$$

It is found that $G \sim 1.1$ over a large range of both T_e and Z. The ratio for the Mg data in Fig. 29 is $G = 1.04$, which establishes the plasma as being in the *low density region* or the *low density limit*. The ratio r is 3.0, which further limits the plasma to be near the *low density limit*. The ratio r approaches zero as the density N_e increases; therefore, the presence of the forbidden line in the spectrum rules out high density plasmas.

As the plasma density increases, the $(1s2s)^1S_0$ state is quenched by electron collisions to the $^1P_1$ level. In this region, the *intermediate density region*, the intensity of the resonance line relative to the intercombination line increases, and the forbidden line R is no longer visible.

As the density is increased even further, the $(1s2p)^3P$ state becomes quenched by electron impact excitations, and the ratio r' of ^1P to ^3P radiation increases. This region is referred to as the *high density region*. The ratio r' is given by

$$r' = \frac{R}{I} = \frac{C(^1P) + C(^3P)B(^3P-^1P)}{C(^3P)B(^3P-^1S)}, \quad (3.2.28)$$

where the two branching ratios are

$$B(^3P-^1P) = \frac{C(^3P-^1P)N_e}{C(^3P-^1P)N_e + N_e S(^3P) + A(^3P-^1S)} \quad (3.2.29)$$

and

$$B(^3P-^1S) = \frac{A(^3P-^1S)}{C(^3P-^1P)N_e + N_e S(^3P) + A(^3P-^1S)}. \quad (3.2.30)$$

Here, $S(^3P)$ refers to direct collisional excitation out of the ^3P level to some higher-energy level that does not cascade through the ^1P level.

Information concerning the plasma can also be obtained from the ratio of intensities of the lithiumlike lines (sometimes referred to as satellites) to the resonance line. The seven 2p–1s x-ray transitions from the $1s2p^2$ and the $1s2s2p$ configurations, as given in the table above, are expected. If we assume that these so-called "satellite lines" are formed by dielectric recombination with the heliumlike ground state [e.g., $(1s^2)^1S_0 + e-(1s2p^2)^2D$] then not all states will be formed, since dielectric recombination, which is the inverse of autoionization, will be parity forbidden for some states. Therefore, x-ray transitions from nonautoionizing states should not be present in plasma x-ray spectra if the dominant mode of excitation is direct electron impact. The three arrows below the abscissa in Fig. 29 indicate the position of x-ray transitions from the nonautoionizing $n = 2$ states of Mg. The absence of these lines in the spectra supports the electron impact hypothesis.

The ratio of the intensity of an observed three-electron satellite line α_s [e.g., $R_3(1s2p\,^1P)2s\,^2P-(1s^22s)^2S$] to the resonance R is given by Gabriel and Jordan[128] as

$$\frac{\alpha_s}{R} = \frac{1.2 \times 10^{-13}}{T_e} \frac{g_s}{g_i} \frac{E(ij)}{Pf(i \to j)} \frac{A_r A_a}{A_a + A_r} \exp\left[(E(ij) - E_s)/T_e\right], \quad (3.2.31)$$

where g_s and g_i are statistical weights of the satellite and ground state recombining ion, respectively; E_s is the energy of the satellite line; A_a and A_r are transition probabilities for decay of the satellite lines by autoionization and radiation, respectively; $E(ij)$ is the transition energy of the resonance line; $f(i \to j)$ is the oscillator strength; and P is a correction factor as tabulated by Van Regermorter.[130] Equation (3.2.31) can be used to determine the plasma temperature if the autoionization and radiation rates are known.

Figure 20 in Section 3.2.3.4 gives the Al spectrum of Nagel et al.[131] from a laser-induced plasma. The heliumlike resonance line ($^1P_1-{}^1S_0$) near 1.6 keV is very strong, and the intercombination line ($^3P_{1,\,2}-{}^1S_0$) at a slightly lower energy is only half as strong. The forbidden line is not present and assumed to be quenched by the plasma. From the preceding arguments, it can be concluded that the plasma is in the medium density region. The line below 1.6 keV labeled S is the lithiumlike satellite line. The intensity ratio S/R can be used in conjunction with Eq. (3.2.31) to obtain an approximate effective temperature for this type of plasma. The ratio of 2p–1s hydrogenlike and heliumlike transitions was used to determine the temperature, using the coronal equilibrium model,[132, 133] and agreed to 20% with the satellite technique.[131] X-ray spectra of laser induced plasma is a relatively new field which will undoubtedly contribute much to our knowledge of x-ray transitions in highly ionized heavy-ion species.

An x-ray spectrum from a low inductance vacuum spark given by Schwob and Fraenkel[134] for Fe is given in Fig. 31. In this spectrum, a very small amount of hydrogenlike Fe XXVI is observed. The heliumlike resonance line is strongly excited and the heliumlike intercombination line is rather weak. The satellite lines are quite prominent but not clearly resolved. The heliumlike forbidden line may in fact be unresolved from the lithiumlike Fe XXIV group. A large number of additional satellite lines are observed in this spec-

[130] H. Van Regemorter, *Astrophys. J.* **136**, 906 (1962).
[131] D. J. Nagel et al., *Phys. Rev. Lett.* **33**, 743 (1974).
[132] G. Elwert, *Z. Naturforsch.* **79**, 432 (1952).
[133] R. C. Elton, in "Plasma Physics" (H. A. Griem and R. H. Loveberg, eds.), Vol. 9A, p. 115. Academic Press, New York, 1970.
[134] B. S. Fraenkel and J. L. Schwob, *Phys. Lett. A* **40**, 83 (1972).

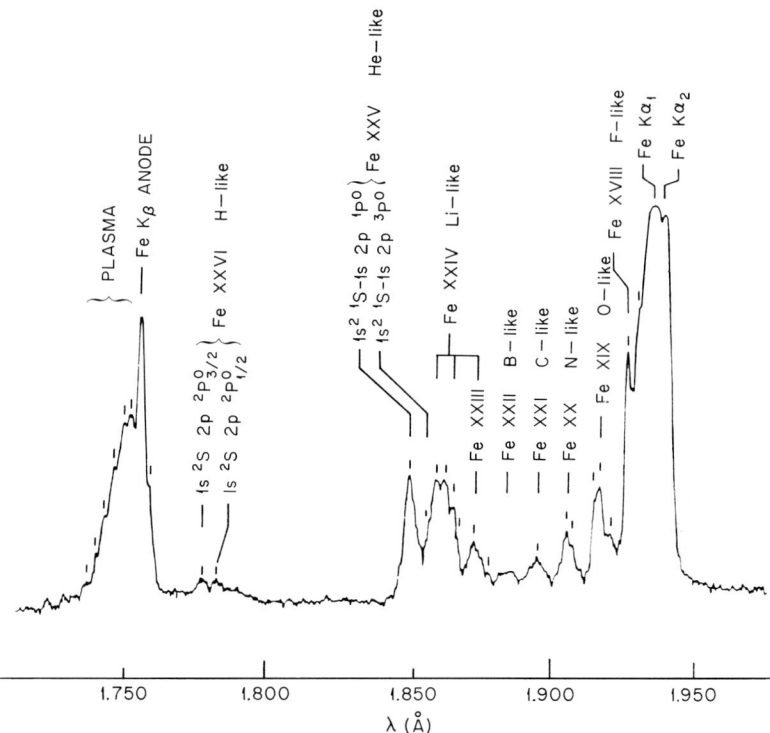

FIG. 31. The few-electron spectrum of Fe as produced in a low inductance vacuum spark. The resonance line ($^1P_1-{}^1S_0$) of Fe XXV is prominent [taken from B. S. Fraenkel and J. L. Schwob, *Phys. Lett. A* **40**, 83 (1972)].

trum. The spectrum is representative of a fairly high-temperature plasma of the order of 10^7 °K.

Figure 32 gives a comparison of Fe K x rays as produced by an accelerator beam[135] and by a high temperature Tokamak environment.[136] The amount of ionization of Fe is much less than in the previous case of a high-temperature spark plasma. The 30-MeV oxygen beam induced spectrum of Burch et al.[135] in the upper part of the figure exhibits predominately two and three inner-shell vacancy states (one K-shell and two or three L-shell vacancies), whereas the Tokamak spectrum of Bretz et al.[136] exhibits a high degree of outer-shell ionization in addition to inner-shell (K and L) ionization. The plasma spectra are produced by multiple electron–ion collisions, whereas the accelerator produced spectra are produced by single collisions between the incident ion

[135] D. Burch, P. Richard, and R. L. Blake, *Phys. Rev. Lett.* **26**, 1355 (1971).
[136] N. Bretz, D. Dimock, A. Greenberger, E. Hinnov, E. Meservey, W. Stodiek, and S. von Godler, Plasma Phys. Lab Rep. MATT-1077, Princeton Univ. (1974).

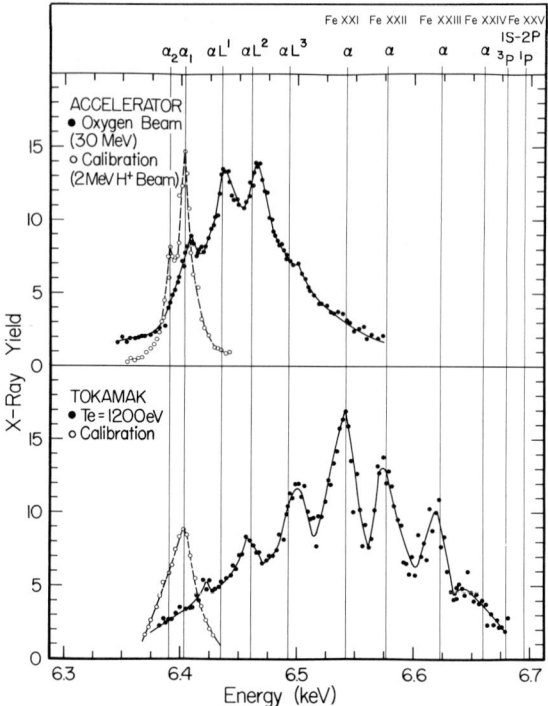

FIG. 32. A comparison of the Kα x-ray spectra of Fe produced by bombarding Fe with a 30-MeV oxygen beam [D. Burch, P. Richard, and R. L. Blake, *Phys. Rev. Lett.* **26**, 1355 (1971)] and as produced in a low temperature Tokamak environment [N. Bretz, D. Dimock, A. Greenberger, E. Hinnov, E. Meservey, W. Stodiek, and S. von Godler, Proceedings of Tokyo Conference of IAEA (1974), paper CN-33/A3-1. The five vertical lines on the left refer to inner-shell ionization of the atom, whereas the six lines on the right refer to outer-shell ionization (see text).

and the target atom. The five vertical lines to the left of the figure correspond to calculated Kα energies of atoms with inner-shell ionization only (e.g., fourth line from the left, αL^2 refers to a Kα transition from an atom with one K-shell and two L-shell vacancies). These energies agree with the observed peaks in the spectrum. The remaining six lines are calculated Kα energies for states with all outer electrons plus one K-shell electron removed (e.g., Fe XXI has 19 outer electrons and one K electron removed). These calculated energies agree fairly well with the observed peaks in the Tokamak spectrum.

In Fig. 33, a comparison is made between beam x rays and target x rays as produced by a monoenergetic high-energy heavy-ion beam.[86] Both spectra are obtained by using the same beam species (Cl) and the same incident energy (105 MeV). Also, the target and beam are selected to have nearly

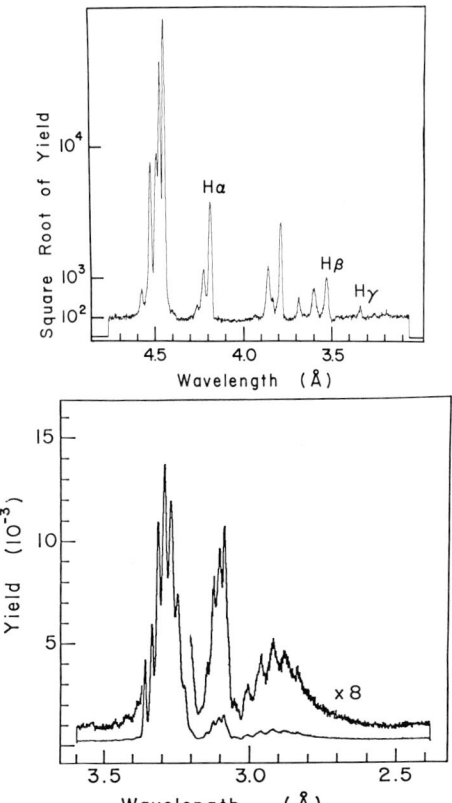

FIG. 33. A comparison of the K x-ray spectra from a calcium target (lower) and from a chlorine beam (upper) for the case of a 105-MeV monoenergetic beam of Cl. The five lines near 4.5 Å in the upper figure are the heliumlike resonance line ($^1P_1$), heliumlike intercombination line (^3P), lithiumlike $n = 2$ ^2P line, lithiumlike ^4P line, and a berylliumlike line, in order of decreasing energy. Lyman α, β, and γ lines are labeled. The target x rays consist of three large groups which are Kα satellites, Kα hypersatellites, and Kβ satellites [P. Richard, C. F. Moore, D. L. Matthews, and F. Hopkins, *Bull. Amer. Phys. Soc.* **19**, 570 (1974)].

equal Z. The lower figure is the target x-ray spectrum which, as stated in the previous paragraph, results from single ion–atom collisions. Three prominent groups of lines can be identified. The lowest-energy group is the Kα satellite group, the next higher-energy group is the Kα hypersatellite group, and the highest-energy group is the Kβ satellite group as discussed in Section 3.2.3. The few electron peaks are not evident. The upper figure contains the Cl beam K x-ray spectrum. This spectrum results from multiple collisions of the beam in the target. The np–1s hydrogenlike lines Hα, Hβ, and Hγ are evident. The five peaks near 4.5 Å are the heliumlike resonance line $(1s2p)^1P_1$–$(1s^2)^1S_0$, the heliumlike intercombination lines $(1s2p)^3P_{1,2}$–$(1s^2)^1S_0$, two lithiumlike satellite lines ^2P and ^4P, and a berylliumlike line, in the order of increasing wavelength. In this spectrum there is no evidence of the peaks seen in the target x-ray spectrum.

3.2.4.2. X-Ray Cross Sections in Ion–Atom Collisions.[137] The cross section σ^I for ionization of atoms by a monoenergetic beam can be deduced from measurements of the x-ray yield per incident particle Y from a target of known number density n by the relation.

$$\sigma^I = \sigma^X/\omega = 4\pi Y/n\Omega\varepsilon A\omega, \qquad (3.2.32)$$

where Ω, ε, and A are the solid angle, detector efficiency, and x-ray absorption correction, respectively. Here, σ^X is the x-ray production cross section and ω is the fluorescence yield of the excited target ion. For the case of proton, deuteron, and alpha particle impact over a wide range of targets and incident energies, the above technique has been successfully applied to the measurement of ionization cross sections. The cross sections compare very well with the theoretical Coulomb ionization theories based on either a plane wave Born approximation (PWBA)[95] or binary encounter approximation (BEA).[138] The cross section peaks at an incident ion velocity near the average orbital electron velocity of the ionized electron shell. The K-shell cross section in the BEA scales as

$$\sigma^I = 2Z^2\sigma_0 G(V)/u^2, \qquad (3.2.33)$$

where Z and u are the projectile Z and target K-shell binding energy, respectively; σ_0 equals 6.56×10^{-14} cm² eV² and $G(V)$ is a function of the scaled velocity $V = v_i/v_e$, where v_i and v_e are incident ion and orbital electron velocities, respectively; and $G(V)$ has a maximum value of ~ 0.6 at $V = 1.0$. At energies such that $V \ll 1$, it has been shown that increased binding and Coulomb deflection can alter the ionization cross sections.[139]

For heavy projectiles, the K x-ray production cross section becomes much more difficult to understand and interpret. Electron capture by the heavy projectile, excitation by many electron effects, and atomic screening of the projectile nucleus may be important in vacancy production, in addition to Coulomb ionization. As we have seen in Section 3.2.3.2, the heavy ion produces multiple ionization so that the value of the fluorescence yield ω used in Eq. (3.2.32) is not well defined. This problem can be overcome by measuring both the Auger electron production cross section (σ^A) and x-ray production cross section (σ^X). The ionization cross section is given by $\sigma^I = \sigma^X + \sigma^A$, and the effective fluorescence yield can be defined as $\bar{\omega} = \sigma^X/(\sigma^X + \sigma^A)$. An

---

[137] For a comprehensive review, see B. Crasemann (ed.)," Atomic Inner-Shell Processes," Vol. 1. Academic Press, New York, 1975.

[138] J. D. Garcia, *Phys. Rev A* **1**, 280, 1402 (1970); **4**, 955 (1971).

[139] G. Basbas, W. Brandt, and R. Laubert, *Phys. Rev. A* **7**, 983 (1973).

additional difficulty arises in that the ionization cross sections depend strongly on the charge state of the projectile.[140, 141] An interesting complication encountered in heavy-ion–atom collisions is the enhancement of cross sections above the Coulomb ionization prediction when the projectile Z (Z_p) becomes comparable to the target Z (Z_T). This enhanced cross section is due to molecular promotion as discussed in Section 3.2.3.8.

Figure 34 gives the ionization cross section[142] for several ion–atom collisions. The data are plotted according to the scaling laws for Coulomb ionization [see Eq. (3.2.33)]. The solid curve is the BEA curve predicted using Eq. (3.2.33). A maximum is observed near $E/\lambda u = V^2 = 1$. For cases where

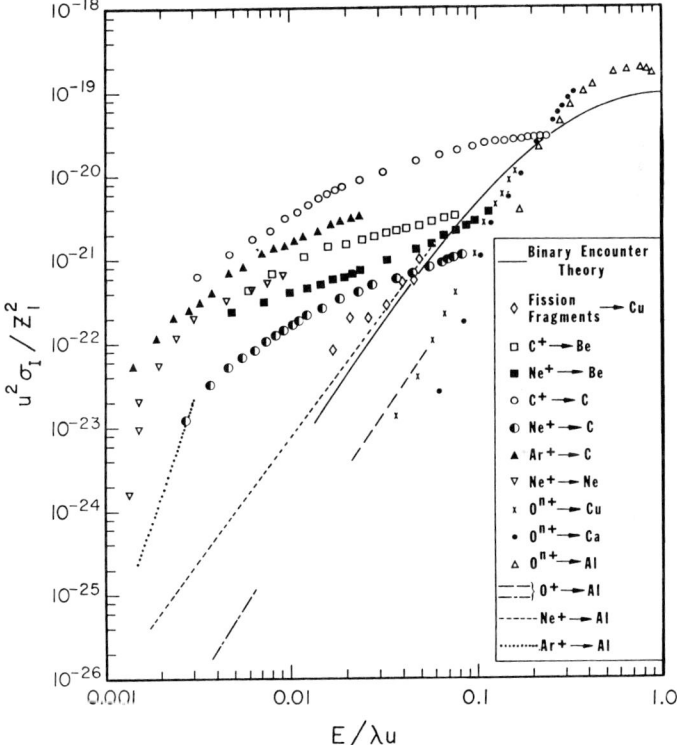

FIG. 34. K-shell ionization cross section for heavy-ion collisions obtained from x-ray measurements [taken from J. D. Garcia, R. J. Fortner, and T. M. Kavanagh, *Rev. Mod. Phys.* **45** 111 (1973)].

[140] J. R. Macdonald, L. M. Winters, M. D. Brown, T. Chiao, and L. D. Ellsworth, *Phys. Rev. Lett.* **29**, 1291 (1972).
[141] J. R. Mowat, D. J. Pegg, R. S. Peterson, P. M. Griffin, and I. A. Sellin, *Phys. Rev. Lett.* **29**, 1577 (1972).
[142] J. D. Garcia, R. J. Fortner, and T. M. Kavanagh, *Rev. Mod. Phys.* **45**, 111 (1973).

$Z_p \sim Z_T$, the cross sections are much larger than the BEA prediction. The variation of the ionization cross section of a given projectile as a function of target Z is given in Fig. 35. A nonmonotonic Z dependence is clearly observed.[143] Meyerhof[144] has explained these data within the framework of the molecular orbital model by using a vacancy sharing mechanism produced by a radial coupling between the $1s\sigma$ and $2p\sigma$ molecular orbitals.

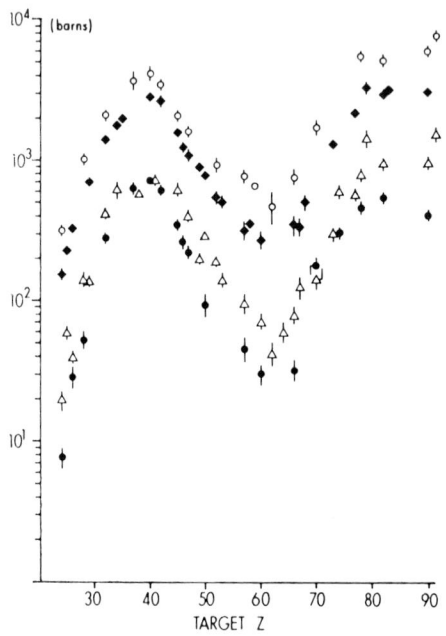

FIG. 35. Bromine $K\alpha$ ionization cross section versus target Z for four incident energies of Br [○ : 110 MeV; ▲ : 85 MeV; △ : 60 MeV; ● : 45 MeV]. A nonmonotonic behavior of the cross section is readily apparent. (Taken from H. Kubo, P. C. Jundt and K. H. Purser, *Phys. Rev. Lett.* **31**, 674 (1973)].

Some additional insight into the ionization mechanism of ion–atom collisions can be obtained by observing the impact parameter dependence of the x-ray production cross section. If the probability for ionization at a given impact parameter b is $P(b)$, then the ionization cross section is

$$\sigma^I = \int 2\pi b P(b) \, db. \quad (3.2.34)$$

Experimentally, one determines $P(b)$ from the ratio of the number of x rays

---

[143] H. Kubo, F. C. Jundt, and K. H. Purser, *Phys. Rev. Lett.* **31**, 764 (1973).
[144] W. E. Meyerhof, *Phys. Rev. Lett.* **31**, 1341 (1973).

3.2. X-RAY REGION

in coincidence with ions scattered at some fixed angle θ to the total number of ions detected at that angle, divided by the fluorescence yield.[145, 146] The impact parameter b is determined from the scattering angle by using a screened Coulomb potential. Figure 36 gives $P(b)$ versus b/r_k, where r_k is the average K-shell radius.[147] The dashed curve is the calculated behavior as given by the semiclassical approximation. For F^{+9} on Ar, the ionization probability has a different b dependence than predicted by Coulomb ionization. This may be due to molecular orbital effects.

For a comprehensive review of the subject of ionization by heavy ions, the reader is referred to Crasemann.[148]

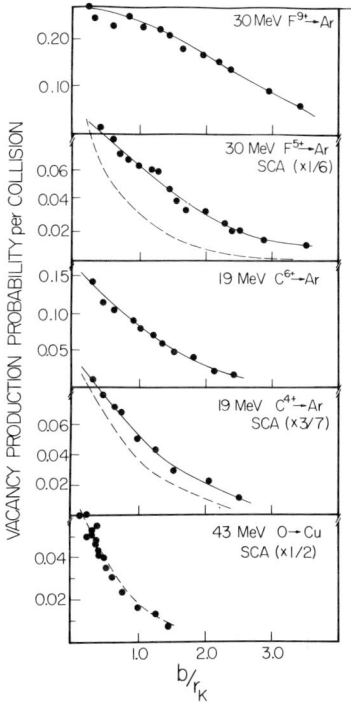

FIG. 36. The impact parameter variation of ionization as obtained from x-ray measurements. Projectiles of F^{9+}, F^{5+}, C^{6+}, and C^{4+} on Ar, and O on Cu, are given. The dashed curves are Coulomb ionization prediction using the semiclassical approximation [R. Randall, C. L. Cocke, B. Curnutte, and J. Bednar, private communications (1965)].

[145] E. Laegsgaard, J. V. Andersen, and L. C. Feldman, *Phys. Rev. Lett.* **29**, 1206 (1972).
[146] N. J. Stein, H. O. Lutz, P. H. Mokler and P. Armbruster, *Phys. Rev. A* **5**, 2126 (1972).
[147] R. Randall, C. L. Cocke, B. Curnutte and J. Bednar, private communications (1975).
[148] B. Crasemann (ed.), "Atomic Inner-Shell Processes," Vol. 1. Academic Press, New York, 1975.

3.3. Far Ultraviolet Region*

3.3.1. Introduction

The far ultraviolet (uv) spectral region discussed in this chapter begins where the radiant energy starts to be absorbed by air. In particular, the absorption is caused by molecular oxygen starting at about 200 nm. At shorter wavelengths (below 100 nm), all atmospheric gases absorb strongly until the hard x-ray region is approached. Consequently, spectrographs for the far uv must be throughly evacuated. This spectral region is aptly termed the *vacuum ultraviolet*, and covers the wavelength range of 0.2–200 nm, an energy range of 6–6000 eV. Thus, this region merges into the x-ray domain. The dividing line is hazy, however, because ultraviolet or optical spectra can be produced from highly ionized atoms at wavelengths as short as 1 nm, whereas conventional x rays, such as the characteristic $L_{2,3}$ emission band of aluminum, can produce long wavelength radiation up to 18 nm. The availability of synchroton radiation further diminishes the dividing lines between the classification of ultraviolet and x radiation.

Work in the far ultraviolet had a very precise beginning. In 1893, Viktor Schumann constructed the first vacuum spectrograph and made the first measurements of vacuum ultraviolet radiation.[1] It was nearly 70 yr later before the first international meeting on vacuum ultraviolet radiation physics was held.[2] During the early part of this century, progress was slow because of the inadequate vacuum technology and nonavailability of light sources, dispersing devices, and detectors. In the 1950's, rapid advances began that were further stimulated by the need for vacuum uv spectroscopy in plasma physics, fusion experiments, and in rocket and satellite flight research in solar physics. Today, the sophisticated instrumentation is available, and the far ultraviolet region is a flourishing and important field. It is the spectral region of photodissociation, photoionization, and photoemission. It has immediate practical applications in aiding our understanding of how solar radiation interacts with the upper atmospheres of the earth and other

---

[1] V. Schumann, *Akad. Weiss. Wien.* **102**, 2A 625 (1893).
[2] G. L. Weissler (ed.), *Proc. 1st Int. Conf. Vacuum Ultraviolet Radiat. Phys.*, 1st [*J. Quant. Spectrosc. Radiat. Transfer* **2**, 313 (1962)].

---

* Chapter 3.3 is by James A. R. Samson.

planets. Recently, an excellent book has been published describing "some aspects of vacuum ultraviolet radiation physics."[2a] This book describes the application of far uv radiation to studies of solids, gases, and hot plasma emissions, and discusses the recent refinements in the design of grating spectrometers.

The instrumentation used in vacuum uv spectroscopy up to 1966 has been thoroughly reviewed by the author[3] and by Zeidel and Shreider.[4] Of historical importance are the works of Lyman[5] and Bomke.[6] The following sections are intended to review the instrumentation and techniques used in the vacuum uv, and to summarize the advances made in the last ten years.

3.3.2. Photon Sources

There are two types of radiation sources that can be used in the far ultraviolet; namely, continuum and discrete line sources. With the availability of synchrotron radiation, continua exist over the entire spectral region. Line spectra, which are characteristic of the radiating atoms, also exist over most of the spectral region. However, the density of the lines in a given spectral interval varies tremendously.

The choice of light source depends on the nature of the investigation. Nanosecond pulses of discrete or continuum radiation may be required, continua may be necessary to observe fine structure in absorption spectroscopy, or a strong line emission may be required for work in photoelectron spectroscopy at high resolution (typical line widths are 0.1–1 mV). It is a fallacy to believe that a continuum source of radiation is always the best source of radiation to measure the absorption spectrum of a gas. Undoubtedly, a continuum source should be used in regions where structure appears in the absorption spectrum, and it is the only source suitable for photographic absorption studies. However, in the continuous photoionization absorption region, more precise measurements of the absolute cross sections can be obtained with a discrete line emission source. The problems of scattered background levels and the determination of overlapping higher-order spectra are easily solved with a line source. Thus, the source should be matched to the application whenever possible.

[2a] N. Damany, B. Vodar, and J. Romand (eds.), "Some Aspects of Vacuum Ultraviolet Radiation Physics." Pergamon, Oxford, 1974.

[3] J. A. R. Samson, "Techniques of Vacuum Ultraviolet Spectroscopy." Wiley, New York, 1967.

[4] A. N. Zeidel and E. Ya. Shreider, "Spektroskopiya Vakuumnogo Ultrafioleta." Izdatel'stvo Nauka, Moscow, 1967.

[5] T. Lyman, "The Spectroscopy of the Extreme Ultraviolet," 2nd ed. Longmans, Green, New York, 1928.

[6] H. Bomke, "Vakuumspektroskopie." Barth, Leipzig, 1937.

Details of light sources have been given by the author.[3] In the following sections, a brief account will be given of the most useful sources available for the far uv.

3.3.2.1. Continuum Sources. 3.3.2.1.1. HYDROGEN CONTINUUM. The H_2 continuum is produced by a dc glow discharge in hydrogen constricted by a capillary of 2–6 mm in diameter. An extensive continuum is produced by electronic transitions from the bound excited $^3\Sigma_g^+$ state to the repulsive $^3\Sigma_u^+$ state, covering the range 500–160 nm. A typical spectrum is shown in Fig. 1 in the region where the continuum merges with the many-lined

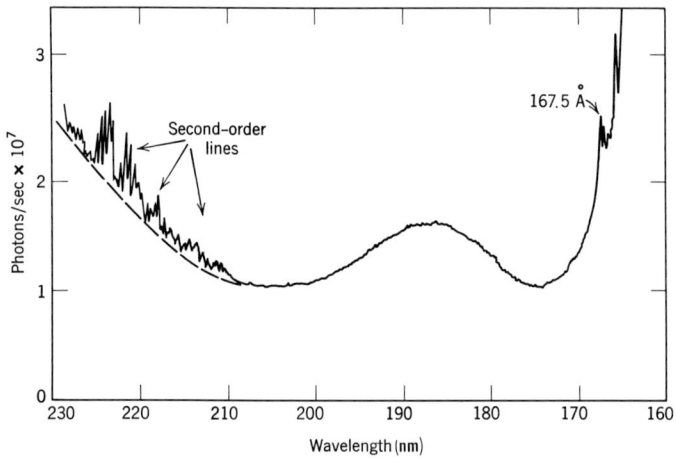

FIG. 1. The H_2 continuum transmitted through a LiF window. The dashed curve indicates the H_2 continuum in the absence of second-order lines [J. A. R. Samson, "Techniques of Vacuum Ultraviolet Spectroscopy." Wiley, New York, 1967].

spectrum. The intensity of the continuum increases approximately linearly as the discharge current increases.

3.3.2.1.2. RARE GAS CONTINUUM. A continuum in helium was first observed by Hopfield[7] in 1930 and later developed by Huffman and co-workers.[8] The useful range of the continuum extends from 60 to 110 nm. The helium continuum has perhaps been the most valuable source in the far uv for the determination of accurate ionization potentials in atoms and molecules. The origin of the continuum requires the formation of excited helium molecules. This is achieved by producing a mildly condensed (0.002 μF) repetitive spark discharge at about 10 kV in pure helium. The purity of the helium is very important in order to avoid numerous impurity lines superimposed on the continuum. A typical spectrum is shown in Fig. 2.

[7] J. J. Hopfield, *Phys. Rev.* **35**, 1133 (1930); **36**, 784 (1930); *Astrophys. J.* **72**, 133 (1930).
[8] R. E. Huffman, Y. Tanaka, and J. C. Larrabee, *Appl. Opt.* **4**, 1145 (1965).

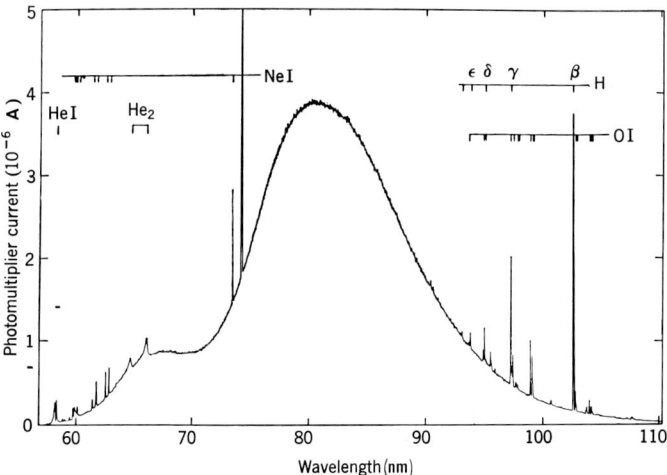

FIG. 2. Molecular helium continuum: pulse repetition rate, 5 kHz; pressure, 44 Torr; power supply, 10 kV and 116 mA [R. E. Huffman, Y. Tanaka, and J. C. Larrabee, *Appl. Opt.* **4**, 1145 (1965)].

Molecular continua can be produced by all the rare gases in a similar manner as the He_2 continuum.[8a] Each continuum has a short wavelength limit starting at the respective resonance line of the rare gas and extending towards longer wavelengths. The optimum operating conditions for each continuum varies with the gas pressure, and are summarized in Table I.

TABLE I. Rare Gas Continua

Gas	Useful range (nm)	Optimum pressure range (Torr)	Maximum intensity[a] (photons/sec)
He	58–110	40–55	$\sim 10^8$
Ne	74–100	>60	
Ar	105–155	150–250	$\sim 10^8$
Kr	125–180	>200	$\sim 10^7$
Xe	148–200	>200	$\sim 10^7$

[a] At the principal maximum with a 0.05 nm bandwidth.

Figure 3 shows typical spectra and how each one slightly overlaps the other. Thus, they provide useful continua from about 60 to 190 nm. Although gas purity is very important, Gedanken and Raz[9] have shown that Xe can be mixed with pure He, Ne, or Ar and still produce the xenon continuum. In fact, they have shown that a similar continuum can be produced by the use

[8a] R. E. Huffman, J. C. Larrabee, and Y. Tanaka, *Appl. Opt.* **4**, 1581 (1965).
[9] A. Gedanken and B. Raz, *Vacuum* **21**, 389 (1971).

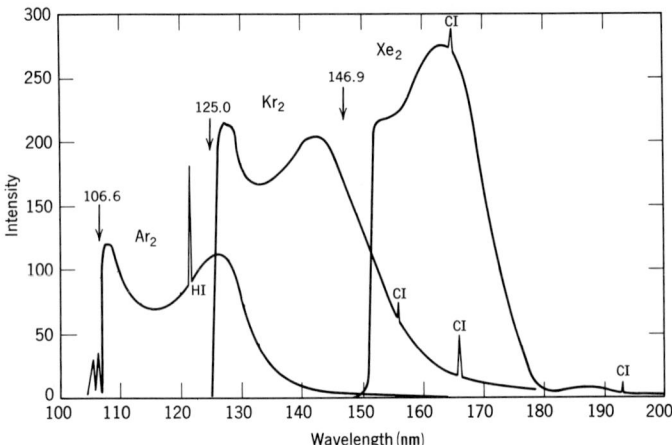

FIG. 3. Molecular argon, krypton, and xenon continua, excited by a 125-W microwave generator [P. G. Wilkinson and E. T. Byram, *Appl. Opt.* **4**, 581 (1965)].

of only 0.25 Torr of Xe with 640 Torr of Ar. This provides a considerable saving of the expensive xenon. Apparently, collisions with the argon atoms are a very effective means for producing excited Xe molecules. The Xe resonance line at 147 nm is present but greatly broadened. A spectrum is shown in Fig. 4.

3.3.2.1.3. PLASMA ARC CONTINUUM. Radiative recombination continuum caused by electron–ion recombination to the ground state would appear to be an obvious source of continuum radiation. However, this type of continuum is usually very weak. Nevertheless, if the number density of the electrons and ions is large enough, a usable continuum is produced. These are the conditions found in the plasma arc source used by Levy and Huffman.[10] Figure 5 shows a schematic of their source. The source operates at about 25 V with an arc current up to 200 A and at a pressure of approximately 1 atm. The continuum starts at the first ionization threshold of the atomic gas used and decreases exponentially with decreasing wavelength. With helium the useful wavelength range is 50.4–30 nm, and with Ne the range covered is 57.5–40 nm.

Caruso and Shardanand[11] have shown that a similar electron–ion recombination continuum occurs at the end of the He II resonance emission lines, starting at 22.9 nm and continuing down to about 14 nm. This continuum is produced by a condensed spark discharge in helium at about

[10] M. E. Levy and R. E. Huffman, *Appl. Opt.* **9**, 41 (1970); *J. Quant. Spectrosc. Radiat. Transfer.* **9**, 1349 (1969).
[11] T. Lyman, *Astrophys. J.* **60**, 1 (1924); *Science* **64**, 89 (1926).

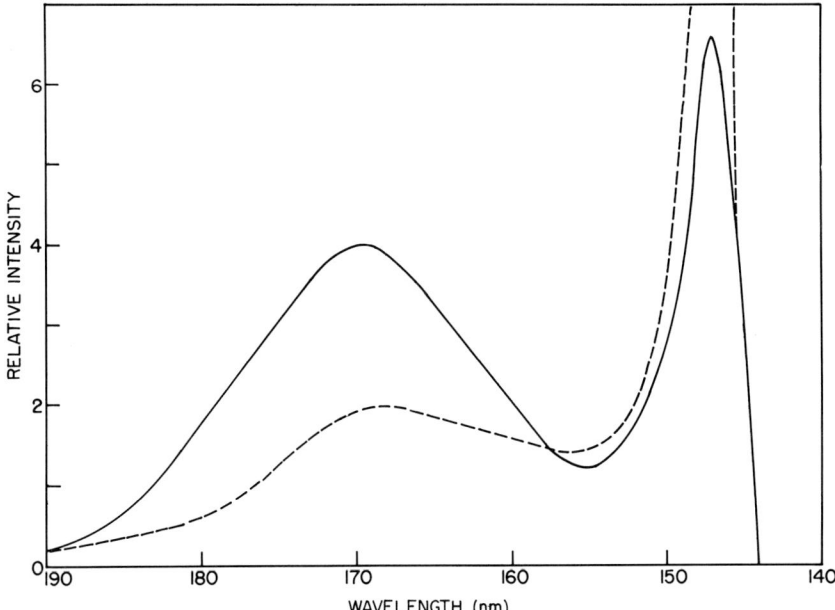

FIG. 4. Emission spectra of Xe–Ar mixtures. The partial pressure of xenon is kept constant at 0.25 Torr with argon at 640 Torr (solid line) and 320 Torr (dashed line) [A. Gedanken and B. Raz, *Vacuum* **21**, 389 (1971), reprinted with permission from Pergamon Press].

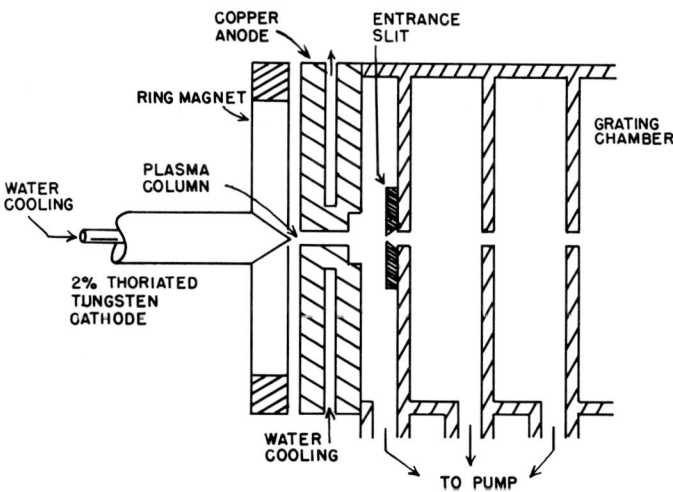

FIG. 5. Schematic diagram of plasma arc radiation source [M. E. Levy and R. E. Huffman, *Appl. Opt.* **9**, 41 (1970); *J. Quant. Spectrosc. Radiat. Transfer* **9**, 1349 (1969)].

0.5 Torr; the source is normally used at lower pressures to produce line spectra (see Section 3.3.2.2.5).

3.3.2.1.4. SPARK SOURCE CONTINUUM. The two main sources of this type are the Lyman[12] flash tube and the Ballofet–Romand–Vodar[13–15] (BRV) source. The Lyman continuum is produced by a condensed, high-voltage spark in a low-pressure gas (usually helium) and confined to a capillary. The continuum originates from the passage of the discharge current through the vaporized wall material but not from the carrier gas. The carrier gas is used only to initiate the discharge. The original Lyman flash tube used a capillary with a 1 mm bore. The source has been greatly improved by Garton,[16,17] who showed that a larger diameter tube could be used (9 mm), provided the current density in the capillary was greater than 30,000 A/cm^2. Typical operating conditions are 20 mTorr of helium, $C = 9$ uF, and $V = 8$ kV. The spark repetition rate is usually less than 1/sec. An excellent continuum is produced from the visible down to about 80 nm. The continuum continues down to about 30 nm but is heavily overlaid with discrete line spectra.

The BRV source operates in a high vacuum. Again, it is a condensed high-voltage spark discharge that is triggered by a sliding spark. There is no capillary, and the discharge is carried by material vaporized from the electrodes. This produces a small volume of plasma at a high temperature and density very close to the surface of the anode, which is made from uranium. No continuum is observed if the anode is made from a light element. The continuum must be viewed very close to the anode surface, and extends from the visible down to about 10 nm. Pulse rates can be as high as 10 pulses/sec. Light duration per pulse depends on the capacitance used but is typically 0.2 μsec. The use of this source for high-resolution spectroscopy has been described by Boursey and Damany.[15]

3.3.2.1.5. SYNCHROTRON RADIATION. The radiation losses experienced by circulating electrons in high-energy machines sets an upper limit to the energies available in a given machine. This drawback to high-energy physics is advantageous to spectroscopy, because the radiation emitted is continuous from the infrared into the x-ray region. Thus, in principle, synchrotron radiation provides the ideal light source.

The first investigation and use of synchrotron radiation was by Tomboulian and Hartman in 1956.[18] Since then, several reviews have appeared

[12] A. J. Caruso and Shardanand, *J. Opt. Soc. Amer.* **59**, 960 (1969).
[13] G. Ballofet, J. Romand, and B. Vodar, *C. R. Acad. Sci.* (Paris) **252**, 4139 (1961).
[14] H. Damany, J.-Y. Roncin, and N. Damany-Astoin, *Appl. Opt.* **5**, 297 (1966).
[15] E. Boursey and H. Damany, *Appl. Opt.* **13**, 589 (1974).
[16] W. R. S. Garton, *J. Sci. Instrum.* **36**, 11 (1959); **30**, 119 (1953).
[17] J. E. G. Wheaton, *Appl. Opt.* **3**, 1247 (1964).
[18] D. H. Tomboulian and P. L. Hartman, *Phys. Rev.* **102**, 1423 (1956).

3.3. FAR ULTRAVIOLET REGION

on the subject.[3,19-21] Synchrotron radiation is produced whenever high-energy electrons are constrained to move in a segment of a circle. In a conventional synchrotron, the electrons are radiating at various rates as their energy is increased from near zero to a maximum in the sinusoidally varying field. In a storage ring, the electrons are brought up to a fixed energy and then allowed to circulate for several hours. Because of their relativistic energies, the radiation pattern of the orbiting electrons is extremely directional. The radiation is emitted in the same direction as the instantaneous velocity vector of the electrons and is contained in a cone of half angle ψ, which is of the order of 1 mrad but varies with the electron energy and the wavelength of the observed radiation. As the electrons circulate, this cone of light sweeps a path in a horizontal plane, as shown in Fig. 6. Typical

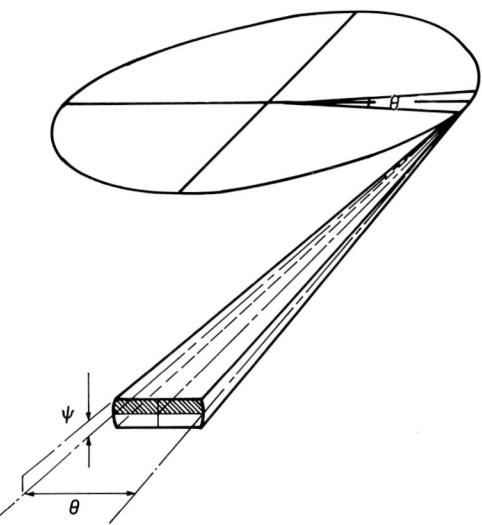

FIG. 6. Cone of synchrotron radiation sweeping a path in the plane of the electron orbit.

angular distributions ψ for a 4-GeV machine with a radius of 20.8 m are shown in Fig. 7 for various wavelengths.

The number N of photons emitted per second per nanometer per milliradian of horizontal distribution (accepting all radiation in the vertical plane) is given by

$$N = 7.9 \times 10^{13} I E^7 \lambda G(y)/R^2, \qquad (3.3.1)$$

where I is the electron current in milliamperes, E the total electron energy

[19] R. P. Godwin, *Springer Tracts Mod. Phys.* **51**, 1 (1969).
[20] K. Codling, *Physica Scripta* **9**, 247 (1974).
[21] K. Codling and R. P. Madden, *J. Appl. Phys.* **36**, 380 (1965).

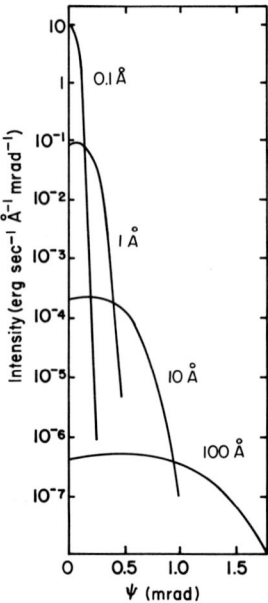

FIG. 7. The variation of angular distribution of synchrotron radiation with wavelength for $E = 4$ GeV and $R = 20.77$ m, the radius of the Daresbury NINA synchrotron [K. Codling, *Physica Scripta* **9**, 247 (1974)].

in gigaelectron volts, R the bending radius in meters of the electrons in the storage ring magnets, λ the wavelength of the radiation measured in nanometers, and $G(y)$ a universal spectral distribution function which is shown in Fig. 8. The variable $y = \lambda_c/\lambda$, where λ_c is given by

$$\lambda_c = (4\pi R/3)(m_0 c^2/E)^3. \qquad (3.3.2)$$

The quantity $m_0 c^2$ is the rest energy of the electron. The expression for the number of photons/sec given in Eq. (3.3.1) is given for a fixed bandwidth ($\Delta\lambda = 1$ nm). It is sometimes useful to express N for a band whose width is always a fixed fraction of the wavelength. For example, in a 10% bandwidth ($\Delta\lambda = 0.1\lambda$), the number of photons is obtained by multiplying Eq. (3.3.1) by 0.1λ (nm).

Figure 9 shows Eq. (3.3.1) plotted as a function of wavelength (nm) and photon energy (eV) for a 10% bandwidth for SPEAR, the storage ring at Standford University. A family of curves are shown for energies of 1.5–4.5 GeV and for a radius $R = 12.7$ m. Because the operating currents vary, the plots are given for $I = 1$ mA. Typical current operation of the SPEAR machine is 50 mA, in which case the flux shown in Fig. 9 would be multiplied by 50. It is interesting to note that, for work in the far uv, there is little

3.3. FAR ULTRAVIOLET REGION

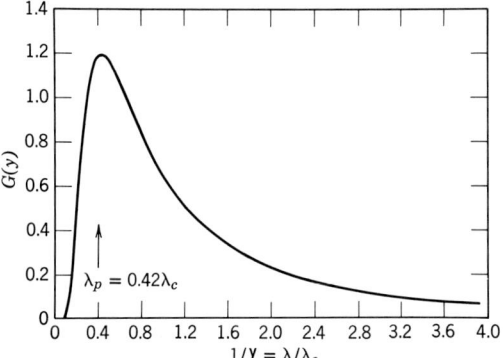

FIG. 8. Universal spectral distribution curve for the radiation from monoenergetic electrons, where $P(\lambda)$ is the instantaneous power radiated per angstrom given by $P(\lambda) = \text{const} \times (E/m_0 c^2)^7 G(y)$, and where $\lambda_c(\text{Å}) = 5.59R/E^3$ (see text) [D. H. Tomboulian and P. L. Hartman, Phys. Rev. **102**, 1423 (1956)].

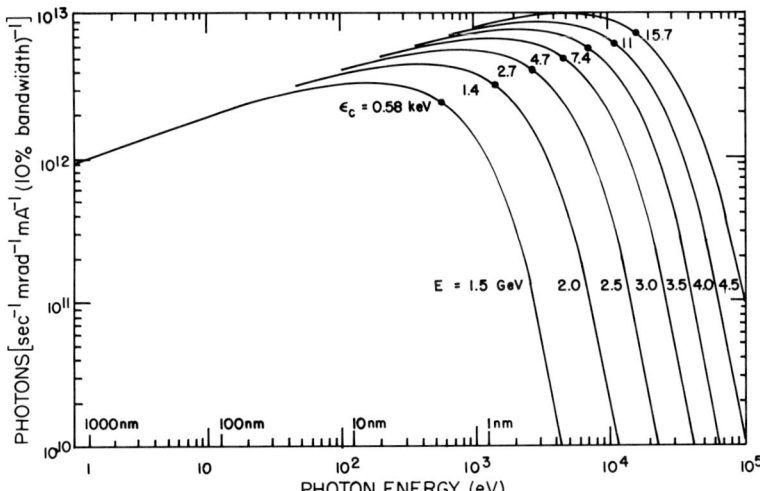

FIG. 9. Spectral distribution of the SPEAR synchrotron radiation at a constant 10% bandwidth ($R = 12.7$ m), where ε_c is λ_c expressed in keV

advantage, in terms of flux output, to increasing the energy of the machine from 1.5 to 4.5 GeV. In fact, it is more troublesome because of the unwanted scattered short wavelengths.

A list of the locations of synchrotrons and storage rings that have, or plan to have, synchrotron radiation programs is given in Table II. Figure 10 compares the output flux of several different accelerators.

TABLE II. Locations of Synchrotron Radiation Facilities

Name	Location	E (GeV)	R (m)	I (mA)	λ_c (nm)
Synchrotrons					
DESY	Hamburg, Germany	7.5	31.70	20	0.042
—	Bonn, Germany	2.3	7.65	30	0.35
—	Bonn, Germany	0.5	1.70	30	7.6
—	Frascati, Italy	1.1	3.60	80	1.5
INS-SOR	Tokyo, Japan	1.3	4.0	30	1.0
FIAN	Moscow, USSR	0.68	2.0	10	3.5
—	Erevan, USSR	6.0	24.7	20	0.064
NINA	Daresburg, UK	5.0	20.8	20	0.093
Storage rings					
TANTALUS I	Stoughton, Wisconsin	0.24	0.65	80	26.0
SPEAR	Palo Alto, California	3.0	12.7	30	2.7
SURF II	Washington, D.C.	0.24		50	
ACO	Orsay, France	0.54	1.1	100	3.9
DCI	Orsay, France	1.8	3.8	250	0.38
INS-SOR	Tokyo, Japan	0.3	1.0	100	20.0
DESY	Hamburg, Germany	1.75	12.1	1000	1.27
DORIS	Hamburg, Germany	3.5	12.1	200	0.16
GINA	Daresburg, UK	2.0	5.5	500	0.39

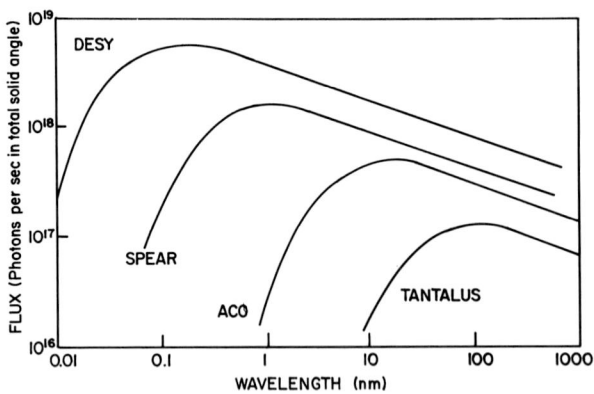

FIG. 10. Spectral distribution of DESY (45 mA), SPEAR (30 mA), ACO (50 mA), and TANTALUS (40 mA) at a constant 10% bandwidth. The characteristic wavelengths λ_c are indicated by vertical lines [M. L. Pearlman, E. M. Rowe, and R. E. Watson, *Phys. Today* **27**, 30 (1974)].

3.3. FAR ULTRAVIOLET REGION

In addition to its continuum nature, synchrotron radiation exhibits several other features not found in conventional laboratory light sources. For instance, it is found that the radiation observed in the plane of the orbiting electrons is essentially 100% plane polarized with the electric vector in the plane of the orbit. As observations are made off the plane, the radiation becomes elliptically polarized.[21-23] Another unusual feature is the extremely short duration of each light pulse. Examples from the SPEAR storage ring show that the duration of a pulse, at present, is 1 nsec, with durations of 0.15 nsec planned. The pulses arrive spaced 781 nsec apart. In TANTALUS I, the pulse length is variable from 1 to 6 nsec, and the pulses are separated by 31 nsec. This unique capability will prove useful for fluorescent and luminescent lifetime studies.

As a far ultraviolet light source, the storage ring is much more attractive than the synchrotron because it is possible to work in the area of the experiments without radiation hazards. In addition, it is possible to monitor the precise beam current and hence any variation in the light intensity. The beam current in the storage ring will be slowly depleted because of electron collisions with residual gas atoms. Typical half-lives for the beams in Tantalus I are about 2 hr for an 80-mA beam and 20 hr for a 1-mA beam.

We have seen that continuum sources are available over the entire far uv spectral region and into the hard x-ray region. The choice of source depends on the experiments planned, and, fortunately, the choices are sufficiently varied to match the most imaginative experiment.

3.3.2.2. Line Sources. Virtually all electrical discharges in gases produce line spectra characteristic of the gas used. Continuum aspects can be enhanced by judicious choice of operating conditions as discussed in the previous sections. A simple dc glow discharge will produce the resonance lines of all neutral atoms, and, under conditions of low gas pressure, can produce the resonance lines of the singly ionized atoms. Thus, in a pure gas, only a few emission lines of good intensity can be expected.

Many more emission lines can be produced if larger concentrations of ions are present in various degrees of ionization. This can be achieved primarily by pulsed or spark discharges. Again, the lower the gas pressure, the higher the degree of ionization present in the gas.

The designs of the various types of light sources are so numerous that no attempt will be made to describe them all. Instead, principles will be discussed with examples of sources and spectra.

3.3.2.2.1. COLD CATHODE DISCHARGE. With this type of discharge, several thousand volts are necessary to initiate a discharge. Once started, the positive ion bombardment at the cathode releases electrons to maintain the

[22] P. Joos, *Phys. Rev. Lett.* **4**, 558 (1960).
[23] K. Westfold, *Astrophys. J.* **130**, 241 (1959).

discharge at lower voltages (300–800 V). To increase the intensity of the radiation, it is necessary to confine the discharge to a capillary with an internal diameter of 1–5 mm and a length that can vary from about 5 to 20 cm. For high-power operation, it is necessary to water cool anode, cathode, and, in some designs, the capillary.

The design of a source should be such that the cathode can be removed for cleaning and polishing. The ion bombardment roughens the surface and leads to excess noise. A large surface area for the cathode is also advantageous for quiet and stable operation.

When resonance lines are produced in a glow discharge, there is a definite relationship between the length of the discharge capillary and the half-width of the resonance lines.[24] In most practical cases, the lines broaden linearly with respect to the increase in length of the capillary. The line width also increases as the pressure increases.

When hydrogen is used in a glow discharge, a profuse many-lined spectrum is produced between 85 and 167 nm. A typical spectrum recorded at the exit slit of a half-meter vacuum monochromator is shown in Fig. 11.

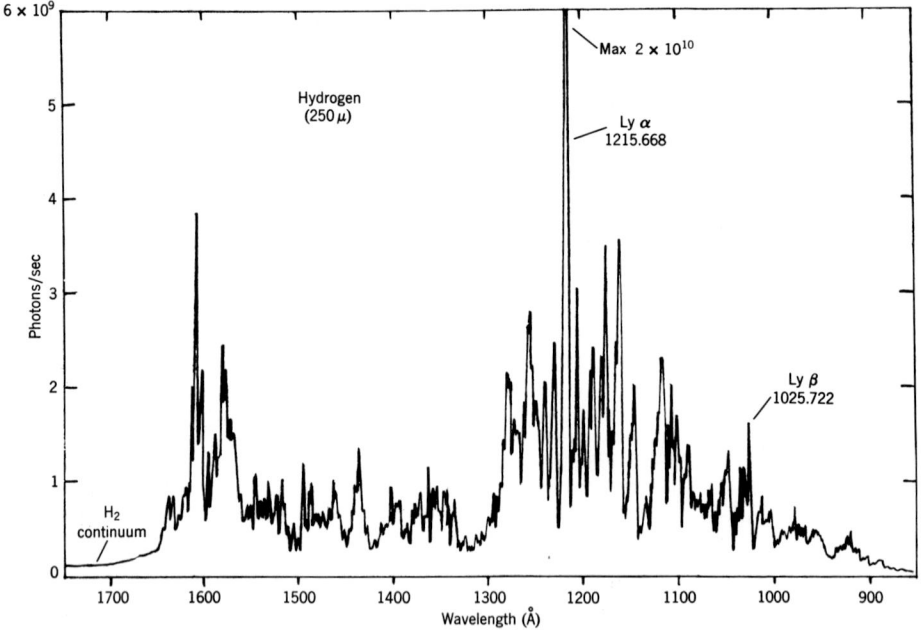

FIG. 11. A H_2 spectrum between 85 and 175 nm. Pressure was 0.25 Torr, and the discharge current was 400 mA [J. A. R. Samson, "Techniques of Vacuum Ultraviolet Spectroscopy." Wiley, New York, 1967].

[24] J. A. R. Samson, *Rev. Sci. Instrum.* **40**, 1174 (1969).

To obtain other line spectra, it is necessary to use the rare gases. Table III lists the most intense lines produced by a dc glow discharge in the rare gases. The ion spectra are very much weaker than those from the excited neutrals. However, they can be enhanced by operating at as low a pressure as possible. Normally, the Kr and Xe discharges are reserved for sources with windows in order to conserve the expensive gases.

TABLE III. The First Resonance Lines of the Rare Gas Atoms and Ions

Gas	First resonance line (nm)	(eV)	Gas	First resonance line (nm)	(eV)
He I	58.4334	21.217	Ar I	104.8219	11.828
He II	30.3782	40.812		106.6659	11.623
Ne I	73.5895	16.848	Ar II	91.9782	13.479
	74.3718	16.670		93.2053	13.302
Ne II	46.0725	26.910	Kr I	116.4867	10.643
	46.2388	26.813		123.5838	10.032
			Xe I	129.5586	9.569
				146.9610	8.436

The rare gas discharges tend to produce only the first resonance lines of the neutrals, and, therefore, constitute essentially monochromatic radiation sources in the far uv. This is particularly advantageous in the field of photoelectron spectroscopy. It is of interest, therefore, to see how the intensity of the doublet lines vary with pressure. Figure 12 shows the intensity

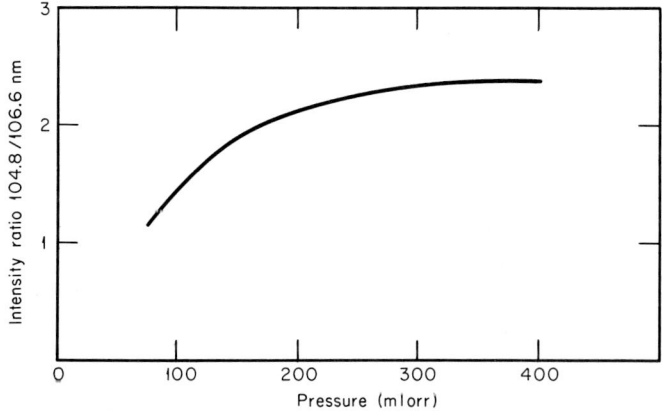

FIG. 12. Ratio of the intensities of the 104.8- and 106.6-nm resonance lines of Ar I as a function of pressure. Discharge current was 400 mA [J. A. R. Samson, "Techniques of Vacuum Ultraviolet Spectroscopy." Wiley, New York, 1967].

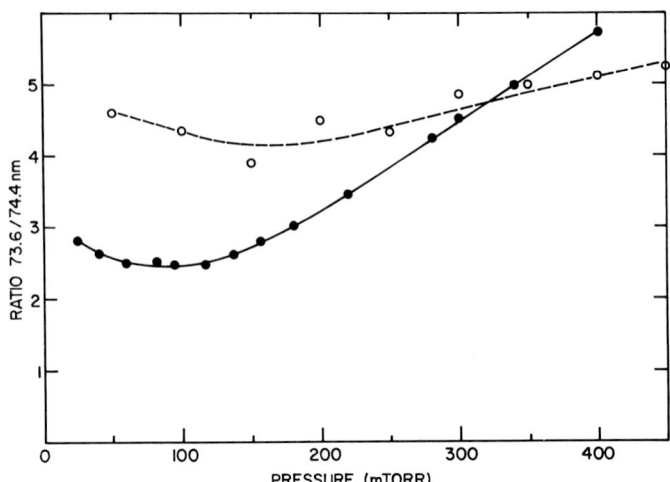

FIG. 13. Ratio of the intensities of the 73.6- and 74.4-nm resonance lines of Ne I as a function of pressure. The discharge currents were 100 mA (dashed curve) and 500 mA (solid curve).

variation of the 104.8/106.6-nm Ar lines with pressure, while Fig. 13 shows similar results for the 73.6/74.4-nm lines of Ne. In photoelectron spectroscopy, the helium discharge has become the most important source. The 58.4-nm line is the most predominant in the spectrum. The second resonance line at 53.7 nm has about 1/50 to 1/100 the intensity of the 58.4-nm line. At pressures of several Torr, the source is extremely monochromatic below 100 nm and is ideal for undispersed studies of photoionization. However, it is often important to produce a more energetic line such as the 30.4-nm He II resonance line. In a glow discharge, the ion spectrum is weak, but at low pressures, the 30.4-nm line increases intensity rapidly. Figure 14 shows the relative intensity of the 30.4-nm line compared with the 58.4-nm He I line as a function of pressure. The absolute intensity of the 30.4-nm line relative to the 58.4-nm line is uncertain because of the unknown efficiency of the monochromator.

3.3.2.2.2. HOLLOW CATHODE DISCHARGE. The hollow cathode discharge is simply another version of the glow discharge, but designed to observe the radiation emitted from the negative glow near the cathode surface. As the gas pressure is reduced, the discharge retreats within the cathode until the negative glow fills the hollow cathode. This is the optimum condition for observing the first ion spectrum. Figure 15 shows a convenient design by Paresce et al.[25] In a similar design, Manson[26] has shown that the hollow

[25] F. Paresce, S. Kumar, and C. S. Bowyer, *Appl. Opt.* **10**, 1904 (1971).
[26] J. E. Manson, *Appl. Opt.* **12**, 1394 (1973).

3.3. FAR ULTRAVIOLET REGION

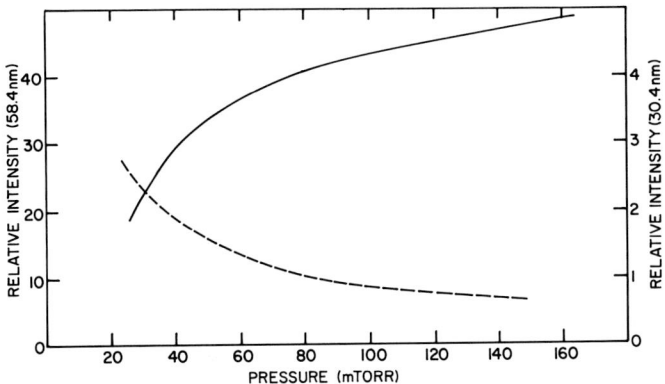

FIG. 14. Relative intensities of the 58.4-nm (solid curve) and 30.4-nm (dashed curve) resonance lines of He I and He II, respectively, as a function of pressure.

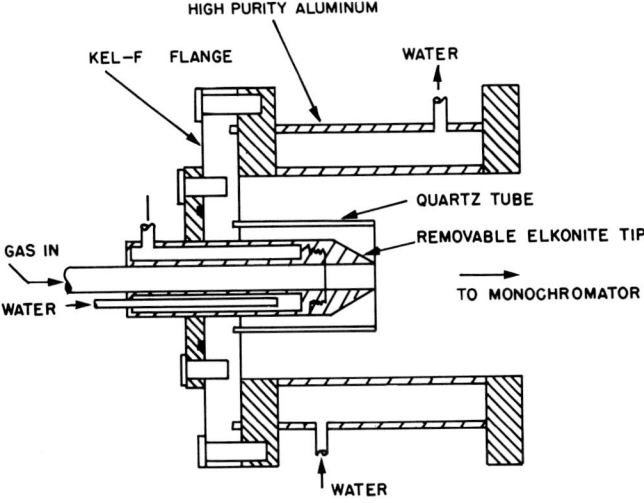

FIG. 15. Schematic diagram of a hollow cathode source [F. Paresce, S. Kumar, and C. S. Bowyer, *Appl. Opt.* **10**, 1904 (1971)].

cathode can produce some Ne III and IV lines between 14 and 28 nm. Although they are weak, the lines can be very useful when detected with a Geiger counter.

3.3.2.2.3. HOT FILAMENT DISCHARGE. In this type of source, the electrons that are necessary to maintain a discharge are produced by a hot filament. Typically, the arc currents are a few amperes and the source voltage is around 100 V. The spectra produced in the rare gases and hydrogen are similar to those produced by a glow discharge. However, one major modification, which uses a magnetic field of a few thousand gauss, is the duoplasmatron.[3] This source effectively increases the ion density in the discharge and hence the ion spectrum.

3.3.2.2.4. MICROWAVE DISCHARGE. One of the main advantages of this source is the absence of electrodes, thus the purity of the radiation produced by the source is improved. It is also very simple to construct. A commercial 125-W, 2450-MHz microwave generator is coupled to an Evenson–Broida cavity[27] that slides over a 12-mm glass or quartz tubing. This type of source has been used with great advantage in photochemistry because of its efficiency in producing atomic resonance lines from many elements.[28, 29] The basic technique is to choose a small amount of the element desired (usually in molecular form) and mix with the main carrier gas, which is usually He or Ar. By this method, almost pure hydrogen Lyman alpha at 1216 Å can be produced.[3] Davis and Braun[29] have shown that isolated atomic lines (undispersed) can be produced with intensities around 10^{15} photons/sec from 110 to 200 nm. Table IV reproduces some of their most intense lines.

TABLE IV. Atomic Lines Produced by a Microwave Excitation Flow Lamp

Emission line (nm)	Atomic specie	Gas mixture % in helium	Emission line (nm)	Atomic specie	Gas mixture % in helium
121.57	H	2.0% H_2	163.36	Br	0.1% Br_2
123.58	Kr	3.0% Kr	157.65		
146.96	Xe	3.0% Xe	193.09	C	1.0% CH_4
130.60	O	10.0% O_2	165.70		
130.49			156.07		
174.52	N	1.0% N_2	182.63	S	0.2% H_2S
174.27			190.03		
138.99	Cl	0.1% Cl_2			

3.3.2.2.5. SPARK DISCHARGE. As with the previous sources, spark light sources come in many varieties. Usually, a capacitor is stored with a given amount of energy at some appropriate voltage and then rapidly discharged in a low-pressure gas or in a high-vacuum sliding spark arrangement. Spectra of some of the most highly ionized atoms have been obtained by a linear theta-pinch discharge used for thermonuclear studies. Recently, Hermansdorfer[30] has reported the spectrum of Ne VIII down to about 6 nm,

[27] F. C. Fehsenfeld, K. M. Evenson, and H. P. Broida, *Rev. Sci. Instrum.* **36**, 294 (1965).
[28] H. Okabe, *J. Opt. Soc. Amer.* **54**, 478 (1964).
[29] D. Davis and W. Braun, *Appl. Opt.* **7**, 2071 (1968).
[30] H. Hermansdorfer, *J. Opt. Soc. Amer.* **62**, 1149 (1972).

3.3. FAR ULTRAVIOLET REGION

and Fawcett et al.[31] have reported spectra from Fe XVIII down to 1.4 nm using a theta-pinch discharge. Another extremely hot spark source is the laser spark; that is, the plasma produced by focusing a high-power laser (400–600 MW) onto a solid target. The spectrum of Fe XVIII has also been achieved by this method,[31] in addition to Al XI.[32]

Probably the most useful source of far uv radiation for general use below 100 nm is the condensed spark discharge in a low-pressure gas ($C \approx 0.2\ \mu F$ and $V \approx 6$ kV).[3] The discharge is confined to a capillary a few centimeters long and a few millimeters in diameter. Repetition rates are in the range 50–100 pulses/sec. The light pulse is about 1 or 2 μsec long. However, this is quite suitable for analog detection. The spectra are characteristic of the gas used in the discharge and are typically from the highly ionized atoms (for example, Ar VI). This type of source produces a very dense line spectrum from at least 10 to 100 nm. Figure 16 shows a spectrum of atomic nitrogen between 20 and 50 nm, when molecular nitrogen is used in the source. Figure 17 shows the longer wavelengths produced by Ar between 40 and 65 nm.

3.3.2.3. Lasers. Laser technology has advanced into the vacuum uv spectral region. The shortest wavelengths produced were announced almost simultaneously in 1972 by Waynant[33] and by Hodgson and Dreyfus.[34] The coherent radiation was produced in the Werner bands of molecular hydrogen extending from about 110 to 127 nm. Previously, lasing had been observed in the Lyman bands extending from 127 to 162 nm.[35–37] Molecular hydrogen has been studied extensively, both from a theoretical[38,39] and experimental point of view.[40] Figure 18 shows the potential energy curves and the transitions involved that emit the Lyman and Werner bands. The pumping of the laser levels is by electron bombardment, of nanosecond duration, either by a fast discharge or a high-intensity, high-voltage electron gun.

In the far uv, the reflectivity of all materials is sufficiently poor that it appears that no lasers in the far uv will operate with optical cavities. It is

---

[31] B. C. Fawcett, A. H. Gabriel, and P. A. H. Saunders, *Proc. Phil. Soc.* **90**, 863 (1966).
[32] M. D. Williams, *J. Opt. Soc. Amer.* **62**, 295 (1972).
[33] R. W. Waynant, *Phys Rev. Lett.* **28**, 253 (1972).
[34] R. T. Hodgson and R. W. Dreyfus, *Phys. Rev. Lett.* **28**, 536 (1972).
[35] R. T. Hodgson, *Phys. Rev. Lett.* **25**, 494 (1970).
[36] R. W. Waynant, J. D. Shipman, Jr., R. C. Elton, and A. W. Ali, *Appl. Phys. Lett.* **17**, 383 (1970).
[37] N. G. Basov, V. A. Danilychev, Yu. M. Popov, and D. D. Khodkevich, Pis'ma Zh. Eksp. Teor. Fiz. **12**, 473 (1970) [*English transl.: JETP Lett.* **12**, 329 (1970)].
[38] A. W. Ali and A. C. Kolb, *Appl. Phys. Lett.* **13**, 259 (1968)
[39] A. W. Ali and P. Kepple, *Appl. Opt.* **11**, 2591 (1972).
[40] R. W. Dreyfus and R. T. Hodgson, *Phys. Rev. A* **9**, 2635 (1974).

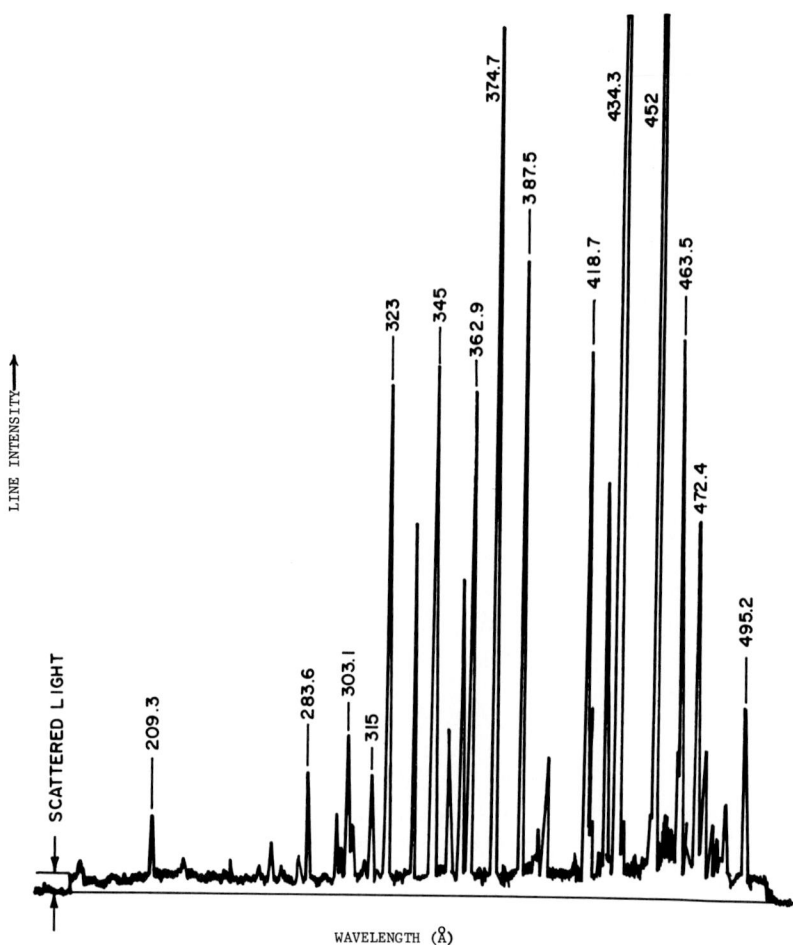

FIG. 16. Nitrogen spectrum between 200 and 500 Å, excited by a condensed spark discharge, where pressure ~ 0.1 Torr and voltage = 5 kV.

therefore necessary to think in terms of high-gain lasers that do not require feedback from mirrors. Shipman[41] has shown that a flat plate Blumlein pulse generator can be made to operate in a traveling-wave mode. Thus, the gas is excited at the speed of light by the traveling-wave, and the inverted population is swept out by stimulation at the same rate. In this manner, Waynant et al.[36] observed the Lyman band lasing with an intensity in the

[41] J. D. Shipman, Jr., *Appl. Phys. Lett.* **10**, 3 (1967).

3.3. FAR ULTRAVIOLET REGION

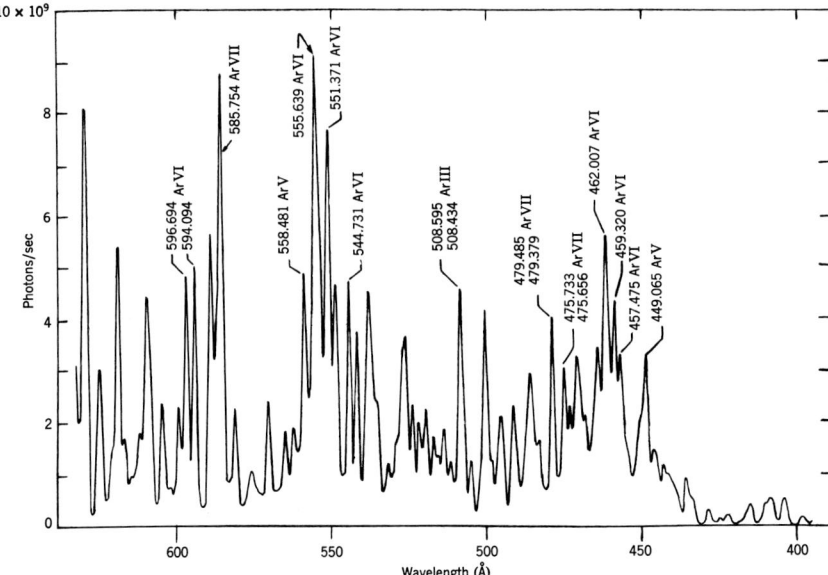

FIG. 17. Argon spectrum between 400 and 650 Å, excited by a condensed spark discharge, where pressure ~ 0.1 Torr and voltage = 5 kV.

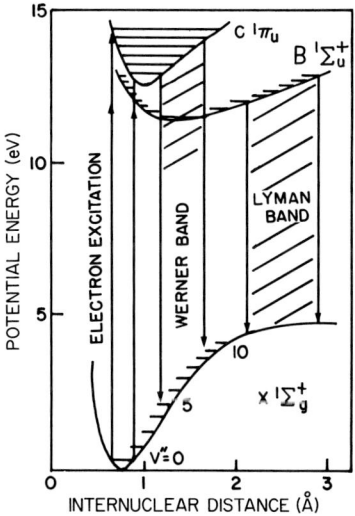

FIG. 18. Potential energy curves for the X $^1\Sigma_g^+$, B $^1\Sigma_u^+$, and C $^1\Pi_u$ states of H$_2$, showing electron excitation of the B and C states and the subsequent emission in the Werner and Lyman bands.

direction of the traveling wave about ten times more intense than in the reverse direction. The power radiated was estimated at about 100 kW.

Extensive studies of the rare gas molecular continua are underway in several laboratories. Lasing action has been observed in the Xe_2 continuum centered on 170 nm.[42–44] Most of these studies involve pumping with 50 kV to 1 MV electrons at several thousand amperes per square centimeter. At present, no success has been obtained in producing lasing action in the He_2 continuum, which would produce the first laser at wavelengths shorter than 100 nm.

Hodgson et al.[45] have produced far uv tunable lasers with the technique of third harmonic generation in Sr vapor from 177.8 to 181.7 nm and from 183.3 to 195.7 nm. They believe the method may be extended to wavelengths around 80 nm.

No commercial far uv lasers are available at present. However, it does seem likely that far uv and x-ray lasers eventually will be available commercially.

3.3.2.4. Standard Sources. Blackbody radiation is, of course, *the* absolute standard source. Tungsten strip lamps and carbon arcs operating at temperatures of 2600 and 3800 K, respectively, have spectral outputs very close to that of a blackbody of the same temperature. These sources are calibrated at the National Bureau of Standards (NBS) and are the main primary standards. However, there is no useful far uv radiation produced at these temperatures. Their shortwave limit is about 200 nm. For blackbody radiation to reach 100 nm, a temperature around 7500 K would be required.

The required high temperatures can only be produced in plasma sources. This technique has been pioneered by Boldt.[46] By use of the wall-stabilized arc plasma, he has produced a number of vacuum uv lines between 110 and 185 nm that reach the blackbody intensity limit at the temperature of the arc. An argon plasma in local thermodynamic equilibrium is used with small additives of H_2, N_2, and CO_2. The strong resonance lines of H, N, C, and O reach the blackbody intensity limit of the local temperature of the plasma.

A source almost equivalent to a blackbody is the hydrogen plasma arc. In a properly constructed source operated in near thermodynamic equilibrium,

---

[42] H. A. Koehler, L. J. Ferderber, D. L. Redhead, and P. J. Ebert, *Appl. Phys. Lett.* **21**, 198 (1972).

[43] N. G. Basov, V. A. Danilychev, and Yu. M. Popov, *Kvantovaya Elektron.* **1**, 29 (1971) [*English transl.: Sov. J. Quantum Electron.* **1**, 18 (1971)].

[44] N. G. Basov, O. V. Bogdankevitch, V. A. Danilychev, A. G. Devyatkov, G. N. Kashnikov, and N. P. Lantsov, *Zh. Eksp. Teor. Fiz. Piz'ma Red* **7**, 404 (1968) [*English transl.: JETP Lett.* **7**, 317 (1968)].

[45] R. T. Hodgson, P. P. Sorokin, and J. J. Wynne, *Phys. Rev. Lett.* **32**, 343 (1974).

[46] G. Boldt, *Space Sci Rev.* **11**, 728 (1970); *J. Quant. Spectrosc. Radiat. Transfer* **5**, 91 (1965).

the spectroscopic properties of the hydrogen plasma are exactly known. Thus, the spectral radiance of the source can be calculated. The plasma is produced in a wall-stabilized arc in pure hydrogen at about 1 atm. Typically, the source operates at about 80 A. Under those conditions, the temperature of the plasma approaches 20,000 K. This source has been developed and studied extensively at NBS,[47,48] and produces a continuum down to 130 nm. The calculated continuum output of the NBS source at 20,000 K is compared in Fig. 19 with the blackbody intensity emitted by a tungsten strip lamp and

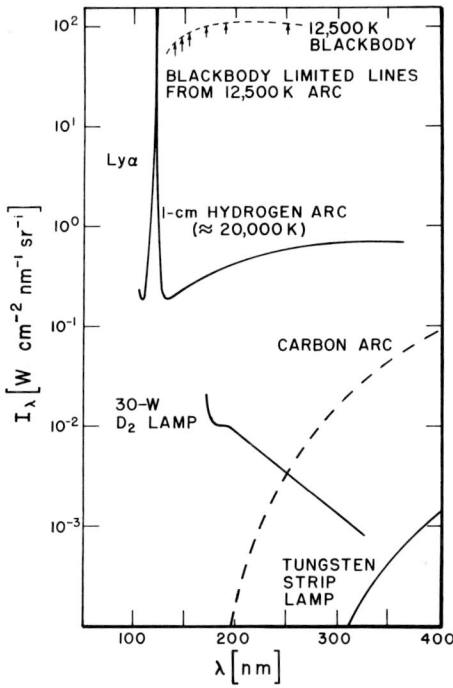

FIG. 19. Spectral radiances of several far uv sources. The tungsten strip lamp output is calculated for a temperature of 2600 K and the carbon arc at 3800 K [W. R. Ott and W. L. Wiese, *Opt. Eng.* **12**, 86 (1973)].

a carbon arc and the output of a deuterium lamp. The continuous emission, at temperatures above 12,000 K, is caused principally by free–bound and free–free transitions of the electrons in the field of the atomic hydrogen nucleus. As the temperature is increased, the continuum emission increases

[47] W. R. Ott, P. Fieffe-Prevost, and W. L. Wiese, *Appl. Opt.* **12**, 1618 (1973)
[48] W. R. Ott and W. L. Wiese, *Opt. Eng.* **12**, 86 (1973).

up to a maximum value and then remains constant during further temperature increases. The maximum is achieved at about 20,000 K. Thus, for a plasma arc temperature of this value, or in excess of this value, no determination of temperature is necessary. This increases the accuracy of the spectral radiance measurement.

The only other absolute standard source of far uv radiation is synchrotron radiation. As described in Section 3.3.2.1.5, the intensity of synchrotron radiation is exactly calculable. In the storage ring, the circulating current can be measured accurately. However, the accuracy in determining the electron energy is a problem. An error of 1 in 170 MeV causes an error of only 1% at 120 nm but 10% at 20 nm. Nevertheless, an absolute standard is available over the entire far uv.

For practical laboratory use, a transfer standard light source is necessary. In general, the output of a gas discharge, producing either a line or continuum spectrum, is quite reproducible, provided the operating conditions are kept constant. These are discharge current, gas pressure, purity (or impurity) conditions, and window transmission. Sources that use LiF or MgF_2 windows may need periodic recalibration because of degradation of these windows. Quartz (or suprasil) is much more stable but will transmit only down to 160 nm. However, the most commonly used transfer standards are deuterium lamps with a short wavelength continuum limit of 165 nm. Several groups have calibrated these lamps against synchrotron radiation, plasma-blackbody radiation, and the hydrogen plasma arc.[49] The National Bureau of Standards now provides deuterium-filled arc lamps with suprasil windows calibrated to 165 nm with estimated uncertainties of 10%.

3.3.3. Dispersive Devices

Although the far uv spectral region was first discovered by the use of prisms, little use is made of them today. Although materials exist for the manufacture of prisms, their usefulness lies only in the Schumann region (the spectral region above 110 nm). For example, prisms may be made from LiF or MgF_2 with shortwave transmission cutoffs at 105 and 115 nm, respectively.

The availability and quality of diffraction gratings capable of covering the entire far uv range make them the only choice for dispersing elements. The grating has a higher resolving power, gives less scattered light, has a nearly linear dispersion (for near normal incidence), and is more useful for precise wavelength measurements than a prism. In this section, we will describe the

---

[49] E. Pitz, *Appl. Opt.* **8**, 255 (1969); D. Stuck and B. Wende, *J. Opt. Soc. Amer.* **62**, 96 (1972); and, at N.B.S., W. R. Ott, and J. D. Bartoe, *J. Opt. Soc. Amer.* **62**, 1372 (1972).

main characteristics of gratings and their use in several different mounts to construct spectrographs and monochromators.

3.3.3.1. Diffraction Gratings. 3.3.3.1.1. THEORY. The grating equation for a plane diffraction grating is

$$\pm m\lambda = d(\sin \alpha + \sin \beta), \tag{3.3.3}$$

where α and β are the angles of incidence and diffraction, respectively, d is the spacing between rulings, λ the wavelength of the diffracted radiation, and m the order of the spectrum. When $m = 0$, we have the *central image* ($\alpha = \beta$). The negative sign applies when the spectrum lies between the central image and the plane of the grating. This is called the *outside order*. When the spectrum lies on the other side of the central image ($+m$), we have the *inside orders*. The signs of α and β in Eq. (3.3.3) are the same when they both lie on the same side of the normal to the plane of the grating.

Because of the lack of materials that transmit far uv radiation, there are no transmission gratings. Further, because of the poor reflectance of most materials at wavelengths shorter than 100 nm, it is necessary to keep the number of reflecting surfaces to a minimum; hence, the development of the concave diffraction grating by Rowland.[50] This single element provides dispersion and focusing. As in the case of a spherical concave mirror, the concave grating will first image a point source into a vertical line (horizontal focus), then into a horizontal line (vertical focus). It can be shown that the best horizontal focal point is given by

$$\frac{\cos^2 \alpha}{r} - \frac{\cos \alpha}{R} + \frac{\cos^2 \beta}{r'} - \frac{\cos \beta}{R} = 0, \tag{3.3.4}$$

where r and r' are the distances of the object and image, respectively, from the center of the grating, and R is the radius of curvature of the grating. Two solutions of this equation are

$$r = R \cos \alpha, \quad r' = R \cos \beta, \tag{3.3.5}$$

and

$$r = \infty, \quad r' = R \cos^2 \beta / (\cos \alpha + \cos \beta). \tag{3.3.6}$$

Equation (3.3.5) is the equation of a circle, with diameter R, and is called the *Rowland circle*. It expresses the fact that diffracted light of all wavelengths will be focused horizontally on the circumference of a circle of diameter R, provided that the entrance slit, or illuminated source, and grating lie on the circle, and that the grating normal lies along a diameter. The arrangement is shown in Fig. 20. This is the normal condition for observing a spectrum.

---

[50] H. A. Rowland, *Phil. Mag.* **13**, 469 (1882).

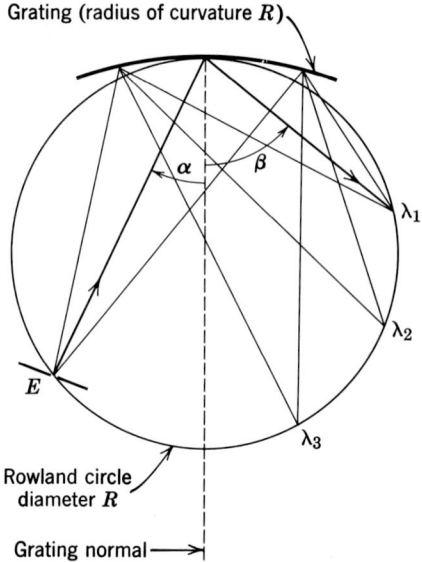

FIG. 20. The Rowland circle. Radiation from the point E is dispersed and focused by the grating at λ_1, λ_2, etc.; α and β are the angles of incidence and diffraction, respectively [J. A. R. Samson, "Techniques of Vacuum Ultraviolet Spectroscopy." Wiley, New York, 1967].

The vertical focal point is given by the equation

$$\frac{1}{r} - \frac{\cos \alpha}{R} + \frac{1}{r'} - \frac{\cos \beta}{R} = 0. \tag{3.3.7}$$

The solutions of this equation are

$$r = R/\cos \alpha, \quad r' = R/\cos \beta, \tag{3.3.8}$$

and

$$r = \infty, \quad r' = R/(\cos \alpha + \cos \beta). \tag{3.3.9}$$

The importance of Eq. (3.3.7) is in the design of instruments to minimize astigmatism; that is, designs which simultaneously satisfy Eqs. (3.3.4) and (3.3.7) for a given r and r'.

Equations (3.3.6) and (3.3.9) represent the case where the incident light is parallel. When the spectrum is viewed near the normal, $\cos \beta \sim 1$, and the two equations become identical. That is, the vertical and horizontal foci coincide, and the image will be stigmatic. This is the condition for the Wadsworth mounting. The grating equation for a concave grating is essentially the same as for a plane grating; namely, $\pm m\lambda = d(\sin \alpha + \sin \beta)$, provided the height of the source above the plane of the grating divided by R is small compared with unity.[3]

3.3.3.1.2. ASTIGMATISM. As discussed in the section above, a point source will be focused, in general, into a vertical line by the concave grating. This aberration is termed *astigmatism*. The length of the astigmatic image z is given by[51, 52]

$$z = (l \cos \beta / \cos \alpha) + L(\sin^2 \beta + \sin \alpha \tan \alpha \cos \beta), \quad (3.3.10)$$

where the first term gives the contribution caused by an object slit of finite vertical height l, and the second term is the astigmatism produced by a point on the object slit; L is the length of the ruled lines illuminated. Astigmatism is most severe at grazing angles of incidence. This is illustrated in Fig. 21. Unfortunately, it is absolutely necessary to use grazing angles to obtain sufficiently high reflectance at the shorter wavelengths (1–30 nm).

3.3.3.1.3. ASPHERICAL GRATINGS. For reviews and details of conventional concave diffraction gratings, the reader is referred to Samson,[3] Beutler,[51] Namioka,[52] Welford,[53] and Palmer and Verrill.[54] Several studies have been made of nonconventional concave gratings. Proposals have been made for spherical gratings with constant groove spacing but with curved grooves[55]; spherical gratings with variable spacing and curved grooves[56]; and cylindrical gratings with constant spacing and curved grooves.[57] While the theory for these new gratings has been detailed, no practical grating has been manufactured. However, their prime advantage over the conventional grating appears to be a large reduction in astigmatism.

Gratings on elliptical[58] and toroidal[59–61] blanks have been discussed also from the point of view of reducing astigmatism. However, only the toroidal grating has been tried out in practice and indeed has shown improved performance.[62]

3.3.3.1.4. HOLOGRAPHIC GRATINGS. A laser-generated interference pattern projected onto high-resolution film is the basis of a simple grating. More complex systems using two laser beams can develop concave gratings with reduced astigmatism. Considerable theoretical progress is being made in this

---

[51] H. G. Beutler, *J. Opt. Soc. Amer.* **35**, 311 (1945).
[52] T. Namioka, *J. Opt. Soc. Amer.* **49**, 446 (1959); *in* "Space Astrophysics" (W. Liller, ed.), p. 249. McGraw-Hill, New York, 1961.
[53] W. T. Welford, *Progr. Opt.* **4**, (1965).
[54] E. W. Palmer and J. F. Verrill, *Contemp. Phys.* **9**, 257 (1968).
[55] Y. Sakayangi, *Sci. Light (Tokyo)* **3**, 1 (1954–1955); **3**, 79 (1954–1955).
[56] M. V. R. K. Murty, *J. Opt. Soc. Amer.* **52**, 768 (1962).
[57] M. Singh and K. Majumder, *Sci. Light (Tokyo)* **18**, 57 (1969).
[58] T. Namioka, *J. Opt. Soc. Amer.* **51**, 4 (1961).
[59] H. Haber, *J. Opt. Soc. Amer.* **40**, 153 (1950).
[60] H. Greiner and E. Schaffer, *Optik* **16**, 288, 350 (1959).
[61] I. V. Peisakhson and I. N. Tarnakin, *Zh. Prikl. Spektrosk.* **1**, 289 (1964); **2**, 218 (1965).
[62] E. Schonheit, *Optik* **23**, 305 (1965/66).

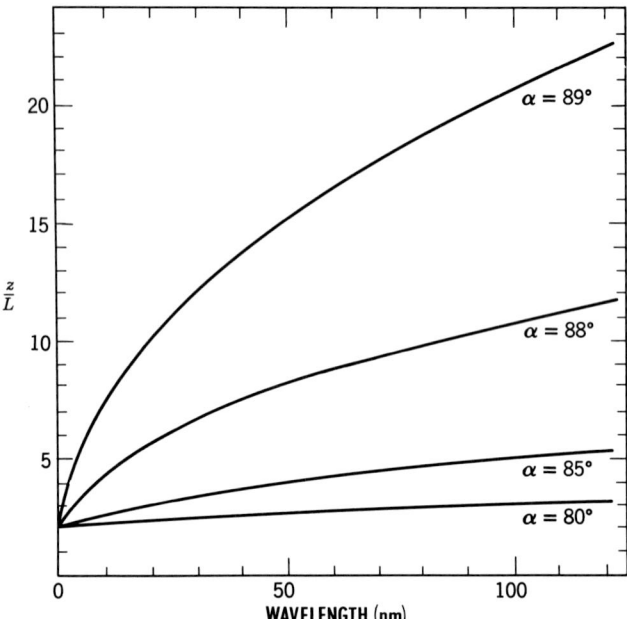

FIG. 21. Length of the astigmatic image z divided by the length L of the rulings illuminated for a 600-line/mm grating as a function of wavelength and angle of incidence α [T. Namioka, *J. Opt. Soc. Amer.* **49**, 446 (1959); *in* "Space Astrophysics" (W. Liller, ed.), p. 249. McGraw-Hill, New York, 1961].

area.[63,64] The Jobin-Yvon Company has developed techniques to produce plane and concave gratings commercially.[64] The main advantages of holographic gratings are the lack of ruling ghosts, the small intensity of stray light, and the capability of reducing astigmatism. However, there is no experimental evidence yet that indicates the holographic grating is more efficient in the far uv. One report on the x-ray performance (4.4 and 2.3 nm) of a plane grating formed holographically describes it as having an efficiency equal to the best mechanically produced grating.[65]

3.3.3.1.5. BLAZE. The efficiency of a grating can be greatly enhanced if the ruled grooves are shaped in such a manner that the surface of each facet is inclined at some angle θ to the surface. This process is called *blazing*. The principle of concentrating radiation into a given wavelength is that this wavelength must be diffracted in a direction that coincides with the direction

---

[63] H. Noda, T. Namioka, and M. Seya, *J. Opt. Soc. Amer.* **64**, 1031, 1037, 1043 (1974).

[64] Jobin-Yvon Optical Systems, Rue du Canal, 91160 Longjumeau, France (in the U.S.A., 20 Highland Ave., Metuchen, New Jersey 08840).

[65] D. Rudolph, G. Schmahl, R. L. Johnson, and R. J. Speer, *Appl. Opt.* **12**, 1731 (1973).

of the specularly reflected beam from the surface of the facet. Figure 22 illustrates the principle. By use of the grating equation (3.3.3) the blazed wavelength is given by

$$m\lambda_{\text{blaze}} = 2d \sin \theta \cos(\alpha - \theta), \tag{3.3.11}$$

where θ is the blaze angle, d the spacing between adjacent rulings, α the angle of incidence, and m the order of the spectrum. All good gratings are now blazed and are useful even for grazing incidence applications where short wavelengths are used (10–30 nm). However, as the wavelengths decrease, a greater premium is placed on the smoothness of the facets if scattered light

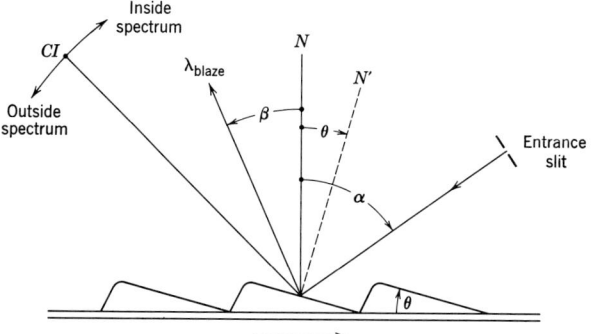

FIG. 22. Cross section of a blazed grating; θ is the blaze angle [J. A. R. Samson, "Techniques of Vacuum Ultraviolet Spectroscopy." Wiley, New York, 1967].

is to be kept to a minimum. A different approach at ruling and producing a blazed condition has been achieved with *phase gratings*. These gratings were developed at the National Physical Laboratories in England by Franks.[66, 67] A profile of the grooves is shown in Fig. 23. If the grooves were simple cuts in a substrate and unable to reflect radiation, the grating would be extremely poor. However, reflection (and diffraction) occurs at both top and bottom surfaces. They are referred to as phase gratings because there is a definite phase relationship between rays diffracted from the top and bottom of the grooves. By a suitable choice of groove depth, beams from the top and bottom of the groove can be made to interfere destructively for the zero-order beam (central image), while, at the same time, interfere constructively for any chosen order at a given wavelength. It is extremely important that the facets of a phase grating be optically flat and smooth. Techniques have been

[66] A. Franks, in "X-Ray Optics and X-Ray Micro-analysis" (H. H. Pattee, V. E. Cosslett, and A. Engstrom, eds.), p. 199. Academic Press, New York, 1963.
[67] L. A. Sayce and A. Franks, *Proc. Roy. Soc. (London)* A **282**, 353 (1964).

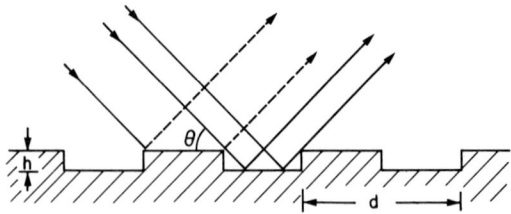

FIG. 23. Cross section of a phase grating. The step height h introduces a phase difference between the diffracted beams indicated by the dashed and solid lines.

developed to assure this condition.[66] The relationship between the groove depth h and the "blazed" wavelength λ_{blaze} is

$$m\lambda_{blaze} = d[2\cos\theta + (d/h)\sin\theta]/[(d^2/4h^2) + 1], \quad (3.3.12)$$

where d is the groove separation, m the order, and θ the grazing angle of incidence.[68] If $2\cos\theta$ is less than $(d/h)\sin\theta$, then Eq. (3.3.12) can be expressed, approximately, as

$$m\lambda_{blaze} \approx 4h\theta. \quad (3.3.13)$$

For a groove depth of 15 nm and a grazing angle of 4°, maximum efficiency should appear in the first order at 4.2 nm. Equation (3.3.12) has been verified qualitatively by Bennett for the 4.45-nm carbon K line.[68] Figure 24 shows his results of the efficiency of a phase grating, with 300 grooves/mm and a groove depth of 15 nm, as a function of the grazing angle.[69] Maximum efficiency occurs at an angle $\theta = 3.2°$ for first-order 4.45-nm radiation. A surprisingly high efficiency of 10% is obtained. As predicted, the central image practically disappears when the first order is a maximum. The minimum wavelength that has been obtained with a phase grating is 0.056 nm.[70]

3.3.3.1.6. REFLECTIVE COATINGS. The selection of the proper reflective coating for a diffraction grating is very important if the highest efficiency is desired. For wavelengths longer than 200 nm, aluminum is unsurpassed. However, although freshly evaporated aluminum has the highest reflectivity of any coating down to about 90 nm, it rapidly oxidizes on exposure to air. By overcoating nonoxidized aluminum by a transparent material, the high reflectivity can be maintained.[71-73] The most useful overcoatings are LiF

[68] J. M. Bennett, Ph.D. Thesis, Univ. of London (1971).
[69] A. Franks, in Proc. Int. Conf. X-Ray Opt. Micro-anal., 6th (G. Shinoda, K. Kohra, and T. Ichinokawa, eds.), p. 57. Univ. of Tokyo Press, Tokyo, 1972.
[70] A. Franks and K. Lindsey, J. Phys. E **1**, 144 (1968).
[71] G. Hass and R. Tousey, J. Opt. Soc. Amer. **49**, 593 (1959).
[72] J. T. Cox, G. Hass, and J. E. Waylonis, Appl. Opt. **7**, 1535 (1968).
[73] A. P. Bradford, G. Hass, J. F. Osantowski, and A. R. Toft, Appl. Opt. **8**, 1183 (1969).

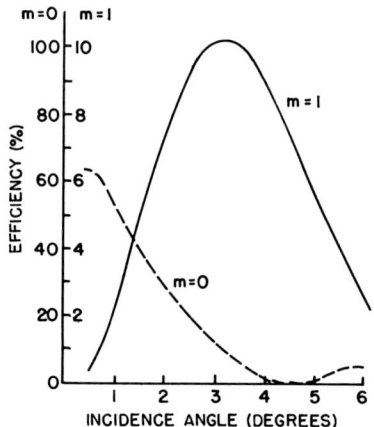

FIG. 24. Absolute efficiency for the zero and first order of a phase grating for 4.4-nm radiation as a function of angle of incidence [A. Franks, *in Proc. Int. Conf. X-Ray Opt. Microanal.*, *6th* (G. Shinoda, K. Kohra and T. Ichinokawa, eds.), p. 57. Univ. of Tokyo Press, Tokyo, 1972, Crown Copyright reserved].

and MgF_2. These materials are transparent down to 104 and 115 nm, respectively. MgF_2 is the most stable overcoating and is little affected by humidity. LiF, on the other hand, is slightly hygroscopic, and this causes some deterioration in a humid atmosphere. Adriaens and Feuerbacher[74] have shown that the reflectivity of Al plus an overcoating of MgF_2 is even greater than that given in Haas and Tousey,[71] Cox et al.,[72] and Bradford et al.,[73] provided the coatings are laid down in an ultrahigh vacuum (10^{-10} Torr) and then annealed at 300°C for about 60 hr in a high vacuum. The effect of overcoating an aluminized grating with LiF and MgF_2 is shown in Fig. 25. The transmission limits of the fluorides dramatically illustrate the dependency of a high efficiency on the original Al coating. The fluorides themselves have poor reflectivities in this spectral region. Thus, all gratings should be aluminized with either a LiF or MgF_2 overcoating for work between 100 and 200 nm.

The normal incidence reflectance of all materials below 100 nm is generally less than 35%. The materials with the highest reflectivities between 40 and 100 nm are those with the greatest densities; namely Pt, Re, Os, and Ir. All have reflectivities greater than 20% at 58.4 nm.[3, 75–79] The highest reported

[74] M. R. Adriaens and B. Feuerbacher, *Appl. Opt.* **10**, 958 (1971).
[75] J. A. R. Samson, J. P. Padur, and A. Sharma, *J. Opt. Soc. Amer.* **57**, 966 (1967).
[76] J. T. Cox, G. Hass, and J. B. Ramsey, *J. Opt. Soc. Amer.* **62**, 781 (1972).
[77] J. T. Cox, G. Hass, J. B. Ramsey, and W. R. Hunter, *J. Opt. Soc. Amer.* **63**, 435 (1973).
[78] G. Hass, G. F. Jacobus, and W. R. Hunter, *J. Opt. Soc. Amer.* **57**, 758 (1967).
[79] G. Hass, J. B. Ramsey, and W. R. Hunter, *Appl. Opt.* **8**, 2255 (1969).

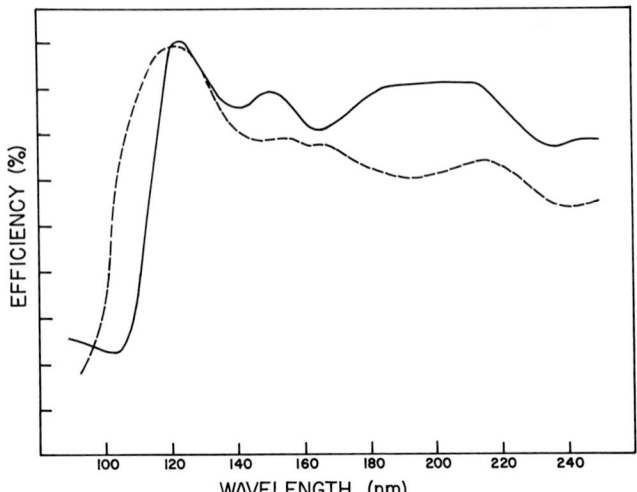

FIG. 25. Absolute efficiency for the first order of a LiF (dashed curve) and MgF$_2$ (solid curve) overcoated blazed grating as a function of wavelength (1200 line/mm plane grating, blazed for 120 nm).

reflectivity at 58.4 nm is 35% for Os deposited on superpolished fused-quartz substrates. For gratings in this wavelength range, it is not possible to have superpolished surfaces. Thus, the choice of coating is probably that which is most readily available. Platinum appears to be the most useful choice.

For wavelengths shorter than 30 or 40 nm, the normal incidence reflectance of all materials is of the order of a few percent. Thus, gratings are normally used with the radiation incident at grazing angles in this spectral range. The grazing angles are typically from about zero to 10°. Very few measurements have been made of the reflectivity of materials versus angle of incidence for wavelengths between about 4 and 40 nm, although considerable data are available from 0.7 to 4 nm.[80, 81] Extrapolating the results of these data, it would appear that Au, Ni, and Cr are among the most satisfactory reflective coatings, considering stability and high reflectivity.

3.3.3.2. Spectrographs and Monochromators. There are two basic types of spectrographs and monochromators, the normal incidence type for work between 30 and 200 nm and the grazing incidence type for work between 0.05 and 30 nm. Although there are exceptions, normal incidence

[80] A. P. Lukirskii, E. P. Savinov, O. A. Ershov, and Yu. F. Shepelev, *Opt. Spektrosk.* **16**, 310 (1964) [*English transl.: Opt. Spectrosc.* **16**, 168 (1964)].

[81] O. A. Ershov, I. A. Brytov, and A. P. Lukirskii, *Opt. Spektrosk.* **22**, 127 (1967) [*English transl.: Opt. Spectrosc.* **22**, 66 (1967)].

3.3. FAR ULTRAVIOLET REGION

usually refers to angles less than 10° and grazing incidence to angles of incidence greater than 80°. The reflectivity of materials is fairly constant for angles of incidence between 0 and 40°, whereas the astigmatism of a concave grating increases rapidly with angle of incidence. Hence, small angles of incidence are preferred for normal incidence instruments. For shorter wavelengths, the normal incidence reflection is so poor that it is necessary to go to grazing angles of incidence to obtain sufficient intensity. A commonly used plane grating spectrograph is the Ebert–Fastie mount shown in Fig. 26.[82–85] With the necessary three reflections, this type of spectrograph

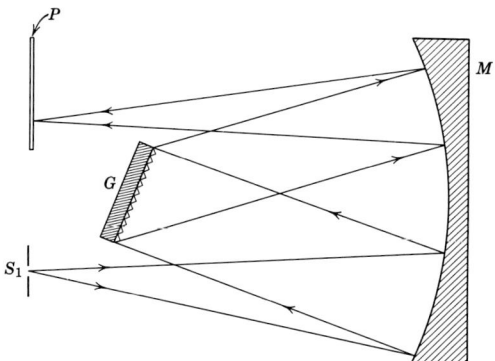

FIG. 26. Ebert–Fastie mount.

is useful only for wavelengths longer than about 110 nm. Its major usefulness is in light gathering power (small f-number). It can readily be made into a monochromator by a simple rotation of the grating. The Czerny–Turner mount has a similar design but uses two separate concave mirrors.

Special photographic film must be used for far uv spectrographs. These are Ilford Q emulsions; Eastman-Kodak SWR, 101-01, and 103-0 uv film; and Kodak-Pathé (France) SC.5 and SC.7 emulsions. The SC.7, SC.5, and 101-01 are by far the most sensitive—at least ten times faster than SWR film at short wavelengths.

By replacing the film plate holder by an exit slit, any spectrograph can be converted to a monochromator. However, to maintain high resolution, the focusing properties of the Rowland circle must be maintained if a concave grating instrument is used. However, resolution is often compromised for

[82] H. Ebert, *Wied. Ann.* **38**, 489 (1889).
[83] W. G. Fastie, *J. Opt. Soc. Amer.* **42**, 641 (1952).
[84] W. G. Fastie, *J. Opt. Soc. Amer.* **42**, 647 (1952).
[85] N. P. Penkin and A. M. Shukhtin, *Izv. Akad. Nauk SSSR, Ser. Fiz.* **12**, 376 (1948).

mechanical convenience or for compatability with a specific research project or light source (e.g., the sun or synchrotron radiation). Many designs exist.[3]

For normal incidence monochromators, off-Rowland circle mountings are commonly used, whereby Eq. (3.3.4) is either satisfied precisely or approximately. The most commonly used approximation to Eq. (3.3.4) is in the Seya–Namioka design.[86-88] This design is the simplest to construct. It has fixed exit and entrance slits, and wavelength scanning is achieved by a simple rotation of the grating about the grating center. To obtain the best approximation to Eq. (3.3.4), the angle subtended by the exit and entrance slits at the center of the grating must be 70°30′, and all the optical elements (slits and grating) should be on the Rowland circle for zero wavelength. For a $\frac{1}{2}$-m grating, the best resolution obtainable is about 0.06 nm.

Lavolee and Robin[89] have shown that it is not necessary to restrict the rotation of a grating to an axis through its vertex. In fact, if a grating is illuminated asymmetrically, it provides an extra degree of freedom in the focusing conditions. Pouey[90] has shown that for a given resolution, the luminosity can be increased by using the optimum width *and* optimum height of a grating. The use of a toroidal grating greatly increases the luminosity by decreasing astigmatism.[91]

For higher resolution, and probably more luminosity, the class of monochromator that involves a rotation and translation of the grating is superior to the Seya–Namioka.[92-94] The mechanical construction is, of course, more complicated. The principle involves the linear translation of the grating along the bisector of the angle subtended by the slits at the center of the grating and a simultaneous rotation of the grating about an axis through its center such that Eq. (3.3.4) is satisfied. Again, fixed exit and entrance slits are achieved with this mount. For best resolution, the Rowland circle mounting should be used.

Grazing incidence monochromators tend to use the Rowland circle mounting. This necessitates either the entrance or exit slits moving. The best design in this class is the Vodar grazing incidence monochromator.[95] In this mounting, the Rowland circle is pivoted about the exit slit, keeping the

---

[86] M. Seya, *Sci. Light* **2**, 8 (1952).
[87] T. Namioka, *Sci. Light* **3**, 15 (1954).
[88] T. Namioka, *J. Opt. Soc. Amer.* **49**, 951 (1959).
[89] M. Lavollee and S. Robin, *J. Opt. Soc. Amer.* **64**, 319 (1974).
[90] M. Pouey, *Opt. Commun.* **3**, 158 (1971).
[91] E. Schönheit, *Optik* **23**, 305 (1965/66).
[92] B. Vodar, *Rev. Opt.* **21**, 97 (1942).
[93] S. Robin, *J. Phys. Radium* **14**, 551 (1953).
[94] McPherson Instrument Corp., Acton, Massachusetts.
[95] M. Salle and B. Vodar, *C. R. Acad. Sci.* (*Paris*) **230**, 380 (1950).

angle of incidence on the grating constant. This construction is shown in Fig. 27. The grating G is rigidly attached to the entrance slit S_1 at a fixed angle of incidence α. The grating is mounted on a turntable constrained to move along the track GS_2, while S_1 is constrained to move along the track $S_1 S_2$. The emerging beam always travels in the same direction.

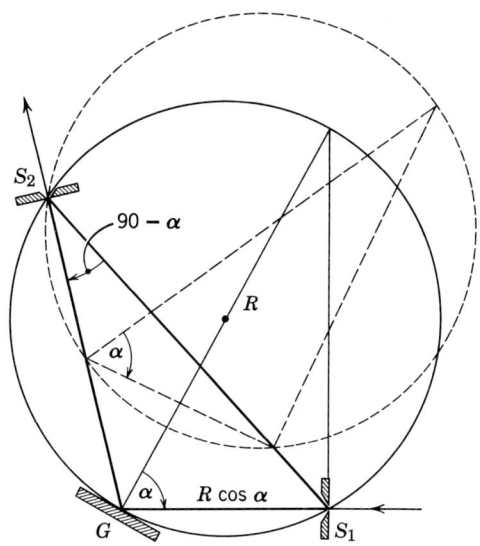

FIG. 27. Vodar-type grazing incidence monochromator. Scanning mechanism maintains the optical elements on the Rowland circle, and maintains a constant angle of incidence. The Rowland circle is pivoted about the exit slit S_2.

Several specialized designs of grazing incidence monochromators have been recently constructed for use with synchrotron radiation facilities. The requirements are, generally, that the exit and entrance slits be fixed, and that reasonably good resolution should be possible. Since synchrotron radiation is extremely directional, it is often possible to omit the entrance slit and also to use something other than the Rowland circle mounting (e.g., plane gratings with concave mirrors). These designs use from two to four optical elements at grazing angles of incidence.[96–99] The extra elements have the added advantage that they can be used to suppress higher-spectral orders by decreasing the angle of incidence on the mirrors as the wavelength increases.

[96] J. B. West, K. Codling, and G.V. Marr, *J. Phys. E.* **7**, 137 (1974).
[97] K. Codling and P. Mitchell, *J. Phys F.* **3**, 685 (1970).
[98] K. P. Miyake, R. Kato, and H. Yamashita, *Sci. Light* **18**, 39 (1959).
[99] H. Dietrich and C. Kunz, *Rev. Sci. Instrum.* **43**, 434 (1972).

3.3.4. Optical Windows and Filters

Stable and rigid window materials exist in the far uv spectral region only down to 104 nm. The most common windows are quartz (180 nm), suprasil (160 nm), sapphire (140 nm), MgF_2 (112 nm), and LiF (104 nm). The wavelengths in parentheses represent their short wavelength transmission limits at room temperature. All of these limits can be reduced about 4 nm at 77 K (liquid N_2 temperature).[100] Carruthers[101] has shown that the transmission of a LiF window, 1 mm thick and cooled to 77 K, was about 10% at 100.8 nm.

In the shorter wavelength range, thin films from 10 to 100 nm thick have been found to transmit radiation in narrow bands.[3, 102] These bands are usually bound by a long wavelength transmission onset that corresponds to the plasma frequency of the valence electrons and a short wavelength cutoff that corresponds to the K, L, M, etc. absorption edges of the material.

Table V lists the available windows in the far uv region, and gives a partial list of the most useful thin films that can be used either as windows or filters.

TABLE V. Transmission Onsets and Bandpasses for Typical Window and Filter Materials

Windows		Filters	
Material	Transmission onset (nm)	Material	Band-pass (nm)
Quartz	200	C	4.4–55
Synthetic fused quartz	160	Be	11–65
Sapphire	140	Al	17–80
BaF_2	134	Mg	25–120
BeO	122	Te	30–70
CaF_2	122	Sb	40–70
MgF_2	112	Sn	50–90
LiF	104	Pb	68–90
		In	70–110

Interference filters have been made successfully between 120 and 200 nm using multiple layers of Al and MgF_2.[103–106] A total of six layers have been used. These filters do not have the narrow widths found at longer wave-

[100] W. R. Hunter and S. A. Malo, *J. Phys. Chem. Solids* **30**, 2739 (1969).
[101] G. R. Carruthers, *Appl. Opt.* **10**, 1461 (1971).
[102] W. R. Hunter, in "Physics of Thin Films" (G. Hass, M. H. Francombe, and R. W. Hoffman, eds.), Vol. 7, p. 43. Academic Press, New York, 1973.
[103] B. Bates and D. J. Bradley, *Appl. Opt.* **5**, 971 (1966).
[104] D. H. Harrison, *Appl. Opt.* **7**, 210 (1968).
[105] E. T. Fairchild, *Appl. Opt.* **12**, 2240 (1973).
[106] A. Malherbe, *Appl. Opt.* **13**, 1275 (1974).

lengths nor have they the same rejection ratio. However, calculations by Spiller[107] suggest that many layers (twelve) of Al and MgF_2 should produce interference filters with a 3-nm band-pass and a rejection ratio of about 30, while still maintaining a transmission of about 27%. The shortest wavelength interference filter produced to date was made by Malherbe.[106] His filter peaked at 121.6 nm and had a 9-nm half-width. The rejection ratio between 121.6 and 160 nm was greater than 10^4, while still maintaining a peak transmission of 15%. This performance is orders of magnitude better than other similar interference filters. However, Malherbe points out that it was necessary to evaporate the thin films in an ultrahigh vacuum (10^{-9} Torr). This is in agreement with the work by Adridens and Feurerbacher[74] (discussed in Section 3.3.3.1.6) for producing maximum reflectance with Al overcoated with MgF_2.

3.3.5. Polarizers

Nearly all birefringent materials are opaque in the far uv region, with the exception of synthetic quartz, sapphire, and magnesium fluoride. Chandrasekharan and Damany[108–110] have tabulated the birefringence of these materials from about 300 nm down to their respective transmission limits. Johnson[111] has shown that a MgF_2 polarizer of the Ronchon or Wollaston type can be constructed[112] for work at wavelengths longer than about 130 nm. For shorter wavelengths, reflection methods must be used. Reflection at Brewster's angle from a dielectric will produce pure plane polarized light, provided the radiation is not absorbed by the dielectric. This imposes a practical short wavelength limit of about 120 nm. When the material is absorbing, partially polarized radiation is produced. Multiple reflections will increase the degree of polarization but will, of course, decrease the intensity of the beam transmitted.

To maintain the original direction of the incident radiation, a reflection type polarizer must be constructed with at least three mirrors, as shown in Fig. 28. The angle of incidence on the first and third mirrors is θ, and $(2\theta - 90°)$ on the second mirror. The mirrors should be optical flats ($\lambda/2$ or better) with a high degree of polish, typical of those used for lasers. This reduces light scattering and maintains the highest transmission. Gold coatings appear to be the most suitable, at least for the range 40–200 nm. Figure 29 shows the calculated degree of polarization P and transmission T

---

[107] E. Spiller, *Appl. Opt.* **13**, 1209 (1974).
[108] V. Chandrasekharan and H. Damany, *Appl. Opt.* **7**, 687 (1968).
[109] V. Chandrasekharan and H. Damany, *Appl. Opt.* **7**, 939 (1968).
[110] V. Chandrasekharan and H. Damany, *Appl. Opt.* **8**, 671 (1969).
[111] W. C. Johnson, JR., *Rev. Sci. Instrum.* **35**, 1375 (1964).
[112] Karl Lambrecht Co., Chicago, Illinois.

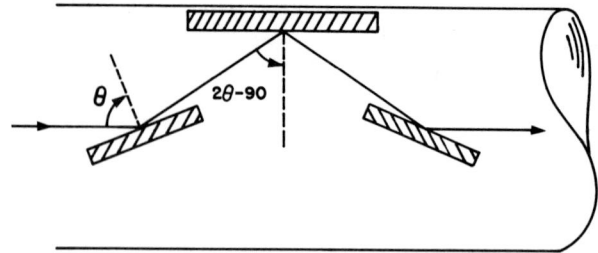

Fig. 28. A three-reflection polarizer to preserve the direction of the incident beam.

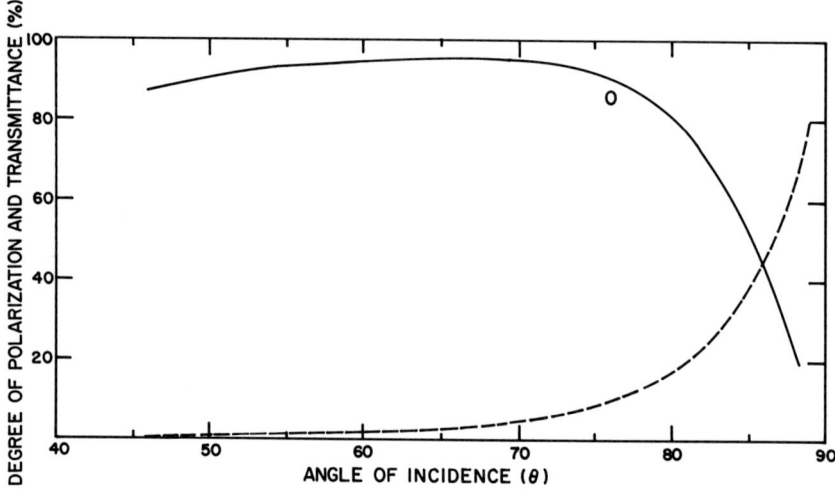

Fig. 29. The degree of polarization (P, solid curve) and the transmittance (T, dashed curve) of a three-reflection polarizer as a function of the angle of incidence θ on the first mirror, for 58.4 nm incident on gold. The single data point O is the experimental value of P for $\theta = 76°$.

of a triple reflection polarizer for 58.4 nm and gold coated mirrors. The values chosen for the optical constants were $n = 1.07$ and $k = 0.85$. Almost identical values are obtained if a value of $k = 0.7$ is used (this represents the typical error spread in k). For the first mirror, P and T are plotted as a function of the angle of incidence θ. The experimental values obtained in our laboratory for $\theta = 76°$ were $P = 85\%$ and $T = 10\%$ when the incident light was unpolarized.

Triple reflection polarizers, of the type shown in Fig. 28, have been evaluated by several investigators.[113,114] Gold coated mirrors and cleaved

---

[113] R. N. Hamm, R. A. MacRae, and E. T. Arakawa, *J. Opt. Soc. Amer.* **55**, 1460 (1965).
[114] V. G. Horton, E. T. Arakawa, and M. W. Williams, *Appl. Opt.* **8**, 667 (1969).

biotite plates have been used. The transmission of the polarizer using biotite was reported to be only 1% in the range 110–200 nm for $\theta = 60°$; however, P was 97–99%.

Another source of polarized radiation is the grating monochromator, especially of the Seya–Namioka design (angle of incidence on grating = 35°). For normal incidence monochromators (for example, McPherson 225), the degree of polarization produced by the instrument is small, varying from 0 to 10%.[115,116] With the Seya design, the output can vary from 30 to 80% as the wavelength is scanned from 200 to 50 nm. Figure 30 shows the degree of polarization produced by a $\frac{1}{2}$-m Seya monochromator with a gold coated grating, blazed for 70 nm and with 1200 lines/nm. Grazing incidence monochromators produce little polarization in the range 0–40 nm. Finally, as described earlier, synchrotron radiation is highly polarized. If observed above or below the plane of the orbiting electron, the radiation is actually elliptically polarized. This may be valuable for special experiments.

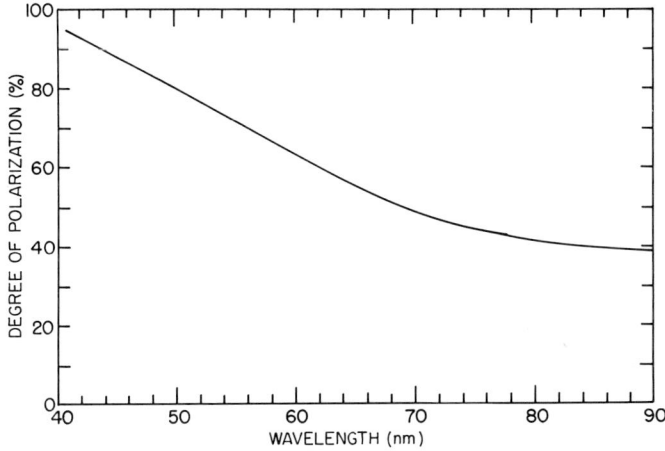

FIG. 30. Degree of polarization of a $\frac{1}{2}$-m Seya–Namioka monochromator with a gold coated grating, blazed for 70 nm, as a function of wavelength.

3.3.6. Detectors

There are many different types of detectors suitable for use in the far uv.[3] The choice is usually dictated by the application. For example, in some applications, the detector should be insensitive to visible radiation (solar blind), and in other applications, detection sensitivity may be more important.

[115] A. Matsu and W. C. Walker, *J. Opt. Soc. Amer.* **60**, 64 (1970).
[116] D. C. Hinson, *J. Opt. Soc. Amer.* **56**, 408 (1966).

However, of great importance is the capability of measuring the absolute photon flux. Thus, in the following sections, a discussion of general detectors will be given, followed by descriptions of absolute standard detectors and secondary or transfer standards.

3.3.6.1. General Detectors. The photomultiplier is one of the most commonly used detectors. It can be used with windows of glass, quartz, sapphire, MgF_2, or LiF to cover the wavelength range from 105 nm to the long wavelength cutoff of the particular photocathode. Photocathodes of cesium or copper iodide are insensitive to radiation of wavelengths longer than about 195 nm, whereas sodium chloride has a cutoff at 148 nm. Photomultipliers with cathodes such as those mentioned above are ideal if there is a large amount of scattered visible light to contend with. For more general applications, a conventional photomultiplier with a standard S11 photocathode and glass envelope can be used with a fluorescent coating (wavelength converter) on the glass window. Sodium salicylate is a very useful wavelength converter. It is sensitive to radiation from the x-ray region to 350 nm, and fluoresces in a band between 370 and 525 nm with a peak that matches the S11 photocathode. Recently, it has been shown to have a relatively flat response down to 10 nm ($\pm 10\%$ variations).[117]

Photodiodes form another type of simple and inexpensive detector for far uv radiation.[3] Virtually any conductor is suitable as a photocathode. However, some materials deteriorate with prolonged contact with air. In these cases, the photodiode must be used with a window. Any metal is suitable for a windowless photodiode. If long term stability of the photodiode is required, then more care must be taken in the selection of the photocathode. Canfield et al.[118] have shown that evaporated aluminum films, subsequently anodized to a thickness of about 15 nm, have about the best long range stability. Figure 31 shows a suitable construction for a photodiode. It is best to insulate both the cathode and anode separately to ground to prevent leakage currents. It is also better to measure the loss of electrons from the photocathode rather than trying to collect all of the electrons. A photoelectric yield curve for aluminum with a natural coating of oxide is shown in Fig. 32. The photoelectric yield of all metals exposed to a laboratory atmosphere are similar to that of Fig. 32. The absolute value of the yield of a particular metal does not vary by more than $\pm 30\%$ of that shown in Fig. 32.[119]

Windowless electron multipliers are about the most sensitive detectors of far uv radiation. They have the advantage that they can be operated either in a photon-counting mode or in a dc mode. The photocathode can be a

[117] J. A. R. Samson and G. N. Haddad, *J. Opt. Soc. Amer.* **64**, 1346 (1974).
[118] L. R. Canfield, R. G. Johnston, and R. P. Madden, *Appl. Opt.* **12**, 1611 (1973).
[119] R. B. Cairns and J. A. R. Samson, *J. Opt. Soc. Amer.* **56**, 1568 (1966).

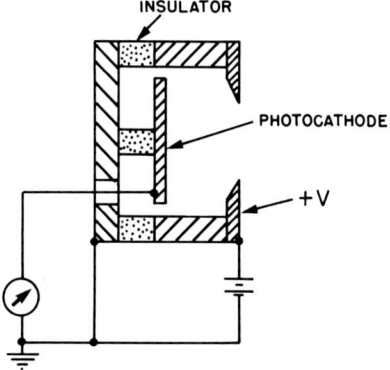

FIG. 31. Photodiode, showing photocathode and anode separately insulated to ground.

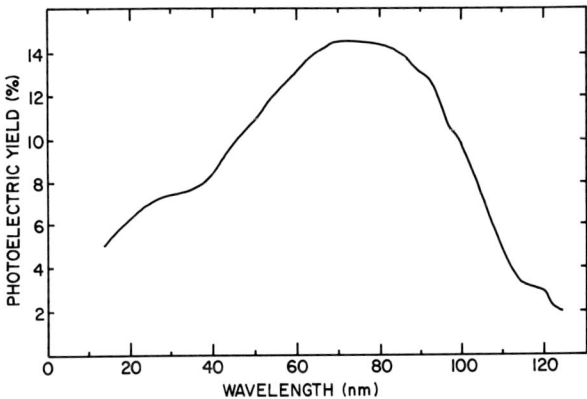

FIG. 32. Photoelectric yield of aluminum (naturally anodized by exposure to air) as a function of wavelength.

metal with essentially zero dark current. The spectral response is then just the same as the metal photodiode. The channel electron multiplier with a continuous dynode structure is ideal for photon counting techniques.[120, 121] For greater sensitivity at shorter wavelengths, the cathode can be overcoated with a thin layer of MgF_2.[122]

Other useful detectors rely on the ionization of gases. These are ion chambers, geiger counters, and proportional counters.[3] The photoionization of a gas is the principle of the absolute photon detector, and will be described in more detail in the following sections.

[120] G. W. Goodrich and W. C. Wiley, *Rev. Sci. Instrum.* **33**, 761 (1962).
[121] J. G. Timothy and L. B. Lapson, *Appl. Opt.* **13**, 1417 (1974).
[122] L. B. Lapson and J. G. Timothy, *Appl. Opt.* **12**, 388 (1974).

Tremendous improvements in the quality of photographic plates and films have been made in recent years. Conventional film can be used when overcoated with a flourescent material (Kodak 103-0 uv sensitized film). However, the more sensitive and commonly used materials are Eastman-Kodak SWR, 104-01, and 101-01; Kodak-Pathé SC.5 and SC.7; and Ilford Q1, Q2, and Q3 films. Recently, the sensitivity of many of these films has been studied.[123, 124] The Kodak 101-01 film appears to have the most constant sensitivity between 80 and 200 nm. It has the added advantage that it is less sensitive to radiation of wavelengths longer than 200 nm.

3.3.6.2. Absolute Standard. The absolute standard detector is the rare gas ionization chamber.[125, 126] The principle of the method is that for radiation more energetic than the first ionization potential of an atom, one ion pair is produced for each photon absorbed by the gas. This holds true for all wavelengths beyond the ionization threshold until the threshold for multiple ionization occurs. However, as long as the ionization efficiency for multiple ionization is known, there is no short wavelength limit to this absolute detector.[127] Thus, the range of the ion chambers starts at the ionization potential of xenon at 102.2 nm and extends, at present, to at least 5 nm. This is the limit of measurements of the multiple ionization of neon. However, the ionization yield of neon only varies from 100% at the threshold of multiple ionization to a maximum of 113.5% at 5 nm. Figure 33 shows the proper design for an absolute standard double ion chamber. When the ion collector plates are of identical lengths, the intensity of the radiation is determined from the relation

$$I_0 = (1/\gamma)[(i_1^2/e)/(i_1 - i_2)], \tag{3.3.14}$$

where γ is the photoionization yield of the gas, i_1 and i_2 are the currents in amperes from the ion collector plates, and e is the electronic charge. The length of the plates, gas pressure, and gas cross section need not be known. A simple but accurate measurement of the two currents is all that is required. The yield γ is unity for measurements down to about 25 nm. To extend the range to shorter wavelengths, the two ion collector plates are connected to form a single ion chamber. The procedure then is as outlined by Samson and Haddad.[127]

Geiger counters and proportional counters can also be used as absolute standards,[128, 129] especially in the soft x-ray region. Provided the photo-

---

[123] W. M. Burton, A. T. Hatter, and A. Ridgeley, *Appl. Opt.* **12**, 1851 (1973).
[124] M. J. B. Fairhead and D. W. O. Heddle, *J. Phys. E.* **4**, 89 (1971).
[125] J. A. R. Samson, *J. Opt. Soc. Amer.* **54**, 6 (1964).
[126] J. A. R. Samson and G. N. Haddad, *J. Opt. Soc. Amer.* **64**, 47 (1974).
[127] J. A. R. Samson and G. N. Haddad, *Phys. Rev. Lett.* **33**, 875 (1974).
[128] D. L. Ederer and D. H. Tomboulian, *Appl. Opt.* **3**, 1073 (1964).
[129] L. Heroux and M. Cohen, ESRO rep. SP-33, p. 125, 1968.

3.3. FAR ULTRAVIOLET REGION

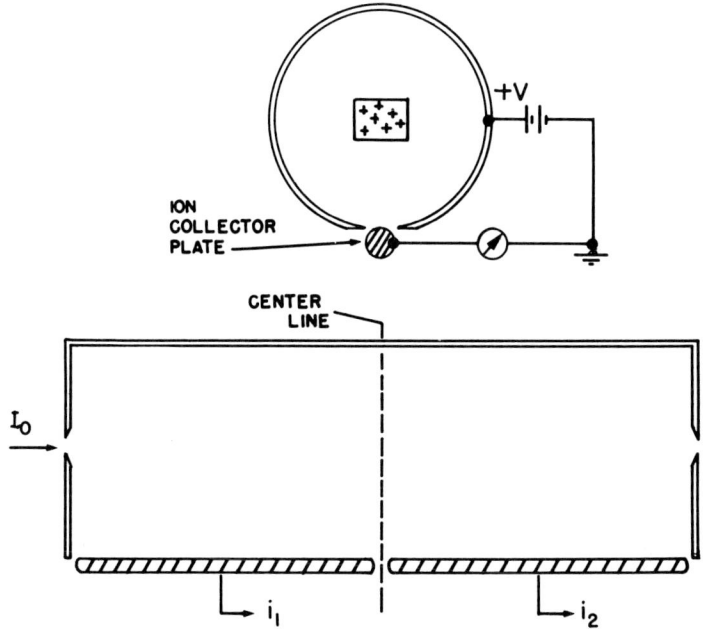

FIG. 33. Double ion chamber for absolute photon flux measurements.

ionization yield of the gas is 100% or greater, one photon will produce one count. However, total absorption of the radiation is required. The need for fragile windows and knowledge of their transmittance in the far uv makes the counters less versatile. They are, of course, much more sensitive than the ion chambers, and find use for very weak sources of radiation.

3.3.6.3. Primary Standard. The thermocouple has long been the primary standard detector for visible radiation. It is not an absolute standard, but must be calibrated with an absolute or primary source of radiation. The response of the thermocouple is proportional to the energy of the photons striking the receiving area. This is true in the visible and infrared spectral regions, but does this "flat" response continue into the far uv, especially since the primary calibration is conducted in a spectral region far from that to be used in the far uv? Canfield et al.,[130] after a very careful study, showed that a calibrated thermocouple detector and a double ion chamber gave similar results at 58.4 nm to within the 3% estimated errors in the measurements. Thus, we have confidence that a thermocouple, once calibrated, can be used over an extended wavelength range well outside that in which it was calibrated.

[130] L. R. Canfield, R. G. Johnson, K. Codling, and R. P. Madden, *Appl. Opt.* **6**, 1886 (1967).

3.3.6.4. Secondary Standards. It is often desirable to have simple detectors calibrated specifically for the spectral range under investigation. These would then be secondary or transfer standards. Although these detectors do not need to have the "flat" response of a thermocouple, they must, of course, be as stable as possible. The National Bureau of Standards has chosen photodiodes as transfer standards.[118] They provide windowless diodes with Al_2O_3 cathodes calibrated from 58.4 to 121.6 nm. For longer wavelengths, diodes with MgF_2 windows and Cs_2Te cathodes are used, and are calibrated from 116.4 to 253.7 nm. The Cs_2Te diodes were developed by Fisher et al.[131] In our laboratory, we have used aluminum photocathodes, naturally oxidized, and calibrated against the single and double ion chambers from 10 to 100 nm. Although these cathodes are relatively stable, it is desirable to have them recalibrated periodically for maximum accuracy.

3.3.7. Wavelength Standards

The standard of wavelength is the orange-red line of Kr 86, which has a vacuum wavelength of 605.780210 nm. All other wavelengths are measured relative to this line. Wherever possible, the measurements are made interferometrically. In the far uv, this is impossible because of the lack of transmission and poor reflectance of all materials. Thus, secondary and tertiary standards are employed. These involve observing the uv lines in a higher spectral order such that they would be in close proximity to first-order standard lines. The grating equation is then used to interpolate between the standard lines. The other technique to obtain far uv standards is by the Ritz combination principle. This is the most precise method. The method consists of determining the relative values of atomic energy levels from observed secondary standards, and then calculating the wavelength of the standard far uv line from the difference between the two levels involved in the transition. In addition, if an accurate measurement can be made of the wavelength produced by a transition from any of the excited levels into the ground level, a whole series of accurate wavelengths can be calculated for transitions from all the known excited levels into the ground state. As an example, Kaufman and Minnhagen[132] have accurately measured the first resonance line of NeI to be $74.37195_5 \pm 0.00002$ nm. From this measurement, they have calculated 23 Ritz standards for the resonance series of neutral neon between 58.2 and 74.3 nm. In a similar fashion, Minnhagen[133] has measured some ground-term combinations of Ar II, resulting in 100

---

[131] G. B. Fisher, W. E. Spicer, P. C. McKernan, V. F. Pereskok, and S. J. Wanner, *Appl. Opt.* **12**, 799 (1973).

[132] V. Kaufman and L. Minnhagen, *J. Opt. Soc. Amer.* **62**, 92 (1972).

[133] L. Minnhagen, *J. Opt. Soc. Amer.* **61**, 1257 (1971).

accurate Ritz standards between 76.2 and 46.5 nm.[134] However, very few standards exist below 50 nm.

Of great help to the workers in this spectral region are the most recent compilations of wavelength put together by Kelly and Palumbo[135] and by Striganov and Sventitskii[136] which cover from 18 to 22 of the elements. A more complete coverage of the elements is given by Moore.[137,138]

3.3.8. Experimental Applications

3.3.8.1. Gases. 3.3.8.1.1. EMISSION SPECTROSCOPY. Measurements of the precise wavelength of the radiation emitted by electrical discharges provide the data necessary to determine atomic energy levels. Traditionally, these discharges were glow, arc, and spark type discharges that were used with different techniques to suppress various orders of ionization. This allowed correct classification of the spectra. The work of Edlen[139] pioneered the study of emission lines in the far uv. The collection of Tables by Moore[137,138] of emission lines and atomic energy levels has proved invaluable to research in this spectral region. However, many lines remain to be classified. The status of our knowledge in the far uv (for a few elements) can be summed up in the monumental collection of wavelengths tabulated by Kelly and Palumbo[135] which covers the literature up to May 1972, but only includes the elements H through Kr. The volume by Striganov and Sventitskii[136] lists the spectra of 22 of the elements most commonly found in plasma physics, and covers a wavelength range from the infrared into the x-ray region.

The interest in solar and thermonuclear physics has stimulated studies of the emission spectra of very hot plasmas. Examples are the theta pinch and stellarator devices, hot laser sparks, and the sun itself. This work has advanced our knowledge of the energy levels of highly ionized atoms. In addition, the emission lines can be used as diagnostic tools to understand the excitation processes taking place in the sun or a laboratory plasma.

---

[134] See also V. Kaufman and J. F. Ward, *Appl. Opt.* **6**, 43 (1967); L. J. Radziemski, Jr., and V. Kaufman, *J. Opt. Soc. Amer.* **59**, 424 (1969).

[135] R. L. Kelly with L. J. Palumbo, "Atomic Emission Lines Below 2000 Angstroms: Hydrogen through Krypton." U. S. Govt. Printing Office, Washington, D.C., 1973.

[136] A. R. Striganov and N. S. Sventitskii, "Tables of Spectral Lines of Neutral and Ionized Atoms." Plenum Press, New York, 1968.

[137] C. E. Moore, Atomic Energy Levels, Nat. Bur. Std. Circ. 467, Vol. I, 1949; Vol. II, 1952); Vol. III, 1958. U. S. Govt. Printing Office, Washington D.C.

[138] C. E. Moore, An Ultraviolet Multiplet Table, Nat. Bur. Std. Circ. 488, Sect. 1, 1950; Sect. 2, 1952; Sect. 3, 1962; Sect. 4, 1962; Sect. 5, 1962. U. S. Govt. Printing Office, Washington, D.C.

[139] B. Edlen, *Rep. Progr. Phys.* **26**, 181 (1963).

3.3.8.1.2. ABSORPTION SPECTROSCOPY. Photographic detection of absorption spectra is by far the most accurate method for determining the energy levels of the absorption structure observed in an atom or molecule. For example, analysis of Rydberg absorption lines of atoms and molecules has led to the most precise determination of ionization potentials of the gases. Surprisingly, no Rydberg series has been identified leading to the first ionization potential of molecular oxygen! A portion of the O_2 absorption spectrum in the vicinity of the ionization threshold is shown in Fig. 34.

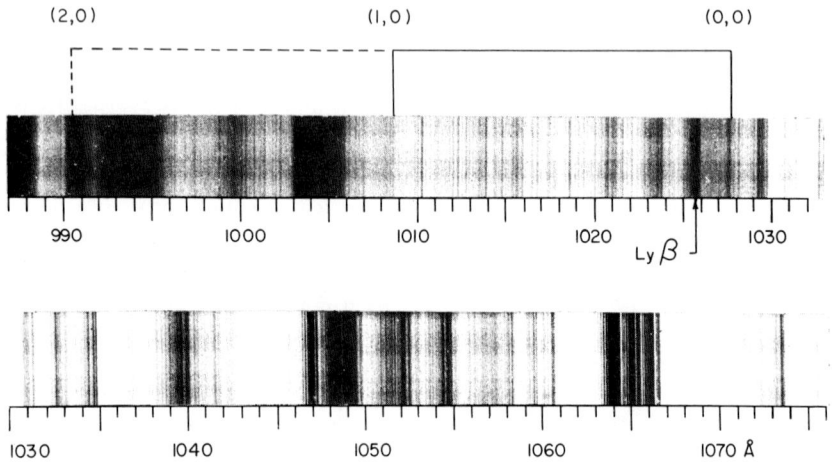

FIG. 34. An O_2 absorption spectrum produced by a Garton type flash tube. The ionization threshold is indicated by the position of the (0, 0) transition. No Rydberg series has been identified [J. A. R. Samson and R. B. Cairns, *J. Opt. Soc. Amer.* **56**, 769 (1966)].

For absorption studies in the region of discrete structure, it is necessary to have a continuum light source. With the availability of synchrotron radiation, it is now possible to photograph the absorption spectrum of any gas in any wavelength region from the infra-red into the x-ray region. Of interest in the far uv is the presence of autoionizing lines and their particular shapes, which give information on the interaction processes between the discrete energy levels and the underlying continuum. Very few atoms in the periodic table have been studied so far at these short wavelengths.

Quantitative absorption cross section measurements are particularly important data in the continuous absorption regions; that is, total absorption cross sections and total photoionization cross sections. Conventional methods for measuring the total absorption cross section σ_t usually involve measuring the intensity of the radiation I_0 before entering an absorption cell and the intensity I transmitted by the cell. Knowledge of the number density of the gas n and the length of the cell L enables the cross section to be determined from the relation

$$\sigma_\tau = (1/nL) \ln(I_0/I). \tag{3.3.15}$$

The detector of the radiation is usually a photomultiplier sensitized to the far uv. Experimental problems are encountered should the absorbing gas fluoresce during the measurements. This is actually quite common in the far uv. Further, if scattered light of longer wavelengths from the monochromator is severe, the photomultiplier may detect this radiation. For accurate work, great care must be taken. However, for measurements at wavelengths shorter than the ionization threshold, the most accurate technique for measuring total absorption cross sections is to use the double ion chamber.[140] The instrument has been described in Section 3.3.6.3, in relation to absolute intensity measurements, and by the author.[3] For use in absorption studies, the cross section is given by

$$\sigma_\tau = (1/nL) \ln(i_1/i_2), \tag{3.3.16}$$

where i_1 and i_2 are the ion currents collected by the first and second collector plates, respectively. The advantages of the double ion chamber are that the detector system is insensitive to the longer wavelength fluorescent and scattered light, and that i_1 and i_2 are measured simultaneously without the usual procedure of filling and emptying the gas cell. Thus, the stability of the light source is not important. The double ion chamber can be used with or without windows since the results are insensitive to spurious absorption of the radiation (caused by gas leaking from the entrance aperture of the ion chamber) before it enters the gas cell. Continuum radiation sources are not so important for absolute measurements of continuum absorption cross sections. In fact, discrete line emission sources allow for a higher overall accuracy in the measurements because there is no uncertainty in determining the scattered light background and the presense of higher-order spectral lines. The total absorption cross sections of the rare gases[140] and most diatomic and triatomic gases[141-143] have been measured down to about 15 nm or less. The total absorption cross section of Xe is shown in Fig. 35. Very little work has been done on polyatomic molecules and the remaining atoms in the periodic table. Accurate experimental data are necessary to check the various theoretical models of the photoionization

[140] J. A. R. Samson, *Advan. At. Mol. Phys.* **2**, 177 (1966).
[141] R. D. Hudson, *Rev. Geophys. Space Phys.* **9**, 305 (1971).
[142] L. C. Lee, R. W. Carlson, D. L. Judge, and M. Ogawa, *J. Quant. Spectrosc. Radiat. Transfer* **13**, 1023 (1973).
[143] L. de Reilhac and N. Damany, *J. Phys.* **32(C4)**, 32 (1971).

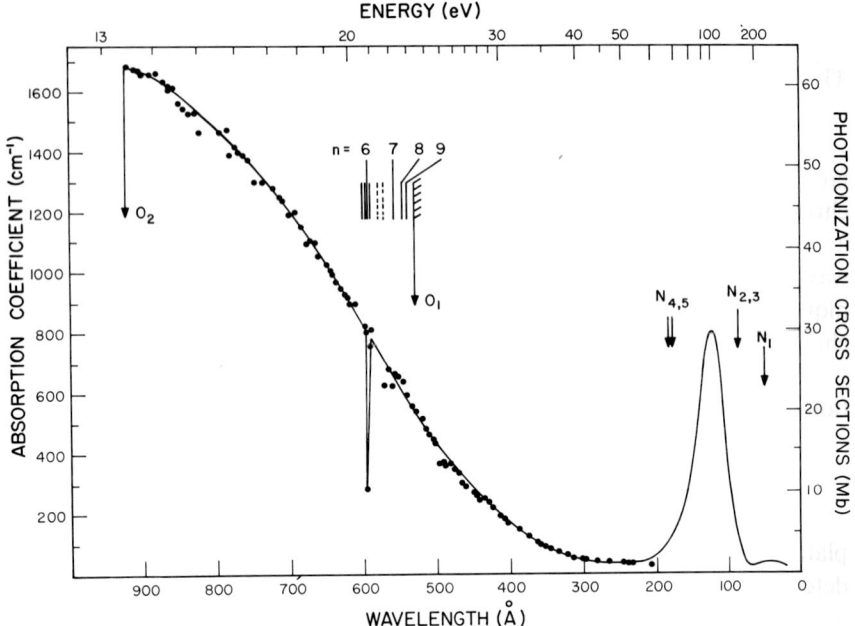

FIG. 35. Total absorption cross section of Xe as a function of wavelength. Structure exists between 50 and 60 nm caused by transitions to states leading to the ejection of an S-shell electron [J. A. R. Samson,. *Advan. At. Mol. Phys.* **2**, 177 (1966)].

process and for practical applications to upper atmospheric physics and astrophysics.

3.3.8.1.3. PHOTOELECTRON SPECTROSCOPY. This section is really a continuation of the discussion on absorption spectroscopy and, in particular, on photoionization. In studying the interaction of far uv radiation with a gas, the ultimate information we desire is the complete description of how the photon energy is shared between all the available absorption channels. For example, what fraction of the absorbed photons produce ionization in a particular vibrational level of a particular electronic state of the ionized molecule? This type of question can be answered by the technique of Photoelectron Spectroscopy. This is a rapidly growing field.[144-147] The principle of the method consists of photoionizing a gas with a monochromatic beam

[144] D. W. Turner, C. Baker, A. D. Baker, and C. R. Brundle, "Molecular Photoelectron Spectroscopy." Wiley (Interscience), New York, 1970.
[145] C. A. McDowell, *Mol. Phys.* **3B**, 847 (1974).
[146] D. A. Shirley (ed.), "Electron Spectroscopy." North-Holland Publ., Amsterdam, 1972.
[147] R. Caudano and J. Verbist (eds.), *Proc. Int. Conf. Elec. Spectrosc.*, *Namur, Belgium* [in *J. Elect. Spectr.* **5**. Elsevier, Amsterdam, 1974].

of photons, and then measuring the kinetic energy spectrum of the ejected photoelectrons. This immediately gives the energy levels of all the states available to radiation up to the energy of the incident photons. Electron energy analyzers with known transmission characteristics are then used to provide accurate transition probabilities for transitions into the specific states of the ion.[148] Figure 36 shows the photoelectron energy spectrum

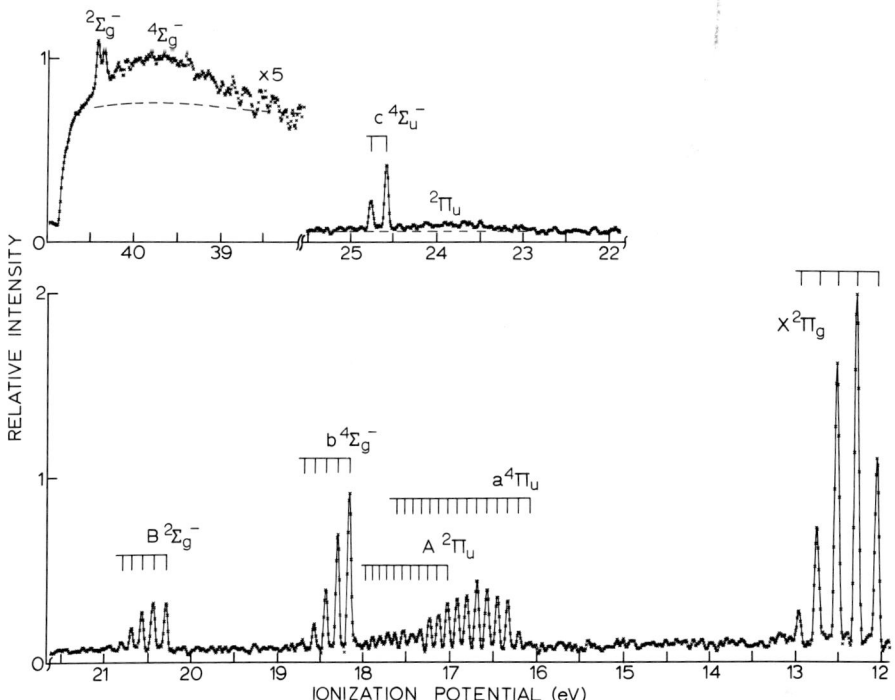

FIG. 36. Photoelectron energy spectrum of O_2 produced by the 30.4-nm line of He II. The spectrum has been corrected for the transmission of the analyzer and was measured at the magic angle and is thus independent of varying angular distributions [J. L. Gardner and J. A. R. Samson, *J. Chem. Phys.* **62**, 4460 (1975)].

obtained from O_2 by 30.4-nm radiation (40.8 eV). In addition to transition probabilities, accurate ionization potentials can be obtained. For example, this method has been used with very high resolution, about 7 mV, to obtain the first ionization potential of O_2 at 12.073 ± 0.001 eV.[149]

3.3.8.2. Solids. Far uv measurements of transmittance, absorptance, and reflectance from solids has increased tremendously of late.[2a] Analysis of the

[148] J. L. Gardner and J. A. R. Samson, *J. Elec. Spectrosc.* **6**, 53 (1975).
[149] J. A. R. Samson and J. L. Gardner, *Can. J. Phys.* **53**, 1948 (1975).

data provide the optical constants of the materials (complex index of refraction) and information on the band structure. In addition, the photoelectric effect has also become a major tool in studying solids. Electron spectroscopy of solids provides a measure of the density of states in the conduction band. For higher photon energies, inner shell electrons can be ejected with energies characteristic of the parent atom, thus providing a tool for chemical analysis of surfaces. This has developed into the field called ESCA, electron spectroscopy for chemical analysis.[150]

[150] K. Siegbahn et al., "ESCA-Atomic, Molecular, and Solid State Structure Studied by Means of Electron Spectroscopy." North-Holland Publ., Amsterdam, 1967.

3.4. Optical Region*

3.4.1. Introduction

This chapter will be concerned with experimental methods of spectroscopy in the near infrared, visible, and ultraviolet wavelength regions; i.e., roughly the regions between 1200 and 200 nm. Above 1200 nm, photographic recording is not applicable, and this has a profound influence on the experimental technique. The borderline is not sharp, though, for other detectors and scanning methods have also found wide application throughout the visible region as well, especially in the past 25 years. The region below 200 nm is different, mostly because the absorption by the atmosphere necessitates the use of high vacua between light source and detector, a requirement that imposes severe restrictions on light sources, spectrographs or spectrometers, and methods of detection. Again, the borderline is not sharp, for conventional spectroscopy can be continued down to approximately 120 nm by using pure nitrogen or helium gas to fill the apparatus (which, by the way, does not relax the requirement of airtightness). Here, the limit is determined by the availability of transparent materials. Occasionally, under favorable circumstances, observations down to about 190 nm have been made in air; on the other hand, absorption sets in above 200 nm, so that vacuum equipment can be of value in the region 240–200 nm as well. As far as the sources are concerned, the borderlines are insignificant. In principle, light sources radiate in all wavelength regions, except that a lower limit is set by the excitation mechanism. Actually, a great deal of the art of experimental spectroscopy is in shifting that limit by constructing light sources and associated energizing equipment suitable to the purpose aimed at.

Roughly speaking, most of the radiation of free neutral and singly ionized atoms is in the visible and ultraviolet regions. This has to do with the fact that the binding energies of the outer electrons in those systems are of the same order of magnitude. More specifically, the $ns-np$ transitions, for the lowest possible value of n, give rise to emission or absorption in the visible and ultraviolet. Incidentally, the same holds for electronic spectra of free molecules ($\sigma-\pi$ transitions). However, there are exceptions to the rule: the resonance lines of the noble gases are in the vacuum region, and those of the

---

* Chapter 3.4 is by P. F. A. Klinkenberg.

heavy alkali atoms are in the infrared. Moreover, the light sources always produce weak spectral radiations in the infrared, many of which tend to be obscured by (pseudo)continua of various origins. The distinction into arc spectra (from neutral atoms) and spark spectra (from singly ionized atoms) is historically connected with the two characteristic sources for the wavelength region in question: the dc arc and the condensed medium-voltage spark discharge. At higher voltages, the second, third, etc., spark spectra appear, but this nomenclature is slightly confusing because the first spark spectrum is sometimes referred to as the second spectrum etc. Therefore, we shall only make use of the unambiguous notation of Roman numerals following the element symbol, I meaning neutral, II singly ionized, etc.

The origin of spectra and the theory of radiation in general are the subjects of Part 2 of this volume. Here we deal with the experimental technique, with an emphasis on the methods used for *research in fundamental spectroscopy*. By this we mean the branch of physics concerned with the internal constitution of atoms, ions, and molecules, including the nuclear effects in electronic states (e.g., hyperfine structure and isotope shift). There are many good textbooks on atomic constitution and spectra, such as those by Herzberg,[1] White,[2] Candler,[3] and Kuhn,[4] while the more advanced theories are covered by the books of Shore and Menzel,[5] Condon and Shortley,[6] Slater,[7] Judd,[8] Wybourne,[9] and others.

3.4.1.1. Definition of Spectral Lines. The great importance of studying spectra in this field lies in the high accuracy with which the wavelengths can be determined, and this is associated with the sharpness of the spectral frequencies as emitted by free particles. According to the Ritz combination principle, each spectral frequency is connected with an electron jump whereby the system falls back from a higher-energy state E_i to a lower one E_f (emission), or is raised from the lower to the higher state (absorption). The validity of Bohr's relation $v = (E_i - E_f)/h$ is not self-evident because it requires the absence of recoil. In the γ region, the effect of recoil is so big that

[1] G. Herzberg, "Atomic Spectra and Atomic Structure." Dover, New York, 1944.

[2] H. E. White, "Introduction to Atomic Spectra." McGraw-Hill, New York, 1934.

[3] A. C. Candler, "Atomic Spectra," Vols. I and II. Cambridge Univ. Press, London and New York, 1937.

[4] H. G. Kuhn, "Atomic Spectra." Longmans, Green, New York, 1962 (2nd ed., 1969).

[5] B. W. Shore and D. H. Menzel, "Principles of Atomic Spectra." Wiley, New York, 1968.

[6] E. U. Condon and G. S. Shortley, "The Theory of Atomic Spectra." Cambridge Univ. Press, London and New York, 1935.

[7] J. C. Slater, "Quantum Theory of Atomic Structure," I and II. McGraw-Hill, New York, 1960.

[8] B. R. Judd, "Operator Techniques in Atomic Spectroscopy." McGraw-Hill, New York, 1963.

[9] B. G. Wybourne, "Spectroscopic Properties of Rare Earths." Wiley, New York, 1965.

3.4. OPTICAL REGION

it delayed the discovery of γ resonance until 1958,[10] but in the optical region it is negligible: The mass of a photon, hv/c^2, is of the order of 10^{-8}–10^{-10} times that of an atom, so that the corresponding frequency change disappears in the natural line width, which is determined in turn by the natural width of the levels. Heisenberg's principle dictates a minimum value for the uncertainty corresponding to the time of observation. This cannot be longer than the natural lifetime of about 10^{-8} sec, except for the ground state and a very small number of excited states, called metastable, which do not take part in the radiation process under laboratory conditions. Now, there is at least one excited state involved in the emission of a spectral radiation, and for that state one has $\Delta E \times \Delta t \geqslant \hbar$, hence, for the line, $h \Delta v \times \Delta t \approx h/2\pi$ or $\Delta v \approx 10^8/2\pi$ Hz. The frequency is connected with the wavelength in vacuo, λ_0, by $v\lambda_0 = c_0$, where c_0 is the vacuum velocity of light. Only the wavelength can be measured directly. It follows that $\Delta v/v = \Delta\lambda/\lambda$ so that, for $\lambda = 500$ nm, one obtains $\Delta\lambda \approx 10^{-5}$–$10^{-4}$ nm for the natural width. It is not much different for other wavelengths because Δt tends to decrease with λ. If E_f is not the ground state, it will be somewhat bigger, but in all light sources there are broadening effects which conceal the natural width completely. Only in very exceptional cases is it possible to reduce the broadenings sufficiently to experimentally verify the prediction of natural width. It might seem that there are no ways to get a sharper line because that is what the unperturbed atoms produce. However, gas lasers can be made to emit at frequencies which are much better defined, without violation of the Heisenberg principle. After all, it is possible, at the expense of intensity, to isolate any spectral frequency with an uncertainty smaller than the natural width; lasers are the only devices to do so at a high-intensity level.

An alternative way to look at artificial line widths is by considering spectrographs and spectrometers as Fourier analyzers. Because the Fourier representation of a wave of finite length contains harmonic components extending to infinity, it is possible to extract a wave train of any desired length from the incident radiation. For all practical purposes, this apparent contradiction can be ignored.

In the optical region, wavelengths are usually measured in air, but the dispersion of standard air is well-known.[11] Until 1960, the basic unit of wavelength was the angstrom, defined by the wavelength of the red cadmium line, 6438.4696 Å in standard air. Because the international unit of length in physics was the Paris standard meter, 1 Å could not be equated to 10^{-10} m which it was intended to be. However, in 1960 it was internationally agreed that the wavelength of an orange krypton line, as emitted by a specified

---

[10] R. L. Mössbauer, *Z. Phys.* **151**, 124–143 (1958).
[11] B. Edlén, *J. Opt. Soc. Amer.* **43**, 339–344 (1953).

light source containing one krypton isotope of mass 86, defines the meter and thus becomes the universal primary standard of length.[12] The reproducibility is about $1:10^9$.[13] Since then, there is no other but historical reason to call 10^{-10} m an angstrom.[14] It is still widely used in spectroscopy and astronomy and will probably continue to be used frequently. Our unit is 10 Å = 1 nm.

It is to be expected that the matter of standard length definition will be reconsidered in the near future. By means of lasers, it has become possible, recently, to determine c_0 as the product of the observed quantities λ_0 and v of one laser line with a precision of $3:10^9$, and with high prospects for improvement to better than $1:10^9$. With both c_0 and v more accurately known than any natural wavelength, it seems likely that the relation $\lambda_0 = c_0/v$ will be used to define the international standard of length.[15] That will not prevent spectroscopists from continuing to calibrate wavelengths against a primary standard wavelength, as they have always done, independent of the definition of standard length.

3.4.1.2. Applications of Optical Spectroscopy. The great importance of spectroscopy to *chemistry* is well-known. In a way, spectrochemistry preceded fundamental spectroscopy. It began on a purely empirical basis, but, in the more advanced stage, it leans heavily on the results derived from the studies of the structures of atoms and ions. The wavelengths are important only for identification purposes, but the emphasis is on the measurement of intensity ratios and, sometimes, absolute intensities. The reader is referred to textbooks in this field such as those of Gerlach *et al.*,[16] Brode,[17] Sawyer,[18] and Harrison *et al.*.[19] Recently, the detection of trace elements has made great progress by the introduction of atomic absorption spectroscopy, which takes advantage of the fact that, in all but very exceptional circumstances, the overwhelming majority of atoms in a discharge find themselves in the ground state. Herewith, the limit of detectability has been improved by a factor of at least 10^3, in suitable cases. The sensitivity of spectroscopic methods combined with their nondestructive character determines their importance in the realm of biology, geology, archeology, medicine, and

[12] *Gen. Conf. Weights Measures, 11th, October 14,* 1960.

[13] E. Engelhard and R. Vieweg, *Z. Angew. Phys.* **13**, 580–596 (1961).

[14] Triple Commission for Spectroscopy, *J. Opt. Soc. Amer.* **53**, 883–893, (1963).

[15] K. M. Baird and G. R. Hanes, *Rep. Progr. Phys.* **37**, 927–950 (1974).

[16] W. Gerlach *et al.*, "Die Chemische Emissions-Spektralanalyse," Vols. I, II, and III. Voss, Leipzig, 1930, 1933, 1936.

[17] W. R. Brode, "Chemical Spectroscopy." Wiley, New York, 1939.

[18] R. A. Sawyer, "Experimental Spectroscopy." Prentice-Hall, Englewood Cliffs, New Jersey, 1946.

[19] G. R. Harrison, R. C. Lord, and J. R. Loofbourow, "Practical Spectroscopy." Prentice-Hall, Englewood Cliffs, New Jersey, 1948.

criminology. Molecules can also be detected by optical band spectra,[20] but in organic chemistry, the most characteristic information is in the infrared region (by absorption spectroscopy) which is outside the scope of this chapter. In industrial applications, very advanced automatic equipment has been developed for checking and controlling the constitution of alloys during the manufacturing process. Conversely, the need for spectroscopic equipment in thousands of industrial plants has been of tremendous importance for fundamental spectroscopy: it created a market for gratings, monochromators, all kinds of auxiliary optics, detectors, and light source accessories, which brought them within easy reach of even small university institutes.

Quite far-reaching has been, and still is, the impact of spectroscopy on *astronomy*. This is based on the facts that stars are light sources, that light travels through space faster than anything else and conserves its spectral character no matter how long the journey takes, that the characteristic frequencies of atoms and ions are constants throughout the universe and over the ages past,[†] and last, but not least, that, through the Doppler shift, light carries information on the radial motion of the sources relative to the observer. And, because the earth's atmosphere is transparent for radiation down to about 300 nm (where absorption by the ozone layer sets in), it so happens that, until recently, *all* information on the chemical composition of the stars, including the sun, and most information regarding the structure of our galaxy, as well as that of the universe, originated in the observation of stellar optical spectra. This is not invalidated by the fact that significant contributions have come from radio astronomy (a branch of spectroscopy too) since about 1950, and that numerous molecular species in interstellar space were discovered by means of observations in the microwave region since 1963.[21] On the contrary, the elimination of the atmospheric obstruction by using space vehicles has opened up the wavelength region below 300 nm to astronomy. Spectacular discoveries have triggered new studies of spectra of highly ionized atoms in laboratories in order to supply the basic knowledge needed to interpret the astronomical spectra. The interest is shifting to the vacuum region because here lies the greatest amount of new information, while space does not favor any particular region.

---

[20] R. W. B. Pearse and A. G. Gaydon, "The Identification of Molecular Spectra." Chapman and Hall, London, 1950.

[21] S. Weinreb, A. H. Barrett, M. L. Meeks, and J. C. Henry, *Nature (London)* **200**, 829–831 (1963).

[†] Actually, these are postulates, but so far no conclusions have been arrived at which cast serious doubts on their validity.

A most significant contribution from optical spectroscopy in ground-based observatories was the discovery, since 1947, of numerous magnetic stars by means of the Zeeman effect, which detects a magnetic field by the characteristic splitting of spectral lines.[22] This has profound implications for the theory of stellar evolution.[†]

The enormous services rendered to astronomy by fundamental spectroscopy have been rewarded by the impetus given to laboratory spectroscopy by astronomers. Some of them grew so impatient to know the results that they started to observe and analyze laboratory spectra and finally became spectroscopists themselves. Outstanding work has been done in this field in a few astrophysical institutes, and astronomers have also contributed to experimental techniques, for instance, by introducing the furnace as a source, improving the art of ruling gratings, and creating a consistent set of secondary wavelength standards. Another gift of astronomy is associated with the fact that, in cosmic sources, matter exists under conditions that are out of reach in a laboratory. So-called forbidden lines, discovered in the solar corona and in nebulae, but also in the aurora, constitute a cornerstone in spectral theory.

In *physics* and *technology*, the use of optical spectroscopy as a diagnostic tool is to be noted. Although in this case it is not such a unique source of information, it has the advantage over other methods that it does not perturb conditions. For instance, the temperature in a hot plasma can be determined most elegantly by measuring the width of spectral lines due to the Doppler effect.[23] The emission of spectral lines is spontaneous so that one only has to provide a window in a strategic place. In cases where there is no spontaneous emission, as in solids at normal temperature, absorption can take over. However, at such high densities, the spectral lines are smeared out into wide regions, except in certain crystals at very low temperatures. Pressure broadening of spectral lines can be used for studying Van der Waals forces and other interactions between atoms and molecules.[24] In the absence of line broadening by external causes, an internal structure can sometimes be observed which is due to nuclear effects in the spectra. This can be employed for determining *nuclear properties*. Practically all ground state spins of stable nuclei have been established by means of optical spectroscopy, while the primary values of most nuclear magnetic moments have also been

---

[22] H. W. Babcock, *Physica* **33**, 102–121 (1967).

[23] T. P. Hughes and R. V. Williams, *Hilger J.* **5**, 19–25 (1958); N. L. Allan *et al.*, *Nature* **181**, 222–224 (1958).

[24] A. Michels and H. de Kluiver, *Physica* **22**, 919–931 (1956); A. Michels, H. de Kluiver, and D. Middelkoop, *Physica* **25**, 163–170 (1959).

[†] For further information, see Vols. 12A, B, and C (Astrophysics) of this treatise.

derived from optical hyperfine structure.[25] These values have been superseded in accuracy by the methods of radio-frequency spectroscopy. On the other hand, many values of nuclear electric quadrupole moments established from optical hyperfine structure and isotope shifts are still the best ones available. The very existence of quadrupole moments and nonspherical nuclei was discovered through optical spectroscopy.[26]

3.4.2. Light Sources

3.4.2.1. Requirements and Limitations. We shall confine ourselves to light sources producing, or vessels absorbing, discrete spectral lines of great sharpness. In practice, this means that the atoms or ions are in a vapor at a pressure of about 1 Torr or lower; i.e., they can be considered as free in so far that the collision probability within the natural lifetime of excited levels is small. Because collisions with atoms of the same kind are much more harmful than those with extraneous atoms, it is often permissible to use quite high foreign gas pressures of 10 Torr or more. In that case, it is the partial pressure of the vapor under study which matters the most. The ideal light source for spectroscopic research would concentrate the energy radiated into narrow spectral lines belonging to the atom or ion of interest, and it would be very bright. These requirements are conflicting, and the great variety of sources reflects the many compromises that can be made in order to satisfy one requirement without greatly jeopardizing others. In certain cases, it is imperative that the relative line intensities are not distorted. This again entails severe restrictions on the source and its operation. The energizing equipment is considered to be a part of the source, but it is usually adjustable, whereby it considerably affects the source conditions. Finally, the nature of the material studied is of great influence.

Basically, the source should perform two functions: (1) it should atomize the material into the vapor state at a pressure of 1 Torr or less, and (2) it should ionize the atoms and excite atoms and ions. The first task is done by heat and/or sputtering; i.e., by the action of colliding heavy particles. The second function is normally carried out by electron bombardment and/or energy transfer in collisions with metastables, the latter being produced again by electron bombardment. In a few arrangements, the two functions are separated spatially, but usually they are performed in the same space, and simultaneously. This is possible as a result of the tremendous difference between the temperature of the radiating discharge plasma and the electron temperature. Even when there is no macroscopic equilibrium, e.g., in intermittent discharges, it is customary to define an effective temperature of the

[25] H. Kopfermann, "Kernmomente." Akad. Verlagsges, Leipzig, 1940. Extended and translated into "Nuclear Moments." Academic Press, New York, 1958.
[26] H. Schüler and Th. Schmidt, Z. Phys. **94**, 457–468 (1935).

vapor which corresponds to the random motion of atoms and ions. The electrons, however, are accelerated by electric fields and lose little energy in collisions with atoms and ions, unless these are excited. Most collisions are elastic and tend to randomize the directions of motion of the electrons only, with the result that the electron gas has an effective temperature which is easily some 10^3 times as high as the plasma temperature. Now, the ionization energies are in the range from 4 to 15 eV for $Z > 10$, while, at $T \approx$ 75,000 K, the mean kinetic energy $3kT/2$ is equivalent to 10 eV. On the other hand, if the effective temperature of the vapor were only 1500 K, we should have serious problems of containment, and so the big difference between vapor and electron temperature is indeed essential for the operation of most light sources. In some exceptional cases, viz., flash pyrolysis and laser produced plasmas, both vaporization and ionization/excitation are caused by light. It will also be clear that vaporization is unnecessary when the material is a gas. The noble gases are particularly easy to handle; the atomic spectra, as well as the band spectra of diatomic molecules such as hydrogen or nitrogen, can also be observed in ordinary Geissler tubes, even at low temperatures, while some metals such as mercury and sodium will develop sufficient vapor when gently heating the tube externally.

For all these reasons, a classification of light sources on the basis of one property is bound to be capricious with regard to other qualities, while the historical order would do no justice to physical considerations. Because it is most important to know what one can do with a source, we shall treat them in the order of increasing excitation.

3.4.2.2. Flame. This is the oldest (artificial) light source in use for spectroscopy. Excitation is mainly by heat, and heat is produced by chemical reactions. The flame is a complicated phenomenon; interest in its regime has revived since it has become possible to use laser light for probing. Metals can be introduced by spraying solutions into the flame, where they are atomized and excited to produce frequencies characteristic of the neutral state. It is used in spectrochemical analysis, but, nowadays, mostly in the form of an absorbing medium (atomic absorption method; cf. Section 3.4.1.2).

3.4.2.3. King's Furnace.[27] This is another example of thermal excitation. It has the advantage over the flame in that the temperature is rather uniform over a large volume and can be varied at will. In the conventional form, a graphite tube is heated by passing a strong current through the thin walls or through a coil around the tube. The material to be studied is placed inside the tube, usually in a noble gas atmosphere to prevent the vapor from diffusing rapidly to the ends. Temperatures near 3000 K can be reached.

[27] A. S. King, *Astrophys. J.* **28**, 300–314 (1908).

This corresponds to an average kinetic energy of the atoms of 0.39 eV only, but owing to the long tail in the Maxwellian velocity distribution, atoms get excited. Even ionizing collisions take place, and subsequent excitation leads to emission of the strongest lines of the singly ionized element. These are completely absent at a temperature of 2000 K. It is the study of relative intensities as a function of temperature, and particularly the determination of "first appearance" of a spectral line when the temperature is gradually increased, which has given rise to the famous temperature classification.[28] This has been of great value, especially for establishing the structure of neutral atoms of low ionization potential such as the rare earths. The temperature classification not only distinguishes, with certainty, between lines from atoms and ions, but also permits the categorization of spectral lines within one stage of ionization.

Because the vapor is in thermal equilibrium, the relative population of the levels obeys a Boltzmann distribution. Hence, the furnace lends itself remarkably to the determination of oscillator strengths, both in absorption[29] and emission experiments.[30] Absolute oscillator strengths can be established when the vapor density is known. Uncertainties arise from the measurement of temperature in the main body of the vapor, the end effects, and the relationship between temperature and saturation density. In most cases, absolute values have been derived from measurements of relative oscillator strengths against a line whose absolute value is known from other experiments. A drawback of the light source is the presence of a strong continuous radiation from the walls, outshining the weaker spectral lines.

The furnace technique for absorption experiments has been modernized recently by introducing induction heating which allows a much more economical use of materials.[31] The containment problems can be solved effectively by exploiting the heat-pipe principle.[32]

3.4.2.4. The Microwave-Excited, Electrodeless Discharge Tube. This is a very important and almost universally applicable source of high-intensity light from neutral and singly ionized atoms. It was invented in 1950 when the discovery was made that the "clean-up" effect becomes negligible at the high frequencies obtained by means of magnetrons.[33] Therefore, its introduction is connected with the development of radar devices during World War II and the commercial applications thereafter.

---

[28] A. S. King, *Astrophys. J.* **37**, 239–281 (1913), followed by numerous articles in the same journal up to 1945.
[29] R. B. King and A. S. King, *Astrophys. J.* **82**, 377–395 (1935).
[30] J. Aarts and G. Bosch, *Physica* **30**, 1673–1681 (1964).
[31] F. S. Tomkins and B. Ercoli, *Appl. Opt.* **6**, 1299–1303 (1967).
[32] C. R. Vidal and J. Cooper, *J. Appl. Phys.* **40**, 3370–3374 (1969).
[33] W. F. Meggers and F. O. Westfall, *J. Res. Nat. Bur. Std.* **44**, 447–455 (1950).

At present, the most useful device is a therapeutic instrument generating an intense electromagnetic field at a frequency of 2450 MHz. Gas contained in a closed quartz tube at a pressure of a few Torr will be strongly excited in this field and, in the absence of "clean-up," continue to radiate intense light for an indefinite time, provided the walls have been thoroughly cleaned before closing the tube. It appears that the correct condition for the electrons to collect energy from the field is reached when the frequency of (elastic) collisions with atoms is comparable to the field frequency. Spectral lines are sharp and can be made sharper by cooling. A mercury lamp can be cooled with running water, while the noble gases will maintain a discharge at lower temperatures. The author succeeded in operating an electrodeless neon lamp immersed in liquid nitrogen, in a 150 MHz field generated by means of a Lecher system connected to a double tetrode oscillator tube; the lines became sharp enough to observe the ^{22}Ne satellite in the natural isotopic mixture by means of an interference instrument. However, most metal spectra require higher temperatures.

An important advance was made in 1953 when halides were introduced for handling refractory metals,[34] which was followed in 1956 by the development of a procedure for preparing sealed-off quartz tubes containing extremely pure halides of rare earths, in quantities of 1 μg to 1 mg, plus carrier gas of a few Torr.[35] In the magnetron field, the tubes first emit the noble gas lines, but metal lines appear in a few minutes and increase in brightness at the expense of the gas lines which finally disappear. If the compounds are not volatile enough, as in the case of rare earth halides, additional heat has to be applied up to about 850°C, and then extreme care has to be taken in cleaning the tubes before sealing off (baking at 1000–1100°C in high vacuum during many hours). This type of source, about 8 cm long and 8 mm o.d., has been of tremendous importance for the study of many spectra of neutral and singly ionized atoms. As an illustration, we mention Hf I and Hf II. Analyses of these spectra had been available prior to 1955, based on a total of 2400 observed lines. The analyses were considered as fairly complete by experts. However, by means of the new light source, the number of observed lines was raised to 6200 (partly due to the absence of obscuring band structures)![36] Obviously, this put the situation on an entirely different basis, and a few years of analysis work were needed to cope with it.

Still greater improvements were obtained for rare earths, and, more specifically, the neutral rare earths. In all light sources, except the furnace, these spectra were at a disadvantage as a result of the low ionization po-

[34] C. H. Corliss, W. R. Bozman, and F. O. Westfall, *J. Opt. Soc. Amer.* **43**, 398–400 (1953).
[35] F. S. Tomkins and M. Fred, *J. Opt. Soc. Amer.* **47**, 1087–1091 (1957).
[36] C. H. Corliss and W. F. Meggers, *J. Res. Nat. Bur. Std.* **61**, 269–324 (1958).

tentials.[37] But in the electrodeless discharge, the excitation can be made just as gentle as in the furnace, while the intensity is at least a thousand times as high and the background is much fainter. As contaminations, one finds Li, Na, and Si from the walls, and the halogen lines, but, as a rule, the metal spectrum is very pure while the lines are as sharp as one can expect at the temperature maintained. A small portion of a Tb spectrogram, obtained in this way, is shown in Fig. 1, together with the iron arc lines, to show the line density difference. Analysis proves that the spectrum is fully developed;

FIG. 1. Spectrum of terbium, between 393.0 and 390.0 nm, from a microwave-excited electrodeless discharge by means of a large grating mounting; iron arc lines are in the central strip.

i.e., that the energy states are occupied according to statistical weight, not according to their energy. If the power is increased, considerable self-absorption leads to broadening of the strong lines and leveling of the observed line intensities. Even self-reversal occurs in extreme conditions.

Correctly prepared tubes can last for dozens of hours under full power, but they finally succumb to various imperfections. Either the clean-up effect causes the metal to disappear, or the halogen pressure builds up too much. The first indication of the latter condition is refusal to ignite without applying a hf leak detector and/or precooling. Once started, the discharge will be constricted into a narrow filament in extreme cases. Then it becomes high time to replace the lamp. Nonetheless, the technique is superior for getting the most out of microquantities of material, as is essential when working with separated natural isotopes or with artificially produced ones.

A smaller size tube (about 3 cm length and 3 mm o.d.) has been developed which, apart from being applicable in a small space (between the poles of a magnet, for instance), is interesting in its own right. First, it does not require

[37] A. G. Shenstone, *Physica* 12, 581–588 (1946).

a noble gas filling; the discharge ignites on a little excess of halogen, usually iodine. Secondly, there is no need for auxiliary heating. Thirdly, the conditions can be varied over a broad range by varying the power. This determines the temperature, thereby the pressure, and consequently the free path. The result is a drastic change of the spectrum, with neutral atoms dominating at high power and singly ionized atoms at low power. However, under no conditions are lines from higher-ionization stages observed in a continuous microwave-excited electrodeless tube. From Fig. 2 it may be seen that a very clear separation of Tb I and Tb II lines can be made from comparison of spectrograms made under widely different conditions. This can never be made with the original large size tube. In addition, at full power the small

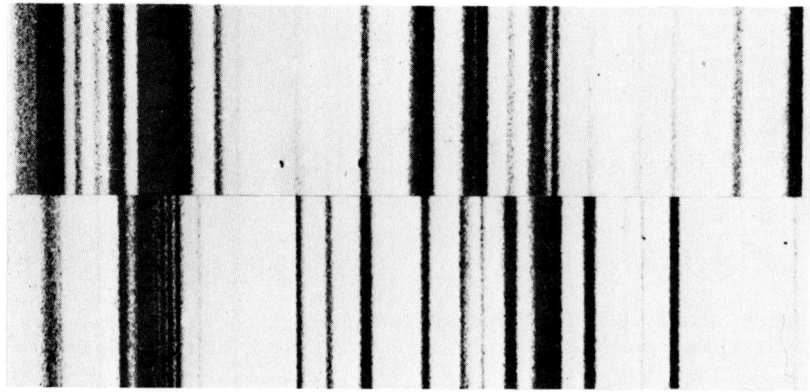

FIG. 2. Spectra of terbium, between 359.45 and 358.93 nm, obtained by means of a large grating mounting under different conditions of excitation in a microwave-excited electrodeless discharge. Top: low-power, Tb II conditions; bottom: high-power, Tb I conditions.

lamp has a considerably higher surface brightness at the expense of line quality, though not dramatically so, while self-reversals become rather frequent. The large noble-gas-filled tubes are superior in line sharpness in any case.

For studies of the Zeeman effect, the small lamps are indispensable. The discharge is perturbed by the magnetic field, and, for stability, it is usually necessary to use a well-adapted microwave cavity. The tubes show a rather capricious behavior, but it is possible to maintain the discharge for hours without interruption, in fields up to 2.4 T.[38] Occasionally, electrodeless tubes have been run at much higher fields in the M.I.T. Bitter solenoid (up to 8 T) when the tuning could be accurately controlled.[39]

[38] P. F. A. Klinkenberg, *Physica* **32**, 1113–1147 (1966).
[39] J. W. Lindner and S. P. Davis, *J. Opt. Soc. Amer.* **48**, 542–547 (1958).

Continuous electrodeless discharges have been in use at much lower frequencies ever since radio tubes became commonplace, especially for gaseous materials. They operate also at extremely low pressures, down to 10^{-3} Torr, and favor the observation of molecular spectra, although several atomic spectra have been thoroughly investigated by means of such sources (e.g., Hg, the noble gases, Br, Zn, etc.), both in the neutral and in the singly ionized state. Still older, originating with Hittorf in 1884, is the *pulsed* electrodeless discharge; the vessel is surrounded by a low-resistance coil through which a battery of condensers is discharged suddenly. In this way, the III and IV spectra are excited along with I and II. After gaining popularity in the twenties, it became almost forgotten, but was rehabilitated as a result of modern electronic equipment, with perfect control of pulse height and duration.[40, 41]

3.4.2.5. Geissler Tubes. These were introduced around 1860 and are still well-known in various forms. They have internal electrodes and are easily excited by means of a 2–10 kV transformer when filled with a permanent gas or a volatile material like Hg at a pressure of a few Torr; the current is 1–50 mA. The part of the discharge observed is the positive column, although it is often referred to as a glow discharge in the literature. A capillary section enhances current density and brightness without undue broadening of lines. With permanent gases, the lines can be sharpened by cooling. This is applied in the ^{86}Kr standard lamp[13] mentioned in Section 3.4.1.; it is operated on dc.

Because of the high conductivity of the gas or vapor, the voltage drop in the positive column is very small. Consequently, the spectrum of neutrals dominates. However, with a capacitor in parallel with the lamp, the II spectrum can be enhanced considerably. Inclusion of metals of low melting point, such as alkalis, causes their I and II spectra to appear, and even the III spectrum when a capacitor is used. It may then be desirable to gently heat part of the tube. Geissler tubes were the forerunners of low pressure discharge tubes now available as spectral lamps which have no constriction and, consequently, do not need such high voltage. When containing metals, they are provided with an enveloping tube to allow the overall temperature to rise sufficiently for maintaining the correct pressure. They are very useful for isolating good quality monochromatic radiations of reasonable luminance and also for purposes of wavelength calibration.

3.4.2.6. DC Hollow Cathode Discharge Tube. This uses the negative glow as a source of light. For different types and modes of operation, the reader is referred to the monograph by Tolansky.[42] The principal advantage

---

[40] L. Minnhagen and L. Stigmark, *Ark. Fys.* **8**, 471–479 (1954); **13**, 21–36 (1957).
[41] L. Minnhagen, B. Petersson, and L. Stigmark, *Ark. Fys.* **16**, 541–544 (1960).
[42] S. Tolansky, "High Resolution Spectroscopy." Methuen, London, 1947.

of this source for hyperfine structure research is the extreme sharpness of the spectral lines. This is due to (1) the absence of electric fields in the hollow cathode plasma, (2) the low partial pressure of the metal vapor, and (3) the intense cooling possible, especially of the all-metal devices introduced by Schüler. The low partial pressure would result in low intensity in any other light source, but, due to the particular cathode shape, the efficiency is exceptionally high so that the lines are not only sharp but also bright. For other purposes, the hollow cathode glow has the advantage of considerable depth which promotes the observation of rare processes. This, in combination with field-free emission, has made it the best source for observing higher series members.

The excitation is brought about partly by electrons and partly by metastable carrier gas atoms; which process prevails depends on conditions and the properties of the material studied. In some cases, the noble gas can be dispensed with,[43] but usually it plays an important role. Because of energy resonance phenomena and also because of the sputtering action, the nature of the noble gas can be critical; the excitation of certain groups of levels may be favored. With helium, the III spectrum is excited, but this was also found occasionally in the regime of pure electron excitation.[43] When argon or a heavier noble gas is used, the excitation can be restricted to the neutral atoms only if the current density is properly adjusted. To the author's knowledge, there is no report of a IV spectrum being observed by means of a dc hollow cathode discharge. Various authors have made extensive studies of I- and II-line intensity ratios as a function of pressure and/or current density with different carrier gases,[44, 45] but there is no unanimity with regard to the results, probably because the number of parameters is so large and some of them are ill-defined.

The hollow cathode source has also been adapted to studies of the Zeeman effect.[46] The discharge is perturbed by the magnetic field, but fields up to 1.2 T have been applied successfully under conditions favorable for observation of lines from neutral rare earth metals[47] (Fig. 3).

Originally, elaborate vacuum systems with gas circulating and purifying devices were needed, although the circulation could be stopped most of the time when the cathode was cooled by means of liquid nitrogen. With the advent of powerful getters, it became possible to use sealed-off tubes. Especially activated uranium was reported to be very effective as such a getter.[48]

[43] A. G. Shenstone and J. T. Pittenger, *J. Opt. Soc. Amer.* **39**, 219–225 (1949).
[44] K. D. Mitchell, *J. Opt. Soc. Amer.* **51**, 846–853 (1961).
[45] H. M. Crosswhite, G. H. Dieke, and C. Salmon Legagneur, *J. Opt. Soc. Amer.* **45**, 270–276 (1955).
[46] R. A. Fisher and A. S. Fry, *Phys. Rev.* **56**, 675–677 (1939).
[47] P. F. A. Klinkenberg, *Physica* **12**, 33–48 (1946).
[48] G. H. Dieke and H. M. Crosswhite, *J. Opt. Soc. Amer.* **42**, 433 (1952).

3.4. OPTICAL REGION

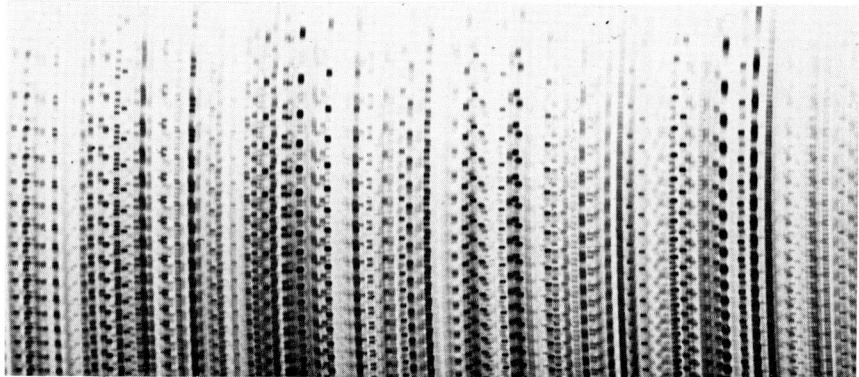

FIG. 3. Zeeman effect in the spectrum of neodymium, between 477.5 and 441.3 nm, obtained by means of a Lummer plate crossed with a three-prism spectrograph from a dc hollow cathode discharge in a magnetic field of 1.109 T; only π-polarized light admitted.

More recently, sealed-off hollow cathode sources have become available commercially for the purpose of atomic absorption spectrometry,[49] but here the only requirement is the emission of sharp *resonance* radiation with sufficient intensity. The hollow cathode itself is manufactured out of the metal to be identified, and a great variety of such cathodes is available. Otherwise, the hollow cathode discharge tube is known for its economy and universality. Even when the material is deposited inside an aluminum cathode cavity in mg quantities, a bright metal spectrum can be obtained for many hours; when oxides or other compounds are used instead, the metal spectrum will develop gradually as the compound is being reduced in the discharge. Special constructions have been made for extracting as much light as possible from small samples.[50] In its original Paschen design, however, it is more wasteful. This type has been used for refractory metals and for metals with a very low sputtering coefficient (like Al), especially for multiplet analysis of II and III spectra. The power consumption is much higher and the (graphite) cathode becomes very hot, so that conditions for observing hyperfine structure are less favorable.

3.4.2.7. The DC Arc. This is one of the oldest and most profoundly studied light sources. The fundamental difference from the gas discharges treated above is that thermal emission of electrons from the cathode is required. This makes it possible to sustain a very high current density which in turn provides the necessary heat to the electrodes through bombardment.

[49] A. Walsh, *Spectrochim. Acta* **7**, 108–117 (1955); R. Lockyer, *Hilger J.* **5**, 55–60 (1959).
[50] H. Schüler and H. Gollnow, *Z. Phys.* **93**, 611–619 (1935); O. H. Arroe and J. E. Mack, *J. Opt. Soc. Amer.* **40**, 386–388 (1950).

The electrode material evaporates quickly, and the vapor emits a reasonably clean spectral radiation in the arc gap, with only some continuous radiation from incandescent particles, mainly in the red and infrared regions. The arc can be operated at atmospheric pressure (open arc) or at reduced pressures down to about 10 Torr, at currents from 2 to 100 A. At low pressures and current, the spectral lines are sharp enough to resolve the widest hyperfine structure patterns. The spectral lines belong to the neutral and the singly ionized state mainly, but, near the cathode, a region of III emission may develop so that, in a stigmatic image, it can be recognized by the pole effect. The intensity distribution over the length of the gap in general is a means for separating the ionization stages. It also depends on pressure and the nature of the surrounding gas. Owing to the radial temperature gradient, an arc source offers favorable conditions for strong self-absorption and self-reversals, even in lines of the singly ionized state. Special high-current devices have been built to exploit this effect.[51] A thorough theoretical and experimental study of line shapes and intensities under the influence of strong self-absorption was published in 1948,[52] and a large amount of information on the arc phenomenon is to be found in the article on electric arcs and thermal plasma in the "Encyclopedia of Physics."[53]

For some metals with low melting points and especially for mercury, the electrodes have to stay cool, and the required emission of electrons is a field emission concentrated on small spots at the cathode. The current may then be reduced to less than 1 A if an inductance in series with the lamp is used to flatten the fluctuations. In order to maintain the pressure needed for arcing, a noble gas can be added at a few Torr, and in the case of mercury, cooling by an air blast or by running water can be applied, whereby the spectral lines become extremely sharp. This kind of source constitutes a transition from the ordinary arcs to the low-pressure gas discharges treated earlier.

For spectrochemical purposes, the arc is popular because it has a high capability of bringing out trace elements, while the materials can be introduced in any chemical compound. Porous graphite electrodes can be used which are impregnated with solutions of the compounds, or the electrodes may contain the material in a drilled hole. Distillation can help to enhance sensitivity; in the case of a uranium matrix, for example (a very unfavorable case because the spectrum possesses thousands of lines), it has been possible, by means of the carrier distillation method, to detect impurities at concentrations down to 0.05 ppm (Ag) and (in the most unfavorable case) 50 ppm (P).[54]

[51] J. Sugar, *J. Res. Nat. Bur. Std.* **66A**, 321–324 (1962).
[52] R. D. Cowan and G. H. Dieke, *Rev. Mod. Phys.* **20**, 418–455 (1948).
[53] W. Finkelnburg and H. Maecker, *Ency. Phys.* **22**, 254–244 (1956).
[54] B. F. Scribner and H. R. Mullin, *J. Res. Nat. Bur. Std.* **37**, 379–389 (1946).

3.4.2.8. The Intermittent Arc (German: *Abreiszbogen*; French: *trembleur*). This is a source operating on dc, but with a voltage too low to maintain a steady arc. The electrodes have to be brought into contact to strike it again, and this is done in rapid succession so that the appearance approaches that of a steady source. The prevailing condition is that of the arc in the stage of ignition which is more violent in excitation and approaches the spark. It can also be run at lower pressure than an arc and the discharge plasma is cooler, giving rise to sharper lines and practically no background. The source requires little space so that it can be mounted between the poles of an electromagnet for Zeeman effect investigations, with the discharge running parallel to the field. In strong fields, the excitation is increased; while, in the zero field situation, the lines of the singly ionized atoms dominate and IV lines are practically absent, in strong fields, the III and IV spectra become outstanding while the I spectrum completely disappears[55] (Fig. 4).

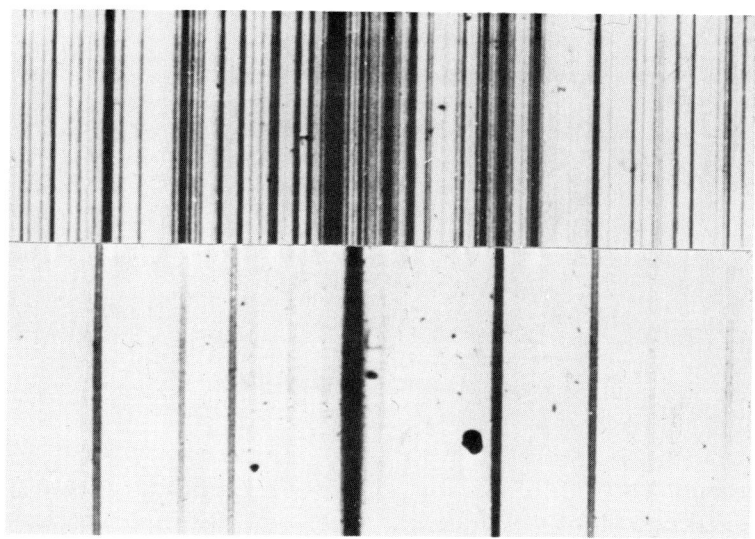

FIG. 4. Spectrum of thorium, between 271.5 and 267.3 nm, obtained by means of a quartz-prism, Littrow spectrograph from an intermittent arc discharge in a magnetic field of 3.8 T (bottom) and in zero field (top). Magnetic enhancement of higher ionizations; in the center is Th IV, $\lambda = 269.397$ nm; the other strong lines in the bottom picture belong to Th III.

The overall intensity is low for many materials; it can be improved when some carrier gas (10 Torr) is admitted, without noticeable loss in sharpness.

3.4.2.9. The Spark. At normal pressure, the spark is the simplest device for obtaining light from ionized atoms. It requires ac high voltage (>10 kV)

[55] P. F. A. Klinkenberg, *Physica* **16**, 185–197 (1950).

and a capacitor parallel to it in order to concentrate the energy into the discharge. This oscillatory circuit may contain an adjustable coil and has a characteristic period $\tau = 2\pi\sqrt{LC}$. In a typical case, $L = 50$ μH and $C = 10^{-2}$ μF, giving $\tau = 4.44$ μsec. If the high voltage is obtained from a transformer whose primary coil is connected to the mains, the frequency is 50–60 Hz and the time interval between successive breakdowns 1/100–1/120 sec. So hundreds of oscillations take place between successive chargings, enough to damp out and restore the initial conditions before every breakdown. If needed, this is promoted by means of an air blast. Usually integrated light is used, but when the spectra are studied as a function of time (time-resolved spectroscopy), excitation is seen to decrease rapidly after the breakdown. In integrated light, the spectrum of neutrals is faint or absent while ionizations up to four times are commonly observed. Because the conditions also depend on the location in the spark gap, a single integrated spectrogram gives information on various ionization stages when the image is stigmatic. In Fig. 5, it is possible to discriminate between Th I, Th II,

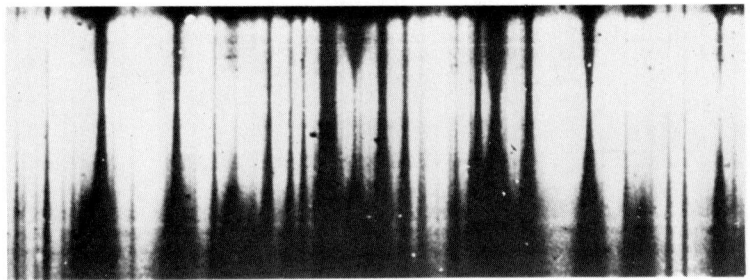

FIG. 5. Spark spectrogram of thorium, showing pole effect, in the same region as Fig. 4. Discrimination of II, III, and IV lines is possible.

Th III, and Th IV. It is also seen that lines are diffuse, especially at the electrodes, and this demonstrates the presence of high electric fields; comparison with the intermittent arc spectrum indeed proves that there are considerable shifts in the spark, due to the quadratic Stark effect.[56]

The electrodes remain relatively cold and the electrons are forced into the gap by the high electric fields, the electron temperature reaches several times 10^4 K, and there is a continuous background in the spectrum due to bremsstrahlung. The mechanism is not fully understood, but thorough experimental and theoretical investigations in the Physics Institute of the USSR Academy of Sciences have revealed important aspects of the spark

[56] P. F. A. Klinkenberg, *Physica* **16**, 618–650 (1950); compare also A. Bardócz, U. M. Vanyek, and T. Voros, *J. Opt. Soc. Amer.* **51**, 283–288 (1961).

plasma; the spark channel is formed in an explosionlike process, causing a shock wave that is responsible for ionization.[57] The breakdown voltage is mainly determined by the gap length, the nature and the pressure of the gas atmosphere, and the electrode shapes, but even under constant conditions it is a very erratic phenomenon. In order to stabilize the spark, e.g., for spectrochemical work, various devices have been developed to trigger it, the simplest of which is a mechanical interruptor in the form of a toothed wheel rotating synchronously with the input voltage and closing an auxiliary gap only at the peak voltage (Feussner), while more sophisticated devices use electronic switching by means of thyratron or ignitron tubes and electron injection into the spark gap by means of an auxiliary electrode.

In time-resolved spectroscopy, the demands on stability are particularly high and, moreover, an exact synchronization of successive observations is necessary so that a constant time-lapse after ignition is ensured. By means of rotating mirror methods,[58] one gets down into the 5μsec range, while the history of each line during the period of observation is recorded on one photographic plate. Synchronization is most conveniently achieved when a lightbeam reflected by the mirror itself in a certain position serves as trigger. If detection is electronic, the sensitivity of the detector can be made to vary synchronously with the spark ignition. In a thorough study of spark conditions by means of Fe lines,[59] a time resolution of 0.5 μsec was reached that way. The adjustable time delay made it possible to pick any desired discharge phase and collect signals for an indefinite period of time.

The initial current in a spark will be proportional to $\sqrt{C/L}$; this will not only determine the total intensity, but also influence the distribution over the various stages of ionization. However, the violence is greatly determined by the peak voltage. There is no use in applying a higher tension if the break-down voltage is not increased. This is about 10 kV in normal air for a gap length of 8 mm, and it is impractical to increase it much above that value. The break-down potential can be increased by using a noble gas, increasing the pressure, or evacuating the spark chamber. The vacuum is an insulator because there are no particles to transport electricity. Consequently, when the pressure is decreased, the spark first becomes less violent; between 40 and 0.01 Torr sparking is not possible at all, but a glow discharge obtains emitting gas lines only, while below 10^{-3} Torr the break-down potential rises so steeply that one has to reduce the gap to about 1mm and still apply

---

[57] S. L. Mandelstam, *Spectrochim. Acta* **11**, 245–251 (1957); *Colloq. Spectrosc. Int. 4th Amsterdam, 1956*.

[58] Á. Bardócz and F. Varsányi, *Nature (London)* **177**, 222–223 (1956); *Z. Naturforsch.* **10a**, 1031–1032 (1955); Á. Bardocz, *Z. Angew. Phys.* **9**, 82–88 (1957).

[59] D. W. Steinhaus, H. M. Crosswhite, and G. H. Dieke, *J. Opt. Soc. Amer.* **43**, 257–270 (1953); *Spectrochim. Acta* **5**, 436–451 (1953).

some 60 kV in order to produce a spark. This is the so-called "hot spark," emitting light from highly ionized atoms only (spectra X–XXV). For a long time, it has been very difficult to cover the ionizations intermediate between those in the ordinary and the hot spark, until the sliding spark was invented; this will be discussed in Section 3.4.2.11. The hot spark is outside the scope of this chapter because it is exclusively used in the vacuum ultraviolet. (The reader is referred to Chapter 3.3.) Finally, it should be mentioned that the spark can also be employed when materials are volatile or rare so that electrodes are not readily available. Spark discharges between a platinum rod and a cup containing a solution of a compound will reveal lines of the elements in the compound, and porous graphite electrodes may be used which are saturated with such a solution. Copper electrodes have been used to hold thin layers formed by evaporation of a solution.[16] The copper spark method was developed into a universal spectrochemical method of high sensitivity and reliability.[60] It was also used to record the spectra of some transuranic elements for the first time.[61]

3.4.2.10. *Pulsed Hollow Cathode Tube.* In order to achieve both excitation of higher ions and good line quality the pulsed hollow cathode tube has come into use.[62] This resembles the oldest type of hollow cathode source, the Paschen tube,[63] much more than the models developed later for hyperfine structure research by Schüler. Pulsed discharges in a hollow cathode were shown earlier to be capable of emitting III lines so sharp that hyperfine structure can be resolved, e.g., in Sc III, La III, Y III, and Bi III[64]; the claim that Pr V was observed in the same source has been disproved by the analysis of Pr V.[65]

In modern applications, the energizing unit is essential. It is similar to the units operating sparks except that a rectifier is needed and a set of coils to prevent oscillations from developing.[66] Much attention is given to problems of controlling the pulse profile, which should be as rectangular as possible in order to create constant conditions during a certain period (see Fig. 6). The study of intensities as a function of the peak current between 50 and 2500 A gives an unequivocal discrimination between the spectra of different ionizations.[67] The cathode glow emits light belonging to ionization stages up to V. Often, they can be discriminated on the basis of the radial

[60] M. Fred, N. H. Nachtrieb, and F. S. Tomkins, *J. Opt. Soc. Amer.* **37**, 279–288 (1947).
[61] F. S. Tomkins and M. Fred, *J. Opt. Soc. Amer.* **39**, 357–363 (1949).
[62] S. Glad, *Ark. Fys.* **10**, 291–340 (1956).
[63] F. Paschen, *Ann. Phys.* **71**, 142–161 (1923).
[64] H. Wittke, *Z. Phys.* **116**, 547–561 (1940).
[65] V. Kaufman and J. Sugar, *J. Res. Nat. Bur. Std.* **71A**, 583–585 (1967).
[66] G. N. Glasoe and J V. Lebacqz, "Pulse Generators." Boston Tech. Publ., Lexington, Massachusetts, 1964.
[67] F. G. Meijer and P. F. A. Klinkenberg, *Physica* **69**, 111–118 (1973).

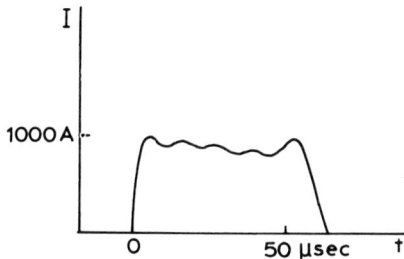

FIG. 6. Pulse profile of pulsed hollow cathode discharge, as observed on oscilloscope screen.

intensity distribution in one single stigmatic spectrogram.[62] The carrier gas is almost exclusively helium.

3.4.2.11. The Sliding Spark. The sliding spark was introduced in 1950.[68] The essential point is that a weakly conducting path between the electrodes causes the breakdown to occur at potential differences of a few kilovolts, even if the electrodes are several mm apart and the spark chamber is evacuated. To this end the electrodes are snugly enclosed in a channel through a ceramic piece which has a bore in the side wall for the light to pass. A strip of graphite on the inner wall of the channel, that can be made simply with a pencil, is sufficient to start with, but during the discharge, sputtered material will be deposited and tend to stabilize it. In principle, the same kind of electronic equipment is applicable here as is used for energizing the pulsed hollow cathode source; ionization similarly depends on the peak current. The line quality is much better than that of the ordinary spark, and the excitation can be made much higher. In 1957, the observation of P V was reported,[69] but much higher ionizations are possible. Indeed, the light source is the ideal link between the normal and the hot spark, and it is being used in the vacuum ultraviolet region as well. It also has advantages over other light sources for observing lower ionization stages. A very complete spectrum of Pr III, for instance, was observed with a helium filled sliding spark chamber (pressure 50 Torr), using a 10 μF capacitor charged to 1000 V, which makes the peak current 10^3 A, while Pr II is strongly present when the peak current is reduced to 50 A by inserting a 500 μH inductor.[70]

3.4.2.12. Special Light Sources. We have not discussed sources made for very special purposes: light sources which have been of some use in the past but are now obsolete, those which are rarely used in the optical region, and those which serve to provide continuous backgrounds for

---

[68] B. Vodar and N. Astoin, *Nature* (*London*) **166**, 1029–1030 (1950).
[69] J. Romand and G. Ballofet, *J. Phys.* (*Paris*) **18**, 641–642 (1957).
[70] J. Sugar, *J. Opt. Soc. Amer.* **53**, 831–839 (1963).

absorption experiments. To mention a few: atomic beam sources constructed for observations of lines practically free of Doppler broadening,[71] electrodeless ring discharge sources for obtaining strong excitation of volatile materials with good discrimination of ionization stages, exploding wire and under-liquid-spark devices for observation of line reversals in ions, and flash pyrolysis and laser-triggered sources that are applied mainly in the vacuum region, although the laser-produced plasma has also been subjected to ordinary discharges for spectrochemical investigations.[72]

Lasers themselves are not spectroscopic light sources in the usual sense. Either they emit extremely coherent light at discrete spectral frequencies (gas lasers, and some crystal lasers at low temperature), or they produce incoherent broad band emission (liquid lasers and semiconductor lasers). However, the latter can be made to concentrate all energy in any arbitrarily narrow interval within the characteristic wavelength region so that tuning becomes possible. Herewith, selective population of atomic levels and effective depopulation of the lowest states can be accomplished, while high-efficiency scanning methods can be applied. The reader is referred to Chapter 5.2 for further details. Recently, successful attempts have been made to use the liquid laser continuum itself by introducing the absorbing vapor *within* the cavity. An extremely weak absorption can be detected because it quenches laser action completely at that particular frequency, which then can be identified by means of a spectrograph.[73]

Finally, the beam-foil technique[74] has been making important contributions to spectroscopy in the last decade, especially with regard to lifetime measurements. For details, the reader is referred to Chapter 5.1.

3.4.3. Spectroscopic Instruments

3.4.3.1. General Considerations. Light can be analyzed into its spectral components (wavelengths, frequencies, etc.) by means of spectroscopes, spectrographs, and spectrometers. With spectroscopes the detection is visual and with spectrographs it is photographic, while, in both cases, the spectral features are separated in space. With spectrometers the separation may be in space or in time, and the detector is any photosensitive device short of the photographic emulsion. If the spectrum is displayed in space, the scanning may be performed either by moving the detector across it or by causing the spectral features to move past the stationary detector (e.g., by rotating a grating). If the spectrum is displayed in time, only the second method is

---

[71] K. W. Meissner, *Rev. Mod. Phys.* **14**, 68–78 (1942).
[72] D. J. Swaine, *Hilger J.* **11**, 69–71 (1968).
[73] A. Dönszelmann, *Comm. E.G.A.S. Conf., 6th, Berlin, 1974*; L. S. Goodman, *Comm. Int. Conf. At. Phys., 4th*, Heidelberg, *1974*.
[74] S. Bashkin, *Nucl. Instrum. Methods* **28**, 88–96 (1964).

applicable. Usually, the detector signal is recorded so that the succession in time is transformed into a spatial succession; it can also be digitized immediately for computer analysis. In Fourier transform spectrometry, there is no display at all, but the integral light is sent through a scanning Michelson interferometer. Here, the analysis is possible only after the scanning is terminated and the complete information on the variation of total flux is available, so that the computer is an indispensable part of the apparatus, though not necessarily on line.

Fourier spectrometry will not be discussed here, because it is dealt with in Chapter 4.2 for the submillimeter region and because its impact on the optical region has been relatively small thus far. In order to compete successfully with standard methods, the theoretical resolving power should be 10^6, and this requires the length of travel to be of the order of a meter. Recently, this has been accomplished; atomic spectra have been measured in the infrared region down to approximately 800 nm, thus securing a comfortable overlap with the photographic region.[75] This is a major technical success; it requires ultraprecise mirror guidance, severe control of conditions for several hours, and, in case of dense spectra, a huge computer memory.[76] However, the advantage of using the integrated light is most valuable in the infrared, where the output of sources is low and detectors have low sensitivity. The big disadvantage inherent in the method is the requirement of source stability. This restricts application to spectra of neutral atoms, and to the stronger II lines which can be excited in stable sources. In conventional scanning of spectra, source instability can be a nuisance for intensity measurements, but it does not lead to errors in wavelength measurements. In photographic work, source instability has no influence on the results. In addition, the advantages of the photographic emulsion are the effect of light accumulation, topographic reliability, high resolution, and image conservation. The biggest disadvantage is the nonlinear response and strong wavelength dependence of sensitivity above 480 nm.

In principle, not only the Fourier transform spectrometer, but *all* spectroscopic instruments perform a Fourier analysis of the incident radiation. But, since no instrument is perfect, the analysis is imperfect; the peculiarities of the instrument will always be superimposed on the real spectrum. In conventional instruments, they take the form of ghost images, continuous background, and distortion of profiles. With regard to line profiles, as given by the apparatus, it will be obvious that they will be convolutions of the true and the instrumental profiles, the latter being the profile of the image if the intrinsic linewidth were infinitely small. Now, we know from Section 3.4.1.1

---

[75] A. Giacchetti and J. Blaise, *Comm. E.G.A.S. Conf.*, 2nd, Hannover, 1970.

[76] J. Connes, H. Delouis, P. Connes, G. Guelachvili, J. P. Maillard, and G. Michel, *Nouv. Rev. Opt.* **1**, 3–22 (1970).

that the natural line width in the optical region is 10^{-5}–10^{-4} nm. However, unless one makes excessive efforts to reduce broadening effects, the intrinsic width of "sharp" lines of atoms in the gaseous state will be of the order of 10^{-3} nm. From this, it follows that the instrumental width (full width at half the maximum intensity, or FWHM) should not exceed 10^{-5} nm in order to record the intrinsic profile; i.e., the resolving power should be at least 5×10^7, a value that can be reached only by means of the Fabry–Perot interferometer, which is a typical high-resolution instrument. In normal situations, the instrumental width is comparable with, or bigger than, the intrinsic line width. Then the resolving power is responsible for the width of the spectral features, and one cannot expect to observe details of the real profile. In addition, in these instruments, the observed spectral lines are the images of an entrance slit in the various wavelengths so that the slit width will also influence the observed profile. It is usual to define the theoretical resolving power, R_{theor} or R, as $\lambda/\delta\lambda$ for an infinitely narrow slit, where $\delta\lambda$ is the smallest wavelength difference between two perfectly monochromatic and equally strong radiations, at the wavelength λ, that can still be seen separated in the image; $\delta\lambda$, or rather the equivalent wavenumber difference $\delta\sigma = \delta\lambda/\lambda^2$, has been termed resolving limit by Tolansky.[42]

3.4.3.2. Spectral Lines as Slit Images. Here the Rayleigh criterion is adopted for defining $\delta\lambda$ in an objective manner: using the diffraction formula $I = I_0(\sin^2 \beta)/\beta^2$, and requiring that the principal maximum of the diffraction image of one of the lines coincides with the first minimum of the image of the other one (Fig. 7), one finds $I = 0.405 I_0$ for the intensity of each line

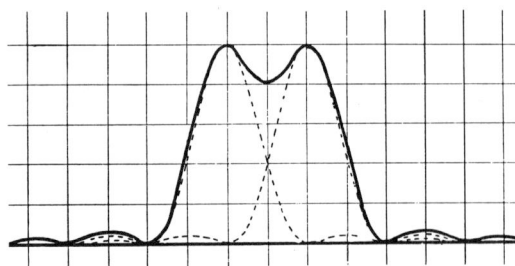

FIG. 7. Definition of resolving power in a diffraction-limited slit image.

at the intersection and hence a total half-way intensity of 81% of the maximum in the summation curve, which is just about sufficient for the human eye to perceive two different images. By means of photometers smaller intensity variations can be measured, and, even when there is no dip, the separation can be evaluated by careful analysis, but such an analysis requires preknowledge about the number of components involved and their intensity ratios. Hence, the results can no longer be considered to follow from one experiment alone. Certainly, the Rayleigh criterion is most adequate for comparing instrumental resolving powers.

The diffraction formula is derived from a discussion of a plane wave with a rectangular aperture, and it describes the intensity distribution at infinity as a function of the diffraction angle θ, measured in a plane through the axis, and the aperture width b: $\beta = \pi b (\sin \theta)/\lambda$ equals half the phase difference between rays arriving from the opposite edges of the aperture. The first minimum obtains for $\beta = \pi$ or $\sin \theta = \lambda/b$, and, since this is always very small, it may be equated to the angle which, for this particular case, shall have the value $\varepsilon = \lambda/b$. In a spectroscope, it is the telescope objective which brings infinity into the focal plane, hence we can identify b with the width w of the parallel beam entering the telescope, measured in the direction perpendicular to the slit: $\varepsilon = \lambda/w$. Now, the angle between two beams of wavelengths λ and $\lambda + \delta\lambda$ is determined by the angular dispersion $D = d\phi/d\lambda$ given by the dispersing element. For $d\phi = \delta\phi = \varepsilon$, we fulfill the Rayleigh condition so that $d\phi/d\lambda$ becomes $\varepsilon/\delta\lambda = \lambda/w\,\delta\lambda$, and

$$R_{\text{theor}} = \lambda/\delta\lambda = wD. \tag{3.4.1}$$

This formula does not imply anything about the dispersing mechanism, but it presupposes uniform intensity distribution over all interfering rays and perfect imagery, except for chromatic errors. A nonuniform intensity distribution will result in a different diffraction pattern. It is possible to suppress the secondary maxima by introducing specially shaped absorbing screens to reduce the aperture. This procedure, called apodization,[77] widens the central maximum and so decreases resolving power; it has not found wide application in the optical region. If the beam has not a rectangular but a circular cross section with diameter w, the effect is that the secondary maxima become much weaker, relatively, and the first minimum is a factor 1.22 farther away from the center. When treating the etalon, we shall find a profile without subsidiary maxima, due to a particular intensity—or rather amplitude—distribution of the interfering rays. However, this is not usually looked upon as apodization, and the image is not considered as a diffraction image, although there is no difference in principle.

Equation (3.4.1) represents the theoretical maximum for an infinitely narrow slit. But the slit has to have a finite width s in order to allow the light to enter. If the light source is imaged on the slit so that the various points in the opening can be considered to vibrate independently, the resulting profile will be the superposition of the intensity curves given by the various points or slit elements; hence, it can be found by integration over the slit

---

[77] P. Jacquinot and B. Roizen-Dossier, *Progr. Opt.* **3**, 29–186 (1963).

width.[78, 79] The first minimum shifts outward when s is increased, so that the effective resolving power diminishes while the intensity increases. Introducing the normal slit width s_n[78] and taking the maximum intensity in the diffraction image at this width to be I_n, assuming $s_n = f_{coll} \times \lambda/4w$, one obtains the relative values for resolving power and intensity as in the tabulation.[78] It is seen that opening the slit wider than $2s_n$ rapidly deteriorates

s	s_n	$2s_n$	$4s_n$	$7.2s_n$
R_{eff}/R_{theor}	0.986	0.943	0.780	0.495
I/I_n	1	2	3.1	4

R_{eff} but may be imperative because of intensity reasons. When R_{eff} has fallen to less than 50% of R_{theor}, the increase of intensity becomes very slow so that a photographic plate will not be better exposed; one simply widens the central maximum and the profile becomes the geometrical image of the slit, with weak diffraction maxima near the edges. However, if a photoelectric or photoconductive detector is used, it is not the illumination (light flux per unit of area) that determines the signal, but the total flux, and this will increase indefinitely so that the slit may be widened, at the expense of R_{eff}, to the limit where adjacent spectral images become entangled. So the best compromise depends on the technical methods and on the purpose of the experiment. The calculation has been repeated by Bousquet[79] with slightly differing numerical results; small corrections for diffraction at the edges of the slit have been computed by Van Cittert.[80]

Actually, there will always be some coherence between the vibrations of points in the slit, and when the light source is *not* imaged on the slit (as is normally the case in the vacuum region, but also when uniform illumination of the slit is preferred), the situation of complete coherence is approached. The normal slit width then corresponds to a diffracted intensity spread over the collimated beam in such a way that the first minima are about 8 times farther apart than the edges of the beam; i.e., the light flux is practically uniform. In practice, because of intensity reasons, wider slits are also used in this case. The effective resolving power then deteriorates more quickly than in the "noncoherence" case. Even when the influence of nonuniform intensity distribution is neglected, it can be shown easily that R_{eff} is 50% at $s = 4s_n$, since the geometrical image then covers half the space between the two first minima for $s = 0$. In addition, the effective value of w decreases because the principal diffraction maximum of the slit is contracted.

---

[78] A. Schuster, *Astrophys. J.* **21**, 197–210 (1905).
[79] P. Bousquet, "Spectroscopie Instrumentale." Dunod, Paris, 1969.
[80] P. H. van Cittert, *Z. Phys.* **65**, 547–563 (1930); **69**, 298–308 (1931).

Both R_{theor} and R_{eff} are instrumental resolving powers; they will have to be corrected for instrumental defects, especially optical defects of the imagery and dispersing elements, in order to obtain the practical resolving power R_p that would be measurable for a sufficiently monochromatic radiation. However, if the proper width of the spectral features themselves approaches or exceeds the instrumental width, the resolution obtained will be still lower than R_p. This has to be taken into account when comparing different instruments with regard to resolving power and luminosity. The traditional viewpoint, for instance, that prism spectrographs produce brighter spectra than grating spectrographs is misleading. The brightness of prism spectra is an apparent one, caused by the overlapping of adjacent wavelengths which are resolved by the grating; moreover, if resolving power is lower, focal distances can be kept smaller. The conclusion that luminosity has to be defined with respect to equal resolving limit was advocated in 1932,[81] but the whole problem was not treated on a quantitative basis until 1954[82]; this culminated in the statement that modern (i.e., blazed) gratings are superior to prisms in all respects. While this is true, it should be taken into account that it really means on the basis of equal resolution. There is no point in using a prism if the resolution required can be attained only by reducing the slit width into the region where serious diffraction losses occur because, for the same purpose, a grating can be obtained that gives higher intensity. Also, it should be realized that the situation where it becomes advantageous to use a grating depends on wavelength and the detection method. Because the photographic emulsion (and the eye) responds to illumination, and a photodetector to flux, the advantage is much more evident in the latter case. Indeed, much wider slits are then generally used, so that the diffraction at the slit plays a subordinate role. Moreover, it is immediately obvious that increasing the focal length for larger linear dispersion does not influence the total flux. Hence, the signal of a photodetector will not change as long as its active area is bigger than the cross section of the beam received, whereas photographic density will decrease with illumination, which is inversely proportional to the square of the focal length. A thorough treatment of the problem can be found in Bousquet[79] (p. 15ff).

3.4.3.3. Prism Instruments. 3.4.3.3.1 RESOLVING POWER. When the dispersing element is a prism, one uses the property of a transparent material which is called dispersion and is defined as $dn/d\lambda$. It can be derived from a Cauchy formula for the index of refraction: $n = n_0 + A/\lambda^2 + B/\lambda^4 + \cdots$. In Fig. 8, the situation is sketched for an isosceles prism in the symmetrical position in air, illuminated by a parallel beam. The two parallel beams

---

[81] E. Pauls, *Phys. Z.* **33**, 405–410 (1932).
[82] P. Jacquinot, *J. Opt. Soc. Amer.* **44**, 761–765 (1954); *Rev. Opt.* **33**, 653–658 (1954).

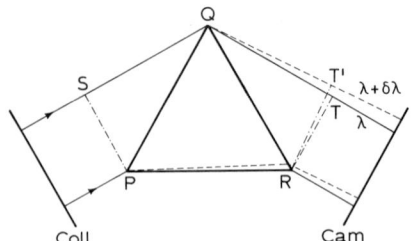

FIG. 8. Prism in minimum deviation.

leaving the prism shall correspond to the wavelengths λ and $\lambda + \delta\lambda$, and the associated wavefronts are RT and $R'T'$. If they are to go to each other's first diffraction minimum in the focal plane of the camera lens, they should include the angle $\varepsilon = \lambda/w$, and so do the wave fronts so that $TT' = \varepsilon \times w = \lambda$ in the limit of small $\delta\lambda$. According to Malus's law, we then have, for λ, $nPR = SQ + QT$, and, for $\lambda + \delta\lambda$, $(n - \delta n)PR = SQ + QT' = SQ + QT - \lambda$, from which $\delta n\ PR = \delta n\ b = \lambda$ or $b\ \delta\lambda \times dn/d\lambda = \lambda$, which yields

$$R_{\text{theor}} = b\ dn/d\lambda. \tag{3.4.2}$$

This is equivalent to Eq. (3.4.1), but it contains quantities characterizing the prism. Of course, Eq. (3.4.2) can be derived from Eq. (3.4.1) by introducing well-known prism relations[18] from geometrical optics to be found in the textbooks.[83-88] Now it is clear that the baselength b can occur in the equation only when the rays are travelling parallel to it and the prism is used to its full extent, because then the path difference, introduced by the prism, between the outermost rays is the longest possible. Consequently, if a stop is used, b is to be replaced by the corresponding maximum path difference within the prism material. It follows also that R_{theor} is highest for the symmetric case, which corresponds with minimum deviation, one of the reasons for preferring the minimum deviation position for the wavelength in the center of the region to be investigated. The attainment of a practical resolving power as close as possible to R_{theor} is subject to all restrictions made in Section 3.4.3.2. Moreover, the technical requirements on the prism are quite high: the material must be optically homogeneous over its entire volume and should not absorb light, while the entrance and exit faces should be flat to a high

[83] R. W. Wood, "Physical Optics." Macmillan, New York, 1923.
[84] F. A. Jenkins and H. E. White, "Fundamentals of Optics." McGraw-Hill, New York, 1937.
[85] B. Rossi, "Optics." Addison-Wesley, Reading, Massachusetts, 1957.
[86] R. S. Longhurst, "Geometrical and Physical Optics," Longmans, Green, New York, 1957.
[87] J. Strong, "Concepts of Classical Optics." Freeman, San Francisco, California, 1958.
[88] M. Born and E. Wolf, "Principles of Optics." Pergamon, Oxford, 1959.

degree of accuracy. Obviously, these requirements are more difficult to satisfy when the prism is bigger, as would be necessary for increasing R_{theor} and desirable for enhancing luminosity. So, there is a technical limit which for glass is of the order of $b = 150$ mm. A more convenient way to increase the path difference is to place two or more prisms in tandem or to make it a double-pass system by using a mirror. Of course, the two methods can be combined in such an autocollimation arrangement in which the light can be admitted by means of a small, totally reflecting prism from a slit mounted on a side tube. Multipass systems have been constructed by introducing additional mirrors, but this reduces the useful region of the spectrum.[89] In the normal case, the refracting angle is close to 60°, but in an autocollimation or Littrow mounting, the prism may be a 30° prism with a reflecting back side, thus acting like a 60° prism with double the size. This is especially advantageous when the material is scarce (e.g., quartz).

The alternative way to increase R_{theor} is choosing materials with large values of $dn/d\lambda$. Here, a paradoxical situation is encountered since it is well-known that high $dn/d\lambda$ is to be expected in regions close to absorption bands.[83–88] However, absorption itself impairs R_{theor} by effectively reducing the aperture. Whether this leads to an optimum value or not when λ is decreased is an academic question because, before this is reached, the absorption reduces the intensity to practically zero. However, around 380 nm, the author was able to resolve Zeeman patterns in a magnetic field of only 2.5 T by means of a three-prism spectrograph with glass optics. One of these patterns, the π pattern of W I, $\lambda = 380.9228$ nm ($g_3 = 1.98$, $g_2 = 1.28$), is shown in Fig. 9. Resolving this pattern requires a resolution

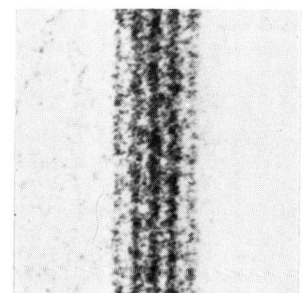

FIG 9. The π pattern of W I, $\lambda = 380.9228$ nm, in a magnetic field of 2.5 T, resolved by means of a spectrograph with three glass prisms.

$R = 35936$; from the appearance on the negative it is concluded that the practical resolving power reached is over 5×10^4. As $D = 0.314 \times 10^{-2}$ rad/nm in this region, we find $w \approx 16$ mm, a relatively low value, due to the large angle of refraction of the rays leaving the third prism which has

[89] A. Walsh, *J. Opt. Soc. Amer.* **42**, 94–95 (1952).

a base of 130 mm. The camera lens focal length is 650 mm, which makes the linear dispersion 2.0 mm/nm.

A possible way to circumvent the problem of size restriction is the use of a liquid enclosed in a prismatic cell with flat glass plates for the faces. A gigantic prism of this type[90] has been employed for studying the character of some forbidden helium lines.[91] An appropriate organic liquid (ethyl cinnamate) served to give a high value of $dn/d\lambda$ in the region of interest where $R > 10^5$ was obtained in a one-prism double-pass arrangement with a total optical path difference of 500 mm for the blue region. The large size had the additional advantage of high luminosity. Extreme care had to be taken to prevent convection currents in the liquid from developing.

From these examples it is seen that the choice of prism material is extremely important; too near the absorption band means a prohibitive loss of light, while too far from it means low resolving power. There is an impressive number of possibilities to meet each special requirement, and if this is combined with the many ways of construction and the various prism shapes in use, the field is almost inexhaustible. For details, the reader is referred to the literature[17–19, 79, 83–88]; we shall confine the discussion to the more common types. Naturally, glass is the most flexible material and, moreover, in the glass region, the imagery can be made most nearly perfect to meet secondary requirements such as achromatism and flatness of focal plane. A very widespread type of glass spectroscope is the constant deviation type which is based on the Pellin–Broca prism shape. This can be thought of as consisting of two 30° and one totally reflecting prism, as shown in Fig. 10.

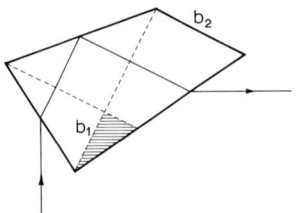

FIG. 10. Constant deviation prism. Length ratios can be different but angles are fixed. There are always useless parts (shaded area in the picture).

As this last prism does not contribute to angular dispersion, the resolving power is determined by the sum of the bases of the two 30° prisms, $b_1 + b_2$. By rotating the prism, any wavelength can be made to be deviated by 90°, which is exactly the position of minimum deviation. This makes it especially useful as a monochromator, but it can also be equipped with an eyepiece to observe a greater part of the spectrum or with a camera for photographing.

[90] A. Couderc, J. Phys. (Paris) **8**, 99S–101S (1937).
[91] J. Brochard and P. Jacquinot, Ann. Phys. (Paris) **20**, 509–534 (1945).

The derivation of R_{theor} only considered the light paths in planes perpendicular to the prism edges, giving rise to images in the median line of the focal plane which all correspond to the center of the slit. Beams arising from other slit points, out of the median plane, will traverse the prism obliquely. Because the optical path differences caused by the prism will then be longer, the rays will behave as if the wavelength were smaller; i.e., they will assemble in images which are shifted towards the shorter wave end, and will have to be deviated more in order to annul the optical path difference in air. This can also be shown geometrically as a result of the fact that the refracting angle is effectively increased. Hence, spectral line images are curved. The excess path length in the prism is optically more serious when the wavelength is smaller, and so the curvature increases when the wavelength decreases; it cannot be corrected for the entire spectrum by introducing a curved slit, as has been done in a few cases. While resolving power should increase a little with inclination, as a result of increased angular dispersion, this is counteracted by the decreasing width of the beam and the decreasing quality of imaging. In practice, one can take advantage of one third of the line length if the full slit is used; the line ends have to be avoided.

3.4.3.3.2. OTHER PROPERTIES. It follows from the above discussion that all prism spectroscopes are stigmatic in principle. Because the prism is traversed by parallel beams and the lenses are axially isotropic, there will be good focusing in any point close to the median plane. However, the collimator lens is not always achromatic, so that the beams going through the prism cannot be parallel for all wavelengths, and, if they are, there are still other imperfections such as spherical aberration and coma to cause small departures farther from the median plane and the wavelength satisfying the minimum deviation condition. But in general they can be neglected; the property of stigmatism is very useful when the distribution of brightness over the surface of the source is an important piece of information. By imaging the source on the slit, one obtains this information straightforwardly in any spectral line (compare Sections 3.4.2.7. and 3.4.2.9., also Fig. 5). The stigmatism is also an asset for combining with an interference instrument. All one has to do is to focus the interference pattern on the slit in order to find it back in the focal plane of the spectroscope (then used with a wide slit), in each spectral image (compare Fig. 3). For accurate wavelength measurements in prism spectra, or for spectrochemical work, it is undesirable to have an irregular intensity distribution over the length of the lines. It can be eliminated by illuminating the slit in a different way so that no image of the source is obtained in the plane of the spectrum.

There is a considerable loss of light at the entrance face of a prism, but this is mainly in one polarization because, for the normal case of a 60° prism, the angle of incidence is not very different from the Brewster angle. Here, the component polarized with the electric vector in the plane of incidence is not reflected at all, while the reflectivity for the oppositely polarized component is about 15%. This situation repeats itself at the exit face. Hence in a train of prisms, originally unpolarized light becomes strongly polarized but the total loss of intensity will not exceed 50%, losses by absorption and scattering being neglected. It may be interesting to note that, due to tarnishing, reflection can be much less than would be expected at an air-glass surface; the transition layer acts as an antireflection coating. A thorough cleaning may then spoil this excellent performance. Special devices have been described in the literature to reduce light loss, but none of them have found general application. Apparently luminosity is not often a limiting factor. With photographic detection, speed can be increased by using short focus lenses, even to the extent that the diffraction image becomes much smaller than the grain, which may be of the order of 10 μm. This has been applied frequently in research of the Raman effect and in stellar spectrography.

If we rewrite the general equation (3.4.1) for the special case of minimum deviation, we obtain the maximum resolving power attainable by means of a prism: $R_{\text{theor}} = w_{\min} \times D_{\min}$, where $D_{\min}$ does not mean minimum angular dispersion but angular dispersion at minimum deviation. Comparing with Eq. (3.4.2), we find

$$D_{\min} = (b/w)\, dn/d\lambda, \tag{3.4.3}$$

a relation which does not contain any element of diffraction theory and therefore can be readily derived from geometrical optics.[79, 84, 86–88]

3.4.3.3.3. PRISM INSTRUMENTS FOR THE ULTRAVIOLET AND THE INFRARED. The use of materials other than glass causes various complications which influence the construction of spectroscopes. First, the optics are not nearly as good as in the glass region, especially when high luminosity is required. Second, some important materials, in particular, crystalline quartz, are double-refracting. In the infrared, materials such as rocksalt, potassium bromide, etc. are in use but mainly for purposes of monochromatization where high resolving power is not the primary requirement. Instead of lenses, one uses mirrors mostly. In the ultraviolet, quartz is the favorite material, while CaF_2 and LiF have been used for the extreme ultraviolet down to 160 and 120 nm, respectively; work in this region also requires the removal of the air in the light path. The discussion will be restricted to quartz instruments. As regards transparency, it has been found that apparently clean natural quartz pieces may be rather different below 220 nm.[92] Specially selected pieces would have a transmissivity of 20–60 times as high as average; they reach down to the limit determined by the absorption in air, approxi-

---

[92] F. Zernike, *Physica* **8**, 81–87 (1928).

3.4. OPTICAL REGION

mately 193.5 nm. Unfortunately, even ordinary quartz becomes rarer, certainly in big pieces. The double refraction is minimized by cutting the pieces so as to have the optical axis parallel to the beam axis. For the prism, this means that it is parallel to the base lines, corresponding to minimum deviation. But even then the double refraction is not quite eliminated, because quartz has the peculiar property of rotating the plane of polarization of light travelling along the optical axis. This is understood to be the result of a small difference in the velocity of propagation of oppositely circular polarizations. The corresponding double refraction would impair the resolving power. Fortunately, nature yields left-handed and right-handed specimen's so that the effect can be compensated for by assembling a prism out of two halves of opposite handedness (Cornu prism). This also helps to minimize the double refraction for rays which travel through the prism in directions making a small angle with the optical axis.

Large quartz lenses cannot be achromatized, so that quartz spectrographs are characterized by the very inclined position of the plate holder. Although this increases linear dispersion, it does not contribute to resolution, as all profiles are enlarged likewise. Actually, it can influence resolution adversely because the near-grazing incidence produces photographic volume effects tending to blur the image. Moreover, because of the other aberrations of the lenses, the spectrum deteriorates sooner when one moves out of the region of minimum deviation and out of the median plane. Usually the focal distances are rather large (0.5–2.50 m) and the apertures relatively small, but even so the practical resolving power may be improved by introducing stops. The focal plane is always curved, but thin glass plates can be bent sufficiently to follow it.

In exceptional cases, large aberrations are tolerated in favor of high luminosity. Figure 11 shows the two quartz (Cornu) prisms, with a 70-mm

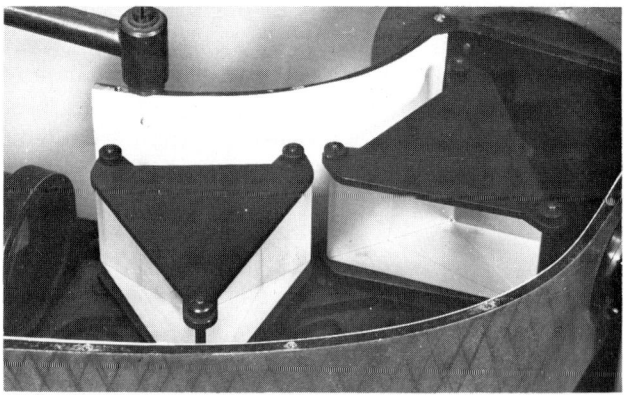

FIG. 11. Quartz (Cornu) prisms of a stellar spectrograph.

base each, of a stellar spectrograph having a nominal aperture $f/5$. Theoretically, they should resolve an interval of 0.012_5 nm at $\lambda = 350$ nm, and of 0.001 nm at $\lambda = 200$ nm. But, as a result of the small focal distance (approximately 0.16 and 0.13 m, respectively), these intervals correspond with distances in the focal plane of about 1.5 μm. In view of photographic graininess, R_{theor} is about 10 times larger than necessary so that some loss of resolving power has no serious effect. In addition, the features in stellar spectra are often broader themselves. The focal plane is strongly curved, but stigmatism is reasonable, as can be seen from Fig. 12, reproducing the extreme ultraviolet part of a spectrogram.

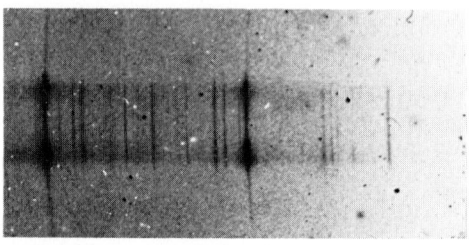

FIG. 12. Far ultraviolet end of a Th-spark spectrogram, taken in air by means of a stellar spectrograph (Fig. 11). Lowest wavelength recorded is $\lambda = 193.0$ nm; strongest line to the left is $\lambda = 200.2$ nm (Th IV).

A different solution of the double-refraction problem is reached by means of the Littrow mounting because automatic compensation is obtained by reversing the direction of light. In addition, it saves material, as already stressed in Section 3.4.3.3.1. The prism is usually one of 30° provided with a reflecting coating on the back side. The totally reflecting prism, sending the light from the slit to the dispersing prism, is placed slightly off the lens axis so that the returning beams will not be obstructed. Since this system is used for large instruments with apertures of $f/20$, or less, aberrations are small. Moreover, the lens may be ground to be aspherical to remove spherical aberration, while coma is minimized by suitably chosen curvatures. The weak point with respect to transmission instruments is the large amount of stray light due to reflection. Reflection from the first lens surface can be prevented from reaching the plate by placing an absorbing strip over the central part of the lens. In front of the focal plane a screen is placed with a long rectangular gap to transmit the spectrum, and care should be taken that from nowhere behind that opening can the reflected slit image be seen. The reflection from the second surface gives rise to a real image which should be inside the spectrograph where it can be stopped by a small screen. However, when making readjustments as needed for observing different parts of the spectrum, the screen has to be displaced, too. With such instruments, a resolving power of

5×10^4 is reached at the short wave end. For 250 nm, this corresponds to a resolving limit of 0.005 nm. With a focal length of about 2 m, the two images can be brought apart some 20 μm (given the strong inclination of the focal plane).

The autocollimation principle has, of course, also been applied to glass optics, in which case an assembly of one 60° and one 30° prism has been quite successful.

It would be advantageous to introduce vitreous quartz because of its isotropy, but, until recently, this was less transparent to the ultraviolet and it could not be manufactured sufficiently homogeneous in large pieces to be useful for this purpose. But some spectrographs of smaller size, equipped with lenses and prisms of the highest quality fused quartz, have now been reported to show good performance. They may well take the lead in the future, especially since purifying procedures have been invented which have proven that the cleanest SiO_2 (Spectrosil B) is *more* transparent than the crystals found in nature.[93,94]

3.4.3.4. Grating Instruments. 3.4.3.4.1. GRATING THEORY. The phenomenon which was originally exploited for producing spectra by means of a grating is diffraction. Since most of the light, then, went to waste into the zeroth order, the grating used to be a notoriously low flux instrument. This is expressed by the well-known formula for the intensity in the diffraction pattern:

$$I = I_0 \frac{\sin^2 \beta}{\beta^2} \frac{\sin^2 N\gamma}{\sin^2 \gamma}, \tag{3.4.4}$$

where I_0 is the zero-order intensity, N the number of diffracting elements, $\beta = \pi b(\sin \theta)/\lambda$, $\gamma = \pi d(\sin \theta)/\lambda$, b the width of each diffracting element, d the constant distance between successive elements measured along a straight line perpendicular to the elements (i.e., the grating constant), and θ the angle of diffraction. In ordinary gratings, b is between $d/2$ and d, while d is of the order of 1 μm. Even the first-order image, then, cannot have an intensity higher than $I_0(4/\pi^2) = 0.405 I_0$, namely, for $b = d/2$; in this case, all even orders vanish because all elements contribute zero intensity, while the third order would have $I = I_0(4/9\pi^2) = 0.045 I_0$. For $b = d$, all diffracted orders are zero; the zeroth order is the direct light either transmitted or reflected: there is virtually no grating any more.

However, right from the start of modern reflection grating technology[95] (Rowland, 1882), it has been found that drastic departures from Eq. (3.4.4)

---

[93] B. Bates and D. J. Bradley, *J. Opt. Soc. Amer.* **57**, 481–485 (1967).

[94] D. J. Bradley, B. Bates, C. O. L. Juulman, and T. Kohno, *J. Phys. Suppl. C2* **28**, 280–286 (1967).

[95] H. A. Rowland, *Phil. Mag.* **13**, 469–474 (1882).

could occur.[†] They were correctly ascribed to the profile of the grooves by Wood; later, the art of ruling gratings was so perfected that groove profiles could be made to specification with great regularity throughout the whole grating. The technical evolution which made this possible was the spectacular development of vacuum coating of large surfaces with highly reflective aluminum layers; the ruling diamond is not subject to wear on this soft material and it permits one to rule deep grooves with smooth faces. While, in primitive gratings, the elements were transparent or reflecting parts of the original surface, the latter disappears completely in blazed gratings, and quasi-specular reflection occurs on the smooth faces of the grooves. As can be seen from Fig. 13, a beam, which is incident at an angle i with the normal to the grating surface S, will favor "diffracted" beams in a direction which

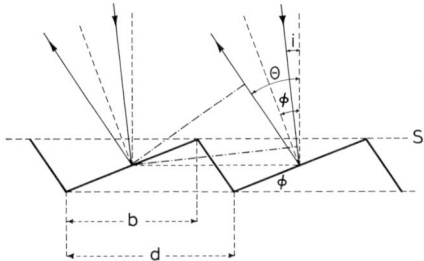

FIG. 13. Blazed grating. If i and θ are on different sides of the normal to S, the sines in Eq. (3.4.5) have to be subtracted, and θ becomes $2\phi + i$. Wavefronts are given by dot-dashed lines.

obeys the law of reflection with regard to the side face enclosing a blaze angle ϕ with the surface. Hence the diffracted beams will show a maximum flux in a diffraction angle $\theta = (2\phi - i)$, and practically no energy will go into the direct image with respect to S. On the contrary, the preferred direction is the zeroth order with respect to the groove face, while it may be any higher order with respect to the grating surface. Hence, blazing a grating may be considered as bringing the principal diffraction maximum of every element into coincidence with higher-order maxima of the complete grating, thus making the factor $\sin^2 \beta/\beta^2$ in Eq. (3.4.4) deliberately equal to unity. The light stolen from the direct image is retrieved in the order desired, corresponding to

$$d(\sin i + \sin \theta) = m\lambda \qquad (3.4.5)$$

in Fig. 13. If the reflection were truly specular, the equation would be

---

[†] Actually, Fraunhofer reported blaze phenomena and had an inkling about their origin, too. His contributions to the ruling art were remarkable; he succeeded in ruling 300 grooves/mm on glass.

satisfied only for particular wavelengths, corresponding to integer values of m, but the slight irregularities of the faces and their small widths (of the order of the wavelengths themselves) result in a blurred reflection image, so that an extended wavelength region around the preferred wavelength can be observed. When the number of grooves per millimeter is very small (for the echelle, around 10), however, the reflection angle is quite well defined so that only a very narrow part of the spectrum can be observed in a particular order. Of course, there is also a possibility for using the opposite groove faces, which adds to the versatility of the grating.

The complete theory of intensity distribution is much too involved to treat here. The dependence on the groove profile has to be evaluated as a boundary value problem in the electromagnetic wave theory which cannot be solved in a general way. An interesting approach has been made by Stroke,[96] who showed, both theoretically and experimentally, that high efficiency requires suppression of polarization of diffracted light, and that the rectangular shape is the worst choice one can make. An efficiency gain from 50–65% to 95% could be reached experimentally by improving the profile.

A very extensive monograph on gratings, including a historical review, was contributed to the "Encyclopedia of Physics" by Stroke,[97] who has been actively engaged himself in the perfecting of the ruling technique and the development of the theory for several years. Basic principles regarding ruling, properties, testing, and use of gratings are treated rigorously in the article quoted.

It is the art of ruling blazed gratings which really placed gratings in a competitive position with prism spectroscopes with regard to speed (compare Section 3.4.3.2.). Just as a prism, the grating needs uniform illumination from an entrance slit, with the same requirements to be met. The theoretical resolving power is found from the Rayleigh criterion. Between two successive maxima of integer orders, Eq. (3.4.4) predicts $N - 2$ secondary maxima, obtained when the last factor vanishes. It is easily seen that the minima adjacent to the mth-order maximum for a particular wavelength λ correspond to a change of path difference of the extreme rays equal to λ, because that makes the path difference between the diffracted rays from any pair of grooves separated by $Nd/2 = W/2$ equal to $(m \pm \frac{1}{2})\lambda$, so that they destroy each other. The order of λ is then $m \pm (1/N)$. According to Rayleigh, we have to place the mth order of $\lambda + \delta\lambda$ in coincidence with the order $m + (1/N)$ of λ, hence $(m + (1/N))\lambda = m(\lambda + \delta\lambda)$, from which

$$R_{\text{theor}} = \lambda/\delta\lambda = mN. \qquad (3.4.6)$$

[96] G. W. Stroke, *Phys. Lett.* **5**, 45–48 (1963).
[97] G. W. Stroke, *Ency. Phys.* **29**, 426–754 (1967).

In large gratings, N is 10^5 or 1.5×10^5, so that R_{theor} can be made several times 10^5 in low orders, independent of wavelength. This property makes the grating superior to the prism, which can rival it only in the small domains where transparent materials with very high dispersion are available. We have already emphasized that large resolving power causes an apparent decrease of flux as a result of less overlapping of adjacent regions within a spectral line.[81] Essentially, the blazed grating is the faster instrument, and this is only enhanced by the fact that it can be made very large. At the same time, this increases R_{theor}, which actually depends on W rather than N, because, in a fixed direction, m is inversely proportional to N if W is constant. We have $m = d(\sin i + \sin \theta)/\lambda$, hence

$$R_{\text{theor}} = mN = W(\sin i + \sin \theta)/\lambda, \qquad (3.4.7)$$

which is equal to the number of wavelengths comprised in the path difference between the two extreme rays; this is exactly the same result as obtained for the prism according to Eq. (3.4.2). The role of N is important in suppressing the subsidiary maxima and obtaining a large "free spectral range."

In order to take advantage of high resolving power, one requires large focal distances. For the angular dispersion, one finds from Eq. (3.4.5):

$$D = \partial\theta/\partial\lambda = m/d \cos \theta, \qquad (3.4.8)$$

which, for $m = 3$, $d = 1/600$ mm, and $\theta = 0$ yields 0.18×10^{-3} rad/nm. If $R_{\text{theor}} = 3 \times 10^5$ ($W \approx 170$ mm), the resolving limit $\delta\lambda$ at 450 nm would be 0.0015 nm, corresponding to an angle $\delta\theta = 27 \times 10^{-7}$ rad. This would require a distance of about 7.5 m to bring the images 20 μm apart, sufficient for resolution on a high-speed photographic emulsion.

As $R_{\text{theor}} = mN$ can be written also as $(m/d \cos \theta) \times Nd \cos \theta$, and $Nd \cos \theta = W \cos \theta =$ the width w of the emerging beam, we again obtain the general formula (3.4.1) derived in Section 3.4.3.2, as expected.

From Eq. (3.4.8) it is evident that the angular dispersion follows a pattern quite different from that in a prismatic spectrum. Around $\theta = 0$ it is almost constant (normal spectrum), and it will slowly decrease on either side away from the normal. Not only in the normal spectrum, but also far out, interpolations are simple in comparison with prismatic spectra, and one does not encounter the crowding of lines in the long wave region characteristic of the latter. Moreover, with a grating, the wavelengths can be measured absolutely by measuring angles, and it is in this way that Ångström established the first system of standard reference lines that gained international acceptance. It should be remarked finally that with gratings the deviation is greater for longer waves, opposite to the situation with prisms.

All equations introduced in this section can be shown to hold for concave gratings as well. A plane grating must be used in a collimated beam, while a concave grating itself has focusing properties.

3.4.3.4.2. CONCAVE GRATING MOUNTINGS. Rowland's great invention, the concave grating,[95] freed the spectroscope from the need of using lenses or mirrors, and opened up the way to high-resolution observations in the infrared and the far ultraviolet, including the vacuum region, where no suitable focusing elements are available. In principle, it is a spherically concave mirror (other concave shapes have not been so successful) on which grooves have been ruled, spaced equidistantly along a chord. All concave grating mountings, except one, are based on the theorem of the Rowland circle, according to which a spectrum is in focus on a circle tangent to the grating surface, having a radius equal to half the grating radius, if the entrance slit is on the circle. This concerns the focusing in the median plane of the circle which is normally horizontal. The place of vertical focusing is outside the circle so that the line image is not stigmatic. The astigmatism of mountings based on the Rowland circle is the principal drawback. At higher angles, it causes a very serious loss of illumination.

The complete theory of the Rowland circle is quite involved, and current treatments of higher-order aberrations are not without errors.[98–100] However, from Fig. 14 it is readily seen that all rays coming from the slit S are incident on the grating G at a constant angle i, while all difffracted rays going to a spectral image L leave the grating under the constant angle θ, to the approximation that the grating surface is coincident with the circle (C is the center of curvature, $OC = r$). This is not strictly true, the grating points being at a distance r from C and the corresponding circle points at $r \cos \phi$, but the difference $r(1 - \cos \phi)$ is extremely small in all practical cases. The effect on focusing is negligble as long as i and θ are not very large; only when the grating is used in grazing incidence must ϕ be reduced considerably (small gratings). However, as the rays actually travel to the grating point, both i and θ are slightly smaller, and this should be compensated for by the larger spacing of the grooves along the grating surface. This makes it plausible that $d(\sin i + \sin \theta) = m\lambda$ remains constant when ϕ is varied; for $\phi = 0$, d is exactly the "real" spacing. The complete computation must also include the rays which move out of the median plane and are inclined to it. The reader is referred to Stroke[97] for the details. As the focal plane is inclined with respect to the incoming rays, the linear dispersion at L is magnified by a factor $1/\cos \theta$, but, at the same time, the distance to the grating is shortened by the same factor so that the net effect is nil. Therefore,

[98] H. G. Beutler, *J. Opt. Soc. Amer.* **35**, 311–350 (1945).
[99] T. Namioka, *J. Opt. Soc. Amer.* **49**, 446–460 (1959).
[100] T. Sai, M. Seya, and T. Namioka, *Sci. Light* **17**, 11–24 (1968).

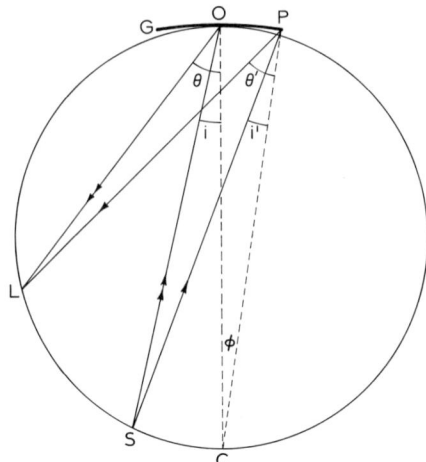

FIG. 14. Theory of the Rowland circle.

one obtains the linear dispersion by multiplying by r:

$$\partial l/\partial \lambda = rm/d \cos \theta. \tag{3.4.9}$$

Gratings with radii up to 13 m have been made, the number of grooves per millimeter varying from 600 to 2400 and the ruled areas having sizes up to about 200 × 100 mm. Smaller gratings with 3600 grooves/mm are also manufactured but are not in use for purposes of research.

The various mountings based on the Rowland circle are that of Rowland himself, which is now obsolete, the Abney mounting, which never acquired popularity, the Paschen–Runge mounting, and the Eagle mounting. We discuss only the latter two. The Paschen–Runge mounting is most directly associated with the description above, because it has a plate holder encompassing a large part of the circle, frequently more than half of it, plus one or more slits in fixed positions. Usually the plate holder is mounted on heavy rails supported by concrete piers in a separate dark room so that plates can be inserted in the open. A system of movable screens in front of the focal circle is needed in order to make it possible to expose different strips of the plates successively. The room is always thermostatically controlled, but barometric pressure changes are usually not compensated. The effect on the spectral lines can be considerable, and exposure may have to be stopped if the change is excessive.[18] The space requirement of the mounting is high, but it has the advantage that it can be made very sturdy and ready for use at any instant at an arbitrary wavelength. Probably the best installation of this type is at the Argonne National Laboratory; its radius is 9.15 m.[101]

---

[101] F. S. Tomkins and M. Fred, *Appl. Opt.* **2**, 715–725 (1963).

The spectrum reproduced in Fig. 1 (Section 3.4.2.4.) was obtained by means of that apparatus.

The Eagle mounting is much more compact because it is an autocollimation system in which one long box can hold slit, grating, and plate holder.[102] The region covered in one exposure is limited by the space inside the box, and it can be varied by rotating the grating and then bringing the Rowland circle and the plate holder into coincidence. This is done by shifting the grating along the line connecting its center with the center of the plate holder and pivoting the latter around the vertical axis in the center. It is practical to admit the light from a side and direct it to the grating by means of a totally reflecting prism mounted a little below the median plane. The virtual image of the slit should be in the vertical axis of the plate holder (off-plane mounting). In this way the prism cannot obstruct the diffracted rays, while the angles with the plane are so small that there is no loss of definition. For the center, one has exactly $i = \theta$, hence

$$2d \sin \theta = m\lambda \qquad (3.4.10)$$

determines the central wavelength.

In the in-plane Eagle mounting, the slit itself is on the Rowland circle adjacent to the plate holder, and it is most practical to rotate the plate holder around the slit for readjustment so that the slit remains the only fixed element of the mounting.

The disadvantage of the limited spectral region with respect to the Paschen–Runge mounting is not as serious as it might seem because, even when the full circle is available, one has to make several exposures. The overlapping of different orders requires the use of different filters or sets of filters and of different photographic emulsions to ascertain the allocation of observed lines to the correct orders. Moreover, the introduction of blazed gratings has rendered observation of extended ranges of spectra less profitable than previously. The additional work of adjustment can be executed swiftly, especially when the instrument is on a stable floor which makes readings reproducible. Requirements for mechanical and temperature stability are the same as for the Paschen–Runge mounting. Exchange or replacement of gratings is simpler since the grating radius is less critical than with the solid circle.

The astigmatism requires the slit to be exactly aligned in both mountings, it reduces the illumination, but is otherwise no disadvantage when wavelength comparisons are to be made. However, when one is interested in the intensity distribution, as for studies of the pole effect for discriminating ionization stages, the source has to be imaged vertically on the plate. To this end the source itself or its image should be in a plane in front of the slit.

[102] A. Eagle, *Astrophys. J.* **31**, 120–142 (1910).

For regions not too far from the normal this plane can be found from the construction of Sirks.[103] In order to achieve full illumination of the grating at the same time, a combination of cylindric lenses can be used. The method can also be convenient for introducing a comparison spectrum whose lines can be easily distinguished from those to be measured. In the Eagle mounting, the plane of vertical focusing is impractically far away when the grating is used at a large angle, as it often is for observing higher orders.

The only *stigmatic* concave grating mounting is based on the work of Wadsworth, who proved that stigmatism is achieved in the grating axis at a distance $r/2$ if the grating is illuminated by a collimated beam.[104] In the original construction, the light was collimated by means of a lens. It was Zeeman who first introduced a concave mirror in the Wadsworth mounting and who thereby realized the compact stigmatic mounting[105] usually attributed to Meggers and Burns[106] or to Fabry and Buisson.[107] Grating and plate holder are rigidly attached to a heavy beam which can pivot around an axis through the grating center. In this, the angle of incidence is varied while observation takes place in a relatively small range of diffraction angles. For focusing, the plate holder can be shifted parallel to itself along the beam. The parabolic focal curve is sufficiently shallow to be approximated by a circle.[108] Zeeman points out the importance of small angles of incidence on the mirror in order to keep the horizontal and vertical foci in the same plane, and the advantages of the mirror with respect to a lens; viz., the absence of chromatic errors and absorption. The largest observable wavelength is about $0.7d$. If this is to be 1200 nm, the grating constant should not be smaller than $1/600$ mm. Because the distance to the focal plane is half that in the other concave grating mountings, the dispersion is also halved, while R_{theor} remains the same. So the actual resolution is often limited by the grain rather than by instrumental properties. However, the solid angle of the light beam is four times as large and the stigmatism still adds to the luminosity so that the mounting is very useful when resolution is not critical. Moreover, just because the flux is higher, it is possible to use slower emulsions which have smaller grain, whereby at least part of the loss is compensated. As with all stigmatic mountings, the Wadsworth–Zeeman mounting is very practical as a monochromator for interferometric measurements. In Zeeman effect investigations, it allows for easy separation of π and σ components in one simultaneous exposure.

[103] J. L. Sirks, *Astron. Astrophys.* **13**, 763–768 (1894).
[104] F. L. O. Wadsworth, *Astrophys. J.* **3**, 47–62 (1896); **2**, 370 (1895).
[105] P. Zeeman, *Arch. Néerl, Sci. Exactes Nat.* **5**, 237–241 (1900).
[106] W. F. Meggers and K. Burns, *Sci. Papers Bur. St.* **18**, 185–199 (1922).
[107] Ch. Fabry and H. Buisson, *J. Phys.* (*Paris*) **9**, 929–961 (1910).
[108] A. G. Shenstone, *J. Opt. Soc. Amer.* **43**, 706 (1953).

3.4.3.4.3. PLANE GRATING MOUNTINGS. The supremacy of concave gratings has lasted about 70 years, and in the vacuum region it is still continuing, for obvious reasons. However, it has been waning in the optical region during the last two decades. Paradoxically, this is due to the same technical improvements which endowed spectroscopy with the blazed grating. We have already seen that this is connected to ruling in soft aluminum coatings. The technique of coating large areas with Al at the same time has made it possible to produce large concave mirrors, with excellent performance throughout the optical region. This was essential for the revival of old plane grating mountings which had been almost forgotten. Moreover, the perfection of the ruling technique reached a point where the curvature of the surface became the principal limit. Highest perfection, both in spacing and in profiling, is possible only on a plane surface.

A plane grating mounting that never entirely disappeared is the Littrow mounting, which is used the same way as with a prism (Section 3.4.3.3.1). The ultraviolet is a problem here because one needs a large aplanatic and sufficiently achromatic lens. This limitation does not exist for the Ebert mounting[109] shown in Fig. 15, which has the plane grating about equally distant from the mirror as the slit, while, in later constructions, it was found

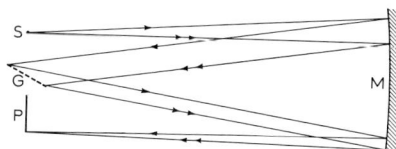

FIG. 15. Schematic arrangement of the Ebert mounting.

advantageous to place it about half way. Because of the symmetry in the light path, the aberrations (coma) can be partly suppressed. Stigmatic images are obtained in a region of reasonable extent around the point of symmetry. With the Littrow mounting, the Ebert mounting has, in common, the ease of wavelength change by rotating the grating. This does not require resetting of the optics and is especially useful for monochromators,[110] and also when recording is nonphotographic, as in the far infrared. By rotating the grating, the spectral lines march across the exit slit behind which a detector; e.g., a photoconductive device (PbS cell), is mounted.[111,112] But the system is also

[109] H. Ebert, *Wiedemann's Ann.* **38**, 489–495 (1889).
[110] W. G. Fastie, *J. Opt. Soc. Amer.* **42**, 641–647, 647–651 (1952).
[111] C. J. Humphreys and H. J. Kostkowski, *J. Res. Nat. Bur. Std.* **49**, 73–84 (1952).
[112] C. J. Humphreys, *J. Opt. Soc. Amer.* **50**, 1171–1173 (1960).

useful in photographic recording, and it has the advantage over the Littrow mounting of the absence of absorption and chromatic errors, while it is not less stigmatic. In the modern, large Ebert–Fastie spectrograph, the rays travel in planes slightly inclined with respect to the median plane through the grating center and perpendicular to the grooves, in such a way that the beam from the slit and the beams returning from the mirror after difffraction by the grating pass over and underneath the grating, respectively. The slit is then mounted centrally and the plate holder covers a long part of the spectrum.[113, 114] Grating exchange is easy and does not affect adjustment. The instrument is also practical for the echelette-type gratings, which allow observation of very high orders without undue ghost intensity.

In a similar way, the old Czerny–Turner mounting[115] was rehabilitated, and it seems to gain importance for the same reasons as the Ebert mounting, particularly for measurements in the infrared.[116] Instead of one it has two concave mirrors, one for collimating and one for focusing. The extra degrees of freedom can be used to minimize residual aberrations.

3.4.3.4.4. MULTIPASS SYSTEMS. The performance of gratings can be improved by using two or more of them in tandem or by using one grating more than once. The first method has been applied to plane gratings by several investigators,[117–120] the second one to plane and concave gratings by others.[121–123] In one case, up to seven diffractions were observed. These methods always involve narrowing of the observable spectral region. A tandem Wadsworth mounting capable of covering an extended region was described recently.[124] A very unusual procedure is to decrease the wavelength by immersing the grating in a transparent oil. In order not to lose the gain in R_{theor} by the effect of refraction, a glass block of suitable shape, in optical contact with the oil, allows a beam to enter at near-grazing incidence. This reduces the improvement to a narrow wavelength region.[125]

---

[113] R. F. Jarrell, *J. Opt. Soc. Amer.* **45**, 259–269 (1955).

[114] S. S. Berman, P. Tymchuk, and D. S. Russell, *Appl. Spectrosc.* **15**, 124–129 (1961).

[115] M. Czerny and A. F. Turner, *Z. Phys.* **61**, 792–797 (1930).

[116] U. Litzén, *Physica Scripta* **2**, 84–87 (1970).

[117] D. H. Rank and T. A. Wiggins, *J. Opt. Soc. Amer.* **42**, 983–984 (1952).

[118] D. H. Rank, A. H. Guenther, C. R. Burnett, and T. A. Wiggins, *J. Opt. Soc. Amer.* **47**, 631–635 (1957).

[119] G. R. Harrison and G. W. Stroke, *J. Opt. Soc. Amer.* **50**, 1153–1158 (1960).

[120] G. W. Stroke and H. H. Stroke, *J. Opt. Soc. Amer.* **53**, 333–338 (1963).

[121] F. A. Jenkins and L. W. Alvarez, *J. Opt. Soc. Amer.* **42**, 699–705 (1952).

[122] W. G. Fastie and W. M. Sinton, *J. Opt. Soc. Amer.* **44**, 103–108 (1954).

[123] E. Hulthén and E. Lind, *Ark. Fys.* **2**, 253–270 (1950); *Proc. Opt. Conf.* 97–111. Chapman and Hall, London, 1952.

[124] J. D. F. Bartoe and G. E. Brueckner, *J. Opt. Soc. Amer.* **65**, 13–21 (1975).

[125] E. Hulthén and H. Neuhaus, *Nature (London)* **173**, 442–443 (1954).

3.4.3.4.5. PRODUCTION OF GRATINGS.† About the production and the testing of gratings, and about grating errors, there is a vast literature. Much information is given by Stroke,[97] but certain procedures have not been fully publicized, for example, the technique of replication. Original gratings are not delivered to the customer any more but they serve as master gratings for replicas. The replicas may be better than the original grating. However, the master will gradually deteriorate as more replicas are made (30–50) so that the latter come in grades warranting certain specifications.

The imperfections in a grating spectrum can include deficient line profiles (affecting resolution), stray light (most troublesome in absorption experiments), and ghost lines. A deficient line profile is due to a number of closely spaced satellites ("grass"), the origin of which does not seem to be quite clear. Ghosts are due to periodic errors in the rulings which reflect the many periodicities of the engine, especially those of the screw. That periodic variations in groove position must lead to ghost images is easily understood by realizing that such grooves constitute a grating themselves with a grating constant $p \times d$, if the error repeats itself on every pth groove. This grating alone would produce $N/p - 2$ equally spaced maxima between two successive principal maxima—on the scale of γ, according to Eq. (3.4.4)—for each wavelength. However, in the perfect grating, the light vector would be zero at these positions owing to the destructive interference of the rays from all grooves. Now that these vectors have the wrong phase for each pth groove, there will be a (negative) resulting amplitude. The same conclusion applies to periodic variations of groove shape because they give rise to amplitude variations. In rich spectra, the ghosts may mask real lines of different wavelength, while other ghost images may be taken for legitimate lines. They can be detected and measured in a cross-dispersion experiment.[126] A great deal of effort has been made to perfect the ruling engines, with increasing success, but no mechanical system can ever be made with the precision required to reduce the small periodic deviations to an optically insignificant level, especially for observation in higher orders, as ghost intensities are proportional to m^2. (A displacement which is metrically the same gives rise to optical path variations, hence to phase variations, which are proportional to m. The resulting amplitudes will therefore also vary in the same way, approximately.)

[126] J. V. Kline and D. W. Steinhaus, *Appl. Opt.* **7**, 2015–2017 (1968).

† For an up-to-date and more detailed review, published after conclusion of this text, the reader is referred to E. W. Palmer, M. C. Hutley, A Franks, J. F. Verrill, and B. Gale, *Rep. Progr. Phys.* **38**, 975–1048 (1975). These authors also thoroughly discuss the manufacture and the properties of *holographic* gratings, a new field of rapidly growing interest.

Consequently, it was a big improvement when interferometric control was introduced. The system was developed at M.I.T. during a decade of studies and trials.[127–131] The advance of the carriage is servo-controlled by the signals from the interference field produced in a Michelson interferometer in which one mirror is mounted on the carriage,[128] and the motion was made continuous to avoid the transient effects accompanying stop and start operation.[129] It was found necessary to employ a second interferometer to provide automatic reduction of rotations of the carriage to 0.01 arc sec. This gave a spectacular improvement in groove straightness, while variations in the control fringe field due to changes of barometric pressure were automatically neutralized.[130] Finally, the influence of varying tilt of the carriage was eliminated by directing the main light beam in the plane of the blank.[131] This proved to reduce satellite intensity considerably. Many plane gratings were ruled over 250 mm length, a distance determined only by the size of available blanks, especially after replacing the original ^{198}Hg lamp by a stabilized helium–neon laser. A typical 300 grooves/mm grating showed almost the theoretical resolving power in the 12th and 16th order (9×10^5 and 12×10^5 respectively, for Hg, $\lambda = 253.7$ nm), while the Rowland ghost intensity in the 11th order of Hg, $\lambda = 546.0$ nm, was found to be 0.05%, which corresponds to only 0.0004% in first order. Satellite intensities were found to be less than 0.3% in the 12th order, while the efficiency in the blaze (i.e., relative to the reflectance of aluminum) was 50% for green light. The departures in the groove positions can be kept within ± 0.003 μm, which is to be compared with the classic Rayleigh condition for maximum path variation of $\lambda/4$ or 0.1 μm. Groove straightness to within 0.025 μm was obtained if the starting stretch of about 5% of the path (the acceleration of the diamond takes place here) is disregarded. An authoritative, nontechnical report is given by Harrison.[132] All these tremendous achievements would have been impossible if the technique of aluminizing had not been invented first. Even now, the wear of the diamond is not quite negligible, and this is one of the reasons why attention was focused on ruling coarser gratings. The other reason is the fact that the blaze can then be made much more effective. Such gratings of approximately 80 grooves/mm or less were termed echelles by Harrison,[133] and several

---

[127] G. R. Harrison, *J. Opt. Soc. Amer.* **39**, 413–426 (1949).

[128] G. R. Harrison and J. E. Archer, *J. Opt. Soc. Amer.* **41**, 495–503 (1951).

[129] G. R. Harrison and G. W. Stroke, *J. Opt. Soc. Amer.* **45**, 112–121 (1955).

[130] G. R. Harrison, N. Sturgis, S. C. Baker, and G. W. Stroke, *J. Opt. Soc. Amer.* **47**, 15–22 (1957).

[131] G. R. Harrison, N. Sturgis, S. P. Davis, and Y. Yamada, *J. Opt. Soc. Amer.* **49**, 205–211 (1959).

[132] G. R. Harrison, *Proc. Amer. Phil. Soc.* **102**, 483–491 (1958).

[133] G. R. Harrison, *J. Opt. Soc. Amer.* **39**, 522–528 (1949).

echelles have been ruled on the M.I.T. engine, down to 27 grooves/mm. Coarser echelles cannot be made in this way because Al coatings cannot be made sufficiently thick. Massive Al-alloy mirrors have been ruled with grating constants down to 2 mm. In common with the old echelon, the diffracting elements of the echelle are the steep sides of the grooves.

Intermediate between the ordinary gratings and the echelles are the echelettes, with about 50 grooves or more per millimeter, which were introduced by Wood in 1910[81, 134] for use in the infrared. This did not require the precision needed for the shorter wave regions, and the reflecting elements were the long sides of the grooves, although Wood compared them with the echelon. The echelette can be considered as the forerunner of blazed gratings. It was not until 1933 that the technique of aluminizing by vacuum evaporation was initiated for the purpose of high reflectivity.[135, 136] In the thirties, efforts were concentrated on the application to large telescope mirrors (up to 2.50 m)[137] and on the production of semitransparent coatings for etalon plates of limited size. Only after 1940 did the procedure become standard practice and high quality coatings up to 25 μm thick were produced, as needed for echelles. Of course, this also extended the possibilities of echelettes. Ruled on glass directly and aluminized afterwards, they became quite useful for observing high orders (30th and higher) in the visible region.[138]

It should be remarked that for *concave* gratings, too, the old speculum metal had been entirely abandoned in favor of aluminized glass or quartz concave blanks, and that the introduction of ruling under interferometric control here also has led to big improvements in performance. In replicas, the grooves are shaped in the substratum so that it is possible to clean the surface and realuminize it. This applies to echelle replicas as well.

3.4.3.4.6. ECHELLE MOUNTINGS. According to Eq. (3.4.7), the resolving power depends on the total width of the ruled surface only, as long as the angles remain constant. In the echelle, the order is high (100–1000), and the total number of grooves N is correspondingly low (e.g., 8000–800). The drawback of high orders is the smallness of the free spectral range, which is

$$\Delta\lambda = \lambda/m, \qquad \Delta\sigma = 1/\lambda m, \qquad (3.4.11)$$

expressed in wavelength or wavenumber units, respectively. For $m = 500$ and $\lambda = 500$ nm, we find $\Delta\lambda = 1$ nm. This means that the instrument has to be used in conjunction with an auxiliary dispersion, capable at least

---

[134] R. W. Wood, *Phil. Mag.* **20**, 770–778 (1910).
[135] J. Strong, *Phys. Rev.* **43**, 498 (1933).
[136] L. Holland, "Vacuum Deposition of thin Films." Chapman and Hall, London, 1956.
[137] J. Strong, *Astrophys. J.* **83**, 401–423 (1936).
[138] R. W. Wood, *J. Opt. Soc. Amer.* **37**, 733–737 (1947).

to resolve that interval. By crossing the dispersions one can obtain an echellegram which is a very compact recording of a large part of the spectrum in two dimensions.[139] An echelle combined with a Wadsworth grating mounting was used extensively for Zeeman effect investigations in rich spectra. The plate holder was constructed to follow the curve of the vertical focus instead of the horizontal one, but the resulting diffuseness of the line ends even far out of the grating axis is harmless. The region 700–200 nm could be covered in a single exposure in two grating orders; the combination was found to be 10 times as fast as a grating mounting of comparable resolution, while requiring 20 times less space.[140] A very practical internal mounting is the Littrow–echelle spectrograph in which the echelle is placed behind a 30° prism and hence, at the same time, takes care of sending the light back through the prism and dispersing it parallel to the direction of the refracting edge of the prism. Obviously, the echelle slit should be perpendicular to that direction while the original spectrograph slit is as wide open as is allowed by the value of $\Delta\lambda$.[139] The echelle can be used further in an external mounting with any stigmatic spectrograph[141]; a special echelle attachment with mirror optics is commercially available for that purpose.

Its high luminosity and compactness have made echelle instruments the ideal spectrographs to send up in space for observations of stellar spectra in the ultraviolet below 300 nm. Efficiency relative to an aluminum mirror has been measured at 5–50% between 110 and 440 nm.[142] A typical rocket-borne Littrow echelle spectrograph with a 30° CaF_2 prism was contained in a case of 133 × 150 × 800 mm. This photographed solar spectra between 400 and 200 nm with a resolution of 0.003 nm, in 60th to 120th order, with plate factors from 0.25 to 0.125 nm/mm.[143] The lens was a CaF_2–quartz–CaF_2 triplet. The weak point was stray light, particularly harmful because the solar spectrum contains outstanding emission lines. A predisperser was necessary to isolate the region of interest, but, even so, the deepest parts of the Fraunhofer absorption maxima will be filled by stray light. Usually, instruments and film can be recovered after parachuting. This kind of research by means of echelles has become standard.[144] Less strain is imposed on the optics in balloon flights, which can also carry the apparatus to heights above the ozone layer. Flights of echelle instruments equipped with a television camera to enhance sensitivity are being prepared.

[139] G. R. Harrison, *Vistas Astron* **1**, 405–413 (1955).
[140] G. R. Harrison, J. E. Archer, and J. Camus, *J. Opt. Soc. Amer.* **42**, 706–712 (1952).
[141] J. R. McNally and P. M. Griffin, *J. Opt. Soc. Amer.* **49**, 162–166 (1959).
[142] W. M. Burton and N. K. Reay, *Appl. Opt.* **9**, 1227–1229 (1970).
[143] R. Tousey, J. D. Purcell, and D. L. Garret, *Appl. Opt.* **6**, 365–372 (1967).
[144] B. C. Boland, B. B. Jones, and S. F. T. Engstrom, *Solar Phys.* **17**, 333–354 (1971).

The echelle can be applied for precision measurements of wavelengths. This is done by identifying the exact order and measuring fractions with respect to a fiducial line corresponding to a constant value of $m\lambda_0$. This reference line is obtained as the image of the echelle slit in all wavelengths by swinging a mirror in front of the echelle,[145] a procedure already well-known with echelons, but simpler with the echelle.

From the remarks made on ghosts in the preceding section it might be inferred that the ghost problem would be prohibitive in echelles. However, the increase of ghost intensity with m^2 applies to one grating with a single grating constant. Actually, the ghost intensity is also inversely proportional to the square of the grating constant when other things are being kept equal. In the echelle the order is high, but the grating constant is correspondingly large, so that the ghost problem is not more serious then in gratings.

3.4.3.4.7. THE ECHELON. This is not manufactured by ruling, but by stacking a number of glass or quartz plates of successively decreasing height. The transmission type was invented in 1898 by Michelson[146] and has been used for hyperfine structure investigations for some 30 years. A coherent beam of light enters the largest plate at normal incidence. The plates are in contact with each other and have the same optical thickness nt everywhere. The diffracted beams leaving the plates then have the constant optical path difference $(n - 1)t$ so that, for $t = 10$ mm and $\lambda = 500$ nm, the order becomes $m \approx 10^4$, Hence, with 30 plates, $R_{\text{theor}} \approx 3 \times 10^5$ can be reached. The free spectral range is only 0.05 nm, and it is not possible to see more than two orders simultaneously with reasonable brightness because, as an essential result of the geometry, the order distance is just half the distance between the two first minima of the diffraction figure due to each element individually. Hence, the echelon can be regarded as the most effective—and the oldest—blazed grating. However, owing to the smallness of the free spectral range, it needs another dispersing instrument to separate neighboring spectral lines.

The reflection type, if it has the same dimensions, yields the order $2t/\lambda = 4 \times 10^4$, so that R_{theor} becomes 1.2×10^6, while the free spectral range decreases to 0.0125 nm. The reflection echelon requires equal *metrical* thickness of the plates and, therefore, it cannot be made by coating the facets of a good transmission echelon. Moreover, in connection with the procedure of optical contacting, it is essential that fused silica be used for the reflection type.[147] It was proposed in 1926[148] and realized as an instrument for

---

[145] G. R. Harrison, S. P. Davis, and H. J. Robertson, *J. Opt. Soc. Amer.* **43**, 853–861 (1953).
[146] A. A. Michelson, *Astrophys. J.* **8**, 37–47 (1898).
[147] W. E. Williams, *Proc. Phys. Soc. (London)* **45**, 699–726 (1933).
[148] W. E. Williams, *Proc. Opt. Convent.* **2**, 982–990 (1926).

hyperfine-structure research in 1930.[149] The mounting was internal, as the back reflector of a Littrow prism spectrograph. An external mounting was used for the same purpose in 1931,[150] but with a 3-m collimating mirror this required an extra path length of 12 m. For observing structures in the most favorable position, the path difference must be variable by regulating the gas pressure in an airtight enclosure. The big advantage of the reflection echelon over other high-order instruments is that there is no restriction on the wavelength, in principle. It also has a special advantage for measuring absolute wavelengths by a method closely related to the one discussed for echelles. For this application, the echelon is equipped with flat side mirrors to provide the necessary fiducial lines.[147] Since the order is much higher than with the echelle, the determination of the exact order number requires greater effort, but the reward is a higher precision of the measurements.[151,152]

From Fig. 16 one obtains, for the path difference between two rays incident on corresponding points of two successive elements,

$$C'A + AC'' = (BC - AB \tan \psi) \cos \psi + (BC - AB \tan \chi) \cos \chi$$
$$= t(\cos \psi + \cos \chi) - h(\sin \psi + \sin \chi),$$

which for small ψ and χ reduces to $2t - h(\psi + \chi)$. In ordinary grating theory, where the grating surface would be the line ACE---, the angle of

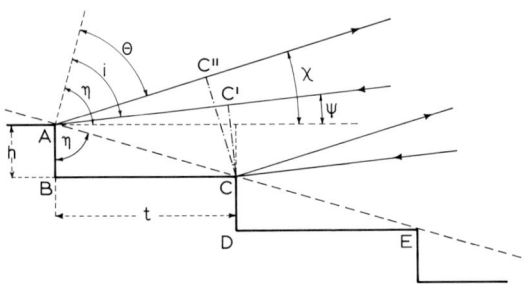

FIG. 16. Principles of the reflection echelon.

incidence would be $i = \eta - \psi$, where η is the blaze angle, and the angle of diffraction would be $\theta = \eta - \chi$. It is easily shown that the grating equation (3.4.5) boils down to the same expression as obtained above, so that in our approximation one has

$$2t - h(\psi + \chi) = m\lambda. \tag{3.4.12}$$

[149] D. A. Jackson, *Proc. Roy. Soc. (London)* **A128**, 508–522 (1930).
[150] R. Ritschl, *Z. Phys.* **79**, 1–25 (1932).
[151] W. E. Williams and A. Middleton, *Proc. Roy. Soc. (London)* **A172**, 159–172 (1939).
[152] T. A. Littlefield and D. T. Turnbull, *Proc. Roy. Soc. (London)* **A218**, 577–586 (1953).

Differentiating for constant ψ and λ, we find $-h\, d\chi = \lambda\, dm$, whence $d\chi/dm = -\lambda/h$. For $dm = 1$ we get the angular distance $\Delta\chi$ between two successive orders, $\Delta\chi = \lambda/h$, which is precisely half the distance between the two first minima of the diffraction figure of an element of width h, as already stated in a more specific way. This is strictly correct only if the light beams make full use of the opening of each element; i.e., there should be no obstruction or "shadowing." It should also be remarked that the derivation assumes the reflection to take place in the plane perpendicular to the edges of the steps, as a small inclination influences the path difference. Because of this, the internal Littrow mounting is at a disadvantage, for the incidence can be correct only for one wavelength.[153]

The intensity envelope given by the diffraction factor $(\sin^2 \beta)/\beta^2$ of Eq. (3.4.4) causes a distortion of the pattern of a hyperfine structure, but it also offers the possibility of measuring intensity ratios of components without having to calibrate plates, or detectors; namely, by determining the gas pressure at which two components have the same apparent intensity as a result of their positions with regard to the fixed envelope. In a transmission echelon the theoretical factor has been found to be unrealistic,[154] whereas in a reflection echelon it holds accurately.[155]

The drawback of the very small spectral range is inherent in using high orders. An interesting device to circumvent it is the compound echelon that was realized as a transmission instrument.[156] In this, the plates had two different thicknesses, in the ratio 10:17, and these were stacked alternatively. Only for angles in which both kinds of plates produce retardations of an integer number of wavelengths will there be maximum intensity for that wavelength. This will occur only at relatively large intervals. Since for constant ψ and m we have $d\chi/d\lambda = -m/h \propto t/h$, this ratio has to be kept constant so that the step heights must be made in the same ratio as the thicknesses in order to get coincidence of all components in the same order. The larger facets were then masked properly in order to make all beams equally strong for maximum resolution. The total number of plates was 56.

In some treatments, echelons are considered on par with interferometers, with which they share the property of small free spectral range and, consequently, the need for auxiliary dispersion. But they are properly to be regarded as gratings, and we have seen that there are intermediate instruments which close the gap between echelons and ordinary gratings. As a diffraction instrument the echelon has to be illuminated with coherent

[153] K. W. Meissner, *J. Opt. Soc. Amer.* **32**, 185–211 (1942); **31**, 405–427 (1941).

[154] H. C. Burger and P. H. van Cittert, *Versl. Kon. Akad. Wetensch. Amsterdam* **29**, 394–400 (1920).

[155] J. H. Gisolf, *Proc. Roy. Acad. (Amsterdam)* **38**, 225–229 (1935).

[156] E. Lau, *Z. Phys.* **80**, 100–104 (1933).

light, while it forms the various interfering beams by division of wave front; hence the necessity of an entrance slit, with all the restrictions as discussed in Section 3.4.3.2, to be accounted for.

3.4.3.5. Interference Instruments. In these, the interfering rays arise from division of amplitude. Because they are coherent by nature, the structure of the incident beam is immaterial, so that no collimation is needed. Of the two competing instruments in this category, the Lummer plate has become obsolete as a result of the tremendous technical improvements of the Fabry–Perot interferometer which, in its rigid form, is best known as an *etalon*. Temporarily, until about 1930, the Lummer plate had been in the lead for short waves, because no nonabsorbing coating materials were available which could procure the high reflectivity needed for etalons. In the Lummer plate, reflectivity does not depend on such coatings, as the reflections within the plane-parallel plate take place at an angle close to the critical angle. By means of a crystalline quartz plate, the ultraviolet down to about 230 nm was accessible for high-resolution work, but there are few top quality plates in existence. The reader is referred to Tolansky[42] for interesting details.

3.4.3.5.1. BASIC FACTS ABOUT THE ETALON. The Fabry–Perot interferometer was invented in 1897,[157] but it has undergone continuous improvements since. Its basic structure and principles are discussed in almost every textbook on optics,[83–88] while more detailed treatments can be found in texts on spectroscopic instrumentation.[42, 79] The reader is also referred to the two articles by Meissner[153] and the report by Jacquinot.[158] The improvements made in the course of time concern the reflecting coatings, the planeness and size of the plates, and the recording technique. The greatest single breakthrough was the invention of dielectric multilayer coatings, as it eliminated the problem of absorption, which had always been the limiting factor with metallic films.[159] A pile of $\lambda/4$ layers of alternately low- and high-index transparent materials, like Na_3AlF_6 and ZnS, deposited on glass or quartz by vacuum evaporation, can make reflectivity arbitrarily high, without undue absorption losses for any wavelength desired, in the infrared and visible regions. In the ultraviolet, other combinations of dielectrics, such as LiF for the low and $PbCl_2$, PbF_2, or Sb_2O_3 for the high index, have to be used.[160] Below 240 nm no suitable materials of high index are available yet. The drawback against metallic layers is the selectivity.

---

[157] Ch. Fabry and A. Perot, *Ann. Chim. Phys.* **12**, 459–501 (1897); *C. R. Acad. Sci. Paris* **126**, 34–36 (1898).
[158] P. Jacquinot, *Rep. Progr. Phys.* **23**, 267–312 (1960).
[159] P. Jacquinot and Ch. Dufour, *J. Phys. (Paris)* **11**, 427–431 (1950).
[160] S. P. Davis, *Appl. Opt.* **2**, 727–733 (1963).

3.4. OPTICAL REGION

Strictly speaking, a $\lambda/4$ (actually $\lambda_0/4n$) layer is correct only for one wavelength. By letting the thickness vary a little over the different layers of the pile, the high-reflectivity region can be spread out to some 10 nm without losing too much on the peak value, but metallic coatings cover hundreds of nm. In the long wave region, it is hard to improve on silver if this is properly deposited,[161] but below 550 nm it becomes increasingly difficult to use metals. The best of them, aluminum, had to wait for application until the evaporation method was developed[136] (compare Section 3.4.3.4.5). In the thirties, the "Hochheim alloy" was successful.[162] Its composition was kept secret but it surely was aluminum skillfully deposited by evaporation, possibly with a little magnesium. Later, it has been found by many investigators that the structure of thin films prepared by evaporation varies with the conditions, and that the most favorable optical properties are obtained when the vacuum is highest and the coating procedure takes the shortest possible time.[163, 164] Recently, the use of aluminum has been extended to 180 nm by applying overcoatings of MgF_2 to prevent oxidation; in this way it was possible to attain 80% reflectivity for the mercury light at 184.9 nm.[93, 94] In addition to the extremely large region covered, Al has the advantage over dielectric coatings of having a small and regular phase dispersion, which makes it preferable for standard wavelength measurements.

Nonetheless, in no region does Al attain the reflectivity values shown by silver about 550 nm, while for problems requiring the highest resolution, the dielectric multilayer coatings have conquered the scene completely. The technique of vacuum deposition has become quite sophisticated. Not only the fact that optical control during the process is necessary,[165] but the condition of uniformity over relatively large areas has given rise to solutions involving complicated rotary motions of the plates, and devices to accommodate more than one plate simultaneously, while the vacuum requirements constantly increased in view of the large distances (over 0.5 m) between source and object which had to be introduced.

All the efforts made towards higher reflectivity would be useless if the plates themselves had not been improved. Actually, as a result of the success of the multilayers, the pressure on the optical firms to manufacture better plates was steadily increasing. While prior to 1940 a flatness of $\lambda/50$ over

---

[161] D. A. Jackson and H. Kuhn, *Nature (London)* **170**, 455–456 (1952); A. H. Jarrett, *Nature (London)* **170**, 456 (1952).

[162] F. Hlučka, *Z. Phys.* **96**, 230–235 (1935).

[163] J. C. Burridge, H. Kuhn, and Anne Pery, *Proc. Phys. Soc. (London)* **B66**, 963–968 (1953).

[164] P. Rouard and P. Bousquet, *Prog. Opt.* **4**, 145–197 (1965).

[165] P. Giacomo and P. Jacquinot, *J. Phys. (Paris)* **13**, 59–64A (1952).

an area of a few hundred square millimeters was considered to be exceptionally good,[166, 167] today a flatness of $\lambda/200$ over a circular area of 70 mm diameter has become standard. Even so, it is always advisable to keep clear of the edges over about 10 mm and still, in many cases, resolution can be improved by using smaller stops.

The interfering rays arise from the light which is multireflected within the plane-parallel space between the two coatings, as illustrated in Fig. 17. The space is usually air-filled or evacuated. Consequently, the etalon plates have no other function than to offer the flat faces for coating. However, they

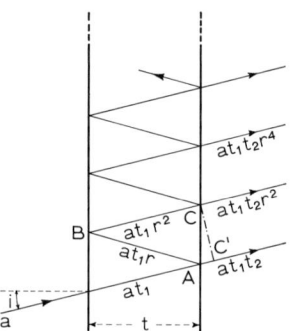

FIG. 17. Formation of fringes by an etalon; in a maximum, the resulting amplitude becomes $at_1t_2(1 + r^2 + r^4 + \cdots) = at_1t_2/(1 - r^2)$.

stand in the path of the light which they should not perturb. If the plates were plane-parallel themselves they would produce weak parasitic fringe systems perturbing the main system. So, the plates are wedge shaped with an angle of up to a few degrees, which throws the parasitic systems off axis, making it possible to block them.[42]

3.4.3.5.2. THEORY OF THE ETALON. With the ideal etalon (used in transmission), the intensity distribution in the interference pattern, formed in the focal plane of a lens, would follow the well-known Airy formula

$$I = I_0/[1 + (4\rho/(1 - \rho)^2) \sin^2(\delta/2)], \quad (3.4.13)$$

where ρ is the reflectivity, δ the phase difference between successive rays, and I_0 contains the factor $t_1t_2/(1 - r^2)$ which is unity only if the absorption is zero. The meaning of the amplitude transmission and reflection coefficients will be clear from Fig. 17, and it follows that $\rho = r^2$, assuming the plate coatings to be equal. The path difference between successive rays is $AB + BC - AC' = 2t \cos i$ in vacuo. If the medium is a gas, refraction

[166] Ch. Dufour and R. Pica, *Rev. Opt.* **24**, 19–34 (1945).
[167] E. Rasmussen, *Kgl. Dan. Vidensk. Selsk, Mat. Fys. Medd.* **23**, No. 3, 1–18 (1945).

3.4. OPTICAL REGION

is negligible, so that the optical path difference becomes $2nt \cos i$ and the corresponding phase difference is $4\pi nt(\cos i)/\lambda_0$. The phase jumps at reflection can be ignored for measurements of narrow structures; they have to be taken into account in comparisons of widely different wavelengths.

Because the condition for maximum intensity, $\sin(\delta/2) = 0$, (or m integer) in

$$2nt \cos i = m\lambda_0 \tag{3.4.14}$$

in a stationary situation depends on i only, the maxima corresponding to slightly divergent or convergent monochromatic light incident along the axis will form a system of concentric circles, with the highest order nearest to the center. These Fabry–Perot fringes (or rings) are sharper for higher values of ρ. If the (extended) light source is at the focus of a lens, each point in the interference pattern corresponds to one point in the source.

Since the curve described by Eq. (3.4.13) and represented by Fig. 18 for $\rho = 90\%$ does not have subsidiary maxima and does not become zero, either, the resolving power cannot be defined by means of the Rayleigh criterion. Instead, one considers two equally strong radiations to be resolved if the two adjacent maxima of the same order intersect at half the peak intensity. This makes the sum intensity in the center equal to I_0; at the position of each individual maximum it becomes $I_0 + (I_0/5)$ when $\sin(\delta/2)$ is small, which, of course, is the only case we are interested in. So, the intensity saddle in the sum curve becomes $\tfrac{5}{6}$ of the peak intensities or 83.3%, which is to be compared with the 81% found for slit instruments (Section 3.4.3.2).

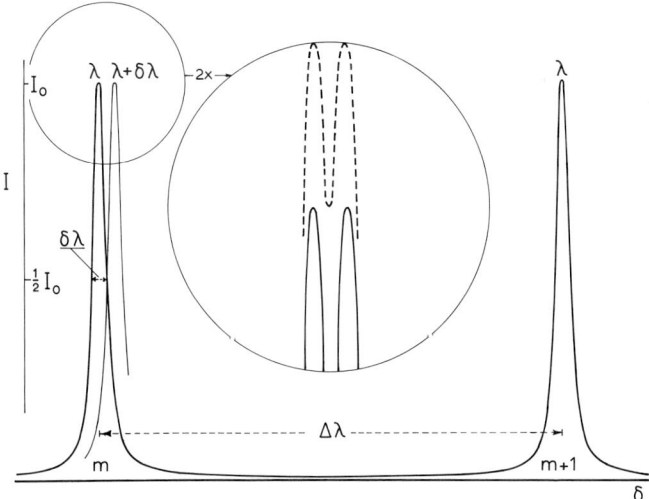

FIG. 18. Instrument profile of an etalon at $\rho = 90\%$, with two just resolved peaks and the sum curve enlarged by a factor of 2 (inset).

There is no use trying to get a better agreement as both criteria are somewhat arbitrary. Now, for small $\sin(\delta/2)$ we have $\delta = m2\pi + \varepsilon$, m being an integer and ε being small. At half-intensity, we have

$$\sin^2(\delta/2) = (1 - \rho)^2/4\rho$$

or

$$\sin(\delta/2) = (1 - \rho)/2\sqrt{\rho} = \sin(m\pi + (\varepsilon/2)) = (\pm)\varepsilon/2.$$

This makes ε, the increase of phase difference for the wavelength λ from the maximum to the point of half-intensity, equal to $(1 - \rho)/\sqrt{\rho}$. For the mth order maximum of the wavelength $\lambda + \delta\lambda$, this phase difference has increased to 2ε, corresponding to an order increase $2\varepsilon/2\pi = (1 - \rho)/\pi\sqrt{\rho}$. This means that the mth order fringe of $\lambda + \delta\lambda$ is coincident with the order $m + [(1 - \rho)/\pi\sqrt{\rho}]$ of λ, which makes $\{m + [(1 - \rho)/\pi\sqrt{\rho}]\}\lambda = m(\lambda + \delta\lambda)$ or

$$R_{\text{theor}} = \lambda/\delta\lambda = m\pi\sqrt{\rho}/(1 - \rho) = mF, \qquad (3.4.15)$$

in which the factor F is called finesse.

The fact that the order can be made indefinitely large by increasing the plate distance t determines the uniqueness of the instrument for investigations of the narrowest structures, but its usefulness also depends on the free spectral range which is found by equating $(m + 1)\lambda$ and $m(\lambda + \Delta\lambda)$, giving $\Delta\lambda = \lambda/m$. Putting $m = 2t/\lambda$, one obtains

$$\Delta\lambda = \lambda^2/2t \quad \text{or} \quad \Delta\sigma = 1/2t. \qquad (3.4.16)$$

For $t = 10$ mm and $\lambda = 500$ nm, we find $\Delta\lambda = 0.0125$ nm only. From $\lambda/\delta\lambda = mF$ and $\Delta\lambda = \lambda/m$ we obtain

$$\Delta\lambda/\delta\lambda = \Delta\sigma/\delta\sigma = F. \qquad (3.4.17)$$

So, in order to utilize the small space between two successive orders most effectively, one requires a high value of finesse, which demonstrates the paramount importance of high reflectivity. For $\rho = 90\%$, we get $F = 30$, which in our example makes $R_{\text{theor}} = mF = 1.2 \times 10^6$. There is a close analogy between Eq. (3.4.15) and Eq. (3.4.6) for gratings, which has led to the designation of F also as the effective number of rays, N_e. We push this idea a little further to find that even Eq. (3.4.1) from diffraction theory is formally applicable to the etalon. The effective width w_e of the emerging beam is $N_e \times 2t \sin i$ (Fig. 17). From $2t \cos i = m\lambda$, we derive $D = di/d\lambda = (-)m/2t \sin i$, hence $w_e D = mN_e = R_{\text{theor}}$. Replacing m by $2t(\cos i)/\lambda$, the angular dispersion becomes

$$D = di/d\lambda = 1/\lambda \tan i, \qquad (3.4.18)$$

which, remarkably enough, is independent of t.

The imperfections of the etalon can be accounted for by redefining the finesse.[79] The one defined by Eq. (3.4.15) is that determined by ρ only and should be termed reflection finesse F_r. Variations of t caused by plate surface irregularities, and eventually also by bad adjustment, can be treated by considering the superposition of the effects of many perfect etalons with different values of t, and be represented by a plate finesse, F_p.[168] This determines another value of R_{theor} and of $\Delta\lambda/\delta\lambda$. From this it will be clear that there is no point in increasing F_r much above F_p.

3.4.3.5.3. THE ETALON IN PRACTICE. The smallness of the free spectral range necessitates the use of an auxiliary dispersion in the same way as discussed for echelons and echelles. On photographic plates, when the dispersions are crossed and the spectrograph is stigmatic, a large region can be recorded on a small area. In order to interpret the observed structures (hyperfine structure or isotope shift patterns), several spectrograms are necessary, with different spacings, preferably not in simple ratios. When dielectric coatings are used, the whole procedure has to be repeated for different wavelength regions. The slit of the spectrograph can be opened as wide as allowed by the distance between adjacent lines, for maximum illumination and for extension of the visible part of the fringes, in behalf of accuracy. The determination of small wavelength differences takes place by measuring the diameters of the rings in several orders and evaluating the readings in terms of the known free spectral range.[42] If one is interested in the structure of a few lines only, it may be more advantageous to use the spectrograph as a monochromator and concentrate on the best possible adjustment for those particular lines, while observing much larger portions of the ring system. This also is the best method for obtaining direct recordings, which are especially important when intensity ratios are to be determined. A very sophisticated instrument has been developed according to this principle in the Laboratoire Aimé Cotton and is marketed under the trade name "Hypeac."[169]

In this apparatus, the Fabry–Perot fringes are forced to pass along a photomultiplier by changing the pressure, while the detector electronically activates a pen recorder. When the pressure change and the paper shift are synchronous, the pen will write a large scale Airy pattern for each component, and if the detector system responds linearly to the flux, and the light source is stable, the resulting curve is a faithful representation of the intensities. The apparatus embodies an Ebert monochromator which allows the convenient isolation of lines by positioning the grating. The output light then passes the etalon which is in an airtight chamber. In principle,

[168] R. Chabbal, *J. Phys.* (*Paris*) **19**, 295–300 (1958).
[169] R. Chabbal and P. Jacquinot, *Rev. Opt.* **40**, 157–170 (1961).

the detector could stand anywhere in the Fabry–Perot pattern, formed in the focal plane of a mirror system, but for reasons of intensity, the center is chosen, where a small circular diaphragm is placed. Its size determines the bandwidth of the radiation admitted, and thus it contributes to F so that a corresponding diaphragm finesse F_d has to be introduced,[79] in addition to F_r and F_p. In order for F_d to exceed F_r, the wavelength change from the center to the edge of the diaphragm should be smaller than $\delta\lambda = \lambda/mF_r$. If at the edge we have $i = \alpha$, which is supposed to be a very small angle, we have $\cos i = 1 - (\alpha^2/2)$, or $\delta(\cos i) = \alpha^2/2$ in $2nt\,\delta(\cos i) = m\,\delta\lambda$, which makes $\alpha = (2/mF_r)^{1/2}$ the maximum permissible value. In the example used before it is $[2/(1.2 \times 10^6)]^{1/2} = 1.291 \times 10^{-3}$. When a is the radius of the central hole and f the focal length of the imaging system, we have $\alpha = a/f$. For $f = 500$ mm, a becomes 0.65 mm to make $F_d = F_r$. Usually, it does not pay to make it much smaller, while the flux decreases quadratically with a. Under optimal conditions, one expects F_r, F_p, and F_d to be of the same magnitude. The profile of a single line depends on all three factors, and the final resolution can be assumed to be $0.6R_{\text{theor}}$.

The index of refraction of a gas increases linearly with density. For standard air it is $1 + (2.7 \times 10^{-4})$, so that about five orders will be recorded, when air is admitted into the evacuated chamber up to about $\frac{1}{2}$ atm. For small values of t, high-index vapors have been used instead of air, and also variation of t has been applied. In most cases, the air flow is assumed to be steady, but it is easy to make corrections for slight variations of the order distance, if any. However, some authors have used devices in which the pressure itself is the primary variable in the recording.[170, 171] Slow fluctuations in the light source can be accounted for when the intensity of the direct light is being recorded simultaneously, to which end a small percentage of the direct beam is deviated by means of a beam splitter to be detected by means of a second photomultiplier.

Pressure scanning is necessarily slow. Mechanical scanning requires extreme caution to maintain parallelism but has been successfully accomplished. This is very fast, covers many orders, and can be made repetitive by controlling t through an electric oscillator.[172] With scanning times of the order of a millisecond or less, the fluctuations in a dc hollow cathode source become negligible. Servo systems have been developed to secure plate parallelism.[173] However, with disruptive light sources, the scanning methods are not appropriate. The photographic method is not subject to

---

[170] H. G. Kuhn and H. J. Lucas-Tooth, *J. Phys. (Paris)* **19**, 293–294 (1958).

[171] J. G. Hirschberg and R. R. Kadesch, *J. Opt. Soc. Amer.* **48**, 177 (1958); J. G. Hirschberg, *Phys. Rev.* **99**, 623A (1955).

[172] D. J. Bradley, *J. Sci. Instrum.* **39**, 41–45 (1962).

[173] J. V. Ramsey, *Appl. Opt.* **1**, 411–413 (1962).

such limitations and, moreover, has the advantage of producing all information on a multitude of lines simultaneously. It is also indispensable for studies of wavelength standards in which case large values of t are often used. The largest value with which a good interference pattern was still observed in a natural source was 1.02 m; namely, in the line at 422.6 nm of CaI in an atomic beam lamp. Its corresponding width was estimated to be less then 5.4×10^{-5} nm![174] However, the photographic method is nonlinear so that intensity evaluation is inaccurate, especially when components perturb each other. Photomultipliers, on the other hand, have a linear response over a wide scale of intensities, when handled properly, so that the recordings can be analyzed easily. The signal can also be digitized immediately, whereupon the data can be fed to a computer for fast processing and analyzing complicated structures with many overlapping components.[175] It is of paramount importance, especially when there are weak components unresolved from strong ones, to know the exact profile of a single line under the conditions of the experiment. If such lines are not found in the spectrum studied, the Th lines lend themselves excellently to that purpose.[176] A very particular application of the central spot scanning method can be made in the study of isotope shifts by means of separated samples. These are in different sources, and the light from each of them is allowed to enter the etalon alternatively during certain periods of time while the scanning proceeds. The interchange can be accomplished by actual exchange of light sources[177,178] or by means of an adjustable mirror system.[175] Figure 19 illustrates the procedure (for Dy); the crosses indicate the switch-over points. In this way, shifts can be measured which are less than $0.1\delta\lambda$ since one isotope component is lying free between adjacent orders of a different isotope.

Apart from the advantage for intensity analysis, the big gain of the scanning methods in comparison with photography is the high sensitivity. This makes it possible, in favorable cases, to decrease the current in a hollow cathode source to about 10 mA, whereby the Doppler broadening is greatly reduced, the gas temperature being nearly equal to that of the coolant.

3.4.3.5.4. LUMINOSITY. It has been proven, generally, that the etalon has a much higher luminosity than the (blazed) grating,[82] even up to a hundredfold. But this is to be understood in the same sense as was mentioned in

---

[174] K. W. Meissner and V. Kaufman, *J. Opt. Soc. Amer.* **49**, 942–943 (1959).

[175] J. W. M. Dekker, P. F. A. Klinkenberg, and J. F. Langkemper, *Physica* **39**, 393–412 (1968).

[176] J. W. M. Dekker, H. F. Bloemhof, J. H. Brouwer, and P. F. A. Klinkenberg, *Physica* **46**, 119–132 (1970).

[177] K. Heilig, *Z. Phys.* **161**, 252–266 (1961).

[178] J. E. Hansen, A. Steudel, and H. Walther, *Z. Phys.* **203**, 296–329 (1967).

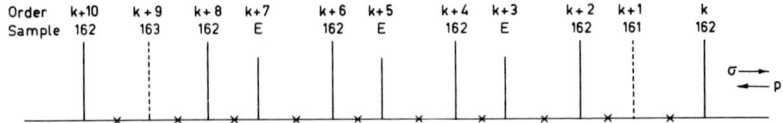

FIG. 19. Measuring isotope shifts by the method of switching light sources; E is the even mass isotope component to be determined against mass 162 (dysprosium), while the barycenters of the odd mass isotope components (hyperfine splitting) are given by the dashed lines. The positions drawn are not true to nature; the shifts are exaggerated for clarity.

Section 3.4.3.1 with regard to grating versus prism. Moreover, the statement is meant for instruments of comparable size, but etalon plates cannot yet be manufactured as large as gratings. One of the principal causes injuring the performance is in the plate imperfections. The theoretical inference, in the absence of absorption, that the resulting amplitude in a maximum is equal to the amplitude of the incident beam (Fig. 17), requires absolute phase agreement. This not being perfect means that maximum transmission is not found in the same direction for different parts of the surface. Apart from this, the intrinsic luminosity lies in the "Jacquinot advantage" of rotational symmetry, which is quite obvious in the case of central spot scanning, and in the principle of division of amplitude, which makes it unnecessary to use an entrance slit. But, again, this cannot be fully exploited if the optical train does not possess equivalent properties. In the Hypeac, loss of light is caused, first, by the Ebert monochromator and, secondly, by the fact that the exit slit is rectangular so that the illuminated etalon area is a strip along a diameter, whose width is limited by the line density in the spectrum. Finally, it has to be borne in mind (Section 3.4.3.1) that the results depend also on the method of detection, because, in photography, the illumination is responsible for the signal, not the flux. Nonetheless, the statement that the etalon is intrinsically the brightest of all dispersive instruments is legitimate.

3.4.3.5.5. COMPOUND SYSTEMS. An elegant method to overcome the drawback of the small spectral range is the compound etalon described by Fabry and Perot in 1899. This consists of two etalons in tandem, with thicknesses in simple ratios. When one etalon has t four times that of the other one, the free spectral range is determined by the thinner one which can be conveniently looked upon as a filter for certain directions, while the resolving power is essentially given by that of the thicker etalon. In the first application, the two etalons were exactly equal and then the gain is only in R_{theor}; viz., a factor of $\sqrt{2}$.[179] The possibility of combining etalons of different t is based on Eq. (3.4.18), which proves that the angular dispersions

---

[179] V. W. Houston, *Phys. Rev.* **29**, 478–484 (1927).

are the same. So, if one component corresponds to beams having the same value of i for both etalons, this is so for all components belonging to the same order numbers. This coincidence position is most easily obtained by pressure adjustment. It does not require the thicknesses to be exactly in an integer ratio, because in orders of 10^4–10^5 a difference of a few is immaterial. In fact, a convenient phase adjustment method makes use of white light fringes occurring only if there is such a difference.[180] Meissner[153] gives an accurate theory of the compound etalon. It was used in the thirties in a series of researches on the Zeeman effect in hyperfine structure, in combination with atomic beam absorption cells.[181–183] The assembly is specially suited for absorption work because this is not hampered by the presence of unsuppressed orders; i.e., those orders transmitted by the bigger gap that should not pass the smaller one. But for intensity reasons, the reflectivities were on the low side so that the suppression was not very effective. Therefore, the invention of multilayer coatings, with negligible absorption, boosted the development.[184] Even three etalons in series have been employed, which, together with an interference filter, can operate without auxiliary dispersion[185] and thus can fully exploit the advantage of rotational symmetry. With such a system, the D_2 line of interstellar sodium in α Cygni has been observed[186] with a resolution of 5×10^5.

A plane-parallel plate clad with reflective coatings on both sides is an etalon in itself, for which the fundamental equation is

$$2nt \cos i' = m\lambda_0, \qquad (3.4.19)$$

where i' is the angle of refraction within the plate. From this, one obtains, under the conditions that i' be small and dispersion be negligible,†

$$\partial i'/\partial \lambda = 1/\lambda \tan i' \quad \text{or} \quad \partial i/\partial \lambda = n^2/\lambda \tan i. \qquad (3.4.20)$$

Comparison with Eq. (3.4.18) shows that the angular dispersion—again independent of t—is much larger than in an air-spaced etalon. The advantage is offset by the high cost and the lack of flexibility, so that this instrument has been used sporadically. However, its advantages for tandem use were

[180] H. G. Kuhn and S. A. Ramsden, *Proc. Phys. Soc. (London)* **B69**, 1309–1311 (1956).
[181] D. A. Jackson and H. G. Kuhn, *Proc. Roy. Soc. (London)* **A167**, 205–216 (1938).
[182] D. A. Jackson and H. G. Kuhn, *Proc. Roy. Soc. (London)* **A173**, 278–285 (1939).
[183] D. A. Jackson, *Physica* **12**, 568–580 (1946).
[184] D. A. Jackson, *J. Phys. (Paris)* **17**, 977–987 (1956); *Phys. Rev.* **103**, 1738–1739 (1956).
[185] J. E. Mack, D. P. McNutt, F. L. Roesler, and R. Chabbal, *Appl. Opt.* **2**, 873–885 (1963).
[186] L. M. Hobbs, *Astrophys. J.* **142**, 160–163 (1965).

† These expressions are only approximately correct because the change of n with wavelength has been ignored with the substitution of $n\lambda$ for λ_0.

soon recognized: only one adjustment is needed. A number of investigations were made after which the method was abandoned.[187-193]

3.4.3.5.6. SPHERICAL ETALON PLATES. A new development came with the invention of the spherical etalon, in the confocal arrangement. This has no dispersion at all so that only the scanning method is applicable.[194] Its luminosity is essentially proportional to R_{theor}, whereas that of a plane etalon is inversely proportional to R_{theor}. For $t > 50$ mm, the latter becomes inferior in throughput; this is equivalent to t (= radius) > 25 mm for the spherical etalon, because the path difference in this instrument is $4t$. In a most interesting application of the spherical etalon, an atomic beam of Ba was used as an absorbing medium *internally*, which enhanced the absorption by a factor of at least 40 and made it possible to study the structure.[195] The trick of introducing emitting or absorbing atoms inside an etalon was independently suggested for a plane etalon.[196] For the case of emission, this led to the conclusion that the light is concentrated in thin layers parallel to the plates because standing waves develop which are associated with one wavelength only, resulting in Doppler narrowing. Actually, only the atoms with the appropriate velocity are in resonance with the electromagnetic field. The phenomenon is closely related to the mechanism of a gas laser, the principal difference being the fact that the atoms in a continuous wave gas laser constitute an *active* medium, a situation in which it is kept artificially by permanent optical pumping or collisional energy transfer. It is here that the concave Fabry–Perot mirrors are superior; t is very large and dispersion is immaterial. But the dielectric coatings are also essential because the amplification factor (the gain) is small and no absorption losses can be tolerated. In practice, various mirror configurations are in use, among which the hemispherical arrangement with one plane and one concave mirror is the most frequent one.

3.4.4. Detection of Optical Radiation

3.4.4.1. General Considerations. Detectors are manufactured in great variety industrially. The choice is made in connection with the problem to solve. If the purpose is precise determination of wavelengths by means of

---

[187] E. Gehrcke and E. Lau, *Z. Tech. Phys.* **8**, 157–159 (1927).
[188] E. Gehrcke and E. Lau, *Phys. Z.* **31**, 973–974 (1930).
[189] E. Lau, *Z. Phys.* **63**, 313–317 (1930).
[190] L. A. Sommer, *Z. Phys.* **75**, 134–136 (1932).
[191] E. Lau and E. Ritter, *Z. Phys.* **76**, 190–200 (1932).
[192] H. C. Burger and P. H. van Cittert, *Physica* **2**, 87–96 (1935).
[193] H. C. Burger and P. H. van Cittert, *Physica* **5**, 177–187 (1938).
[194] P. Connes, *J. Phys.* (*Paris*) **19**, 262–269 (1958).
[195] D. A. Jackson, *Proc. Roy. Soc.* (*London*) **A263**, 289–308 (1961).
[196] A. Kastler, *Appl. Opt.* **1**, 17–24 (1962).

3.4. OPTICAL REGION

a dispersive instrument in the region below 1000 nm, there is no method of detection matching the photographic one. For precise measurements of intensity ratios, let alone of absolute intensities, it is not in the same favorable position, especially not when a stable source or a source of good reproducibility is available. In case the instrument is nondispersive, the photographic method is not applicable, unless one introduces very special procedures as embodied in the SIMAC.[197] We already have pointed out the advantages of photoelectric detection in connection with total flux considerations in Sections 3.4.3.2 and 3.4.3.5.4. In cases of very low flux it may be imperative to use photoelectric devices, and in the ultraviolet, photon counting is applicable. Moreover, in the long wave region, above 800 nm, photographic sensitivity decreases rapidly; although materials are available which can record radiation up to 1250 nm, direct-recording methods are much in use in that region, while they are indispensable in the farther infrared. A very important piece of information, namely, the intensity distribution over the source in each spectral frequency, is immediately available on a spectrogram, albeit on a nonlinear scale, as demonstrated in Fig. 5. This sort of information would be retrievable by any other method of detection only by performing several repeated scans, with a very stable light source under reproducible conditions, such as cannot often be maintained rigorously.

An advantage of direct recording which cannot possibly be imitated by means of the photographic method lies in the possibility of adjusting the immediate response. In a photomultiplier, one can vary the dynode voltage at will and, with photocells or photoconductive devices, one can change the external amplification (this is, of course, also possible with a multiplier). When this is done in synchronism with an intermittent light source, all background radiations in unwanted frequencies, including frequency zero, can be suppressed. In other applications, the principle has been used for obtaining time resolution.[59] When the light source is stationary, it may be useful to interrupt the beam artificially in order to make amplification possible and remove background.

Instead of using the immediate response, it is also possible to integrate the signal over a certain period of time, thereby enhancing the signal-to-noise ratio. When this is done in combination with repeated sweeping over a certain part of the spectrum in a multichannel device, one can get rid of long period source fluctuations in the same way as with the photographic method, but with lesser noise and greater reliability with regard to intensity calibration.

At very low light levels, photography becomes impractical because of an intensity threshold effect. It is possible in that case to intensify the image by focusing it on a specially prepared photocathode, accelerating the electrons

---

[197] R. Chabbal and R. Pelletier, *Jap. J. Appl. Phys.* **4** Suppl. *I*, 445–447 (1965).

released, and bringing them to an image by electron optics.[198] There are various ways to record that image, but the most successful device uses a photographic emulsion specially made for detection of electrons of a few kilovolts. About 1 in 10 electrons will cause activation of a grain, while for photons, this ratio is about 1 in 1000 for normal emulsions. In addition, the response to electrons is practically linear and the intensity threshold effect does not exist for the photoelectric process. As a result, quantum efficiency is increased by a factor of 20–100. Small scale spectra of stellar nebulae have been obtained that way.[199] Because the photoelectric layer has a small size (~ 20 mm) the electronic camera is not adapted to conventional problems. It has been applied in connection with a SIMAC spectrometer, where the information in a wavenumber region of 1500 cm^{-1} (equivalent to 37 nm at 500 nm), with $R_{theor} = 10^6$, could be stored in a small area, having a capacity of 75,000 spectral elements (against 500 in classical unidirectional dispersion spectroscopy); however, the loss of resolution was considerable.[200] In theory, a further increase in sensitivity could be expected from introducing one or more stages of secondary emission, but it has been found that this causes too great a loss in resolution.

Attempts to record spectra conventionally beyond 1250 nm on the basis of the effect of heating have been made several times, but it has not led to convincing practical results. For a review of evaporography, see Lecomte.[201]

We shall not devote much space to discussing the construction and manipulation of detectors; the reader is referred to the literature.[17–19, 42, 202] Only some of the principal features and applications will be treated, in connection with actual results.

3.4.4.2. The Photographic Emulsion. This consists of small AgBr crystals embedded in a gelatin layer with a number of other substances which, although present in minute quantities, play an essential role in the photographic process. Since the mixture is dry and solid, it would be more appropriate to call it a suspension, but we shall follow common usage. Normally, the sensitivity to light stretches from 480 to 200 nm. In order to get beyond 480 nm, the emulsion must be sensitized by means of a suitable organic dye, and each dye has a characteristic wavelength region for which it works optimally. The emulsions and the sensitized emulsions are manufactured industrially; they are fixed either to glass (plates) or celluloid (films). The only operation executed by the user is hypersensitizing, a procedure consisting of bathing in ammonia which enhances sensitivity of sensitized emulsions

[198] A. Lallemand, *C. R. Acad. Sci. Paris* **203**, 243–244, 990–991 (1936).
[199] O. Struve, *Sky Telesc.* **14**, 224–227 (1955).
[200] R. Chabbal, Ph. Bied Charreton, and R. Pelletier, *J. Phys.* (*Paris*) **28**, C2, 209–214 (1967).
[201] J. Lecomte, *J. Phys.* (*Paris*) **10**, 27–31D (1949).
[202] J. Strong, "Procedures in Experimental Physics." Prentice-Hall, Englewood Cliffs, New Jersey, 1938.

for a couple of hours. The slower the emulsion, and the higher the wavelength, the stronger is the enhancement. It easily reaches a factor of 100, but it is insignificant below 650 nm, while it does not improve so much on the fastest emulsions. A certain amount of enhancement seems to be obtained by bathing the plate in distilled water. There are also reports about increasing sensitivity by baking and by treating the emulsion with mercury vapor, but this is seldom done. However, sensitivity is known to increase with temperature so that arrangements for heating during exposure are not so uncommon. Although the main aspects of the photographic process are understood on the basis of modern theories of the solid state,[203] many details are not quite clear so that there is still room for legend and witchcraft.

The lower wavelength limit given has nothing to do with the sensitivity of the AgBr itself, but with the absorption by the gelatin. One remedy lies in admixing the emulsion with a fluorescent oil, converting the short into longer waves; the old spectroscopists used to treat commercial plates with a lubricating oil in order to "uv"-sensitize them. But the more successful procedure is to decrease the amount of gelatin drastically during manufacture (this was also done formerly by etching commercial emulsions with sulfuric acid). In extreme cases, these Schumann plates have been known to contain so little of it that the developed image could be wiped away easily, the silver deposit not being held in position by its environment. Schumann plates can be used throughout the vacuum region down to 1.5 nm, but they are more sensitive than normal plates also in the region between 250 and 200 nm.

Apart from the wavelength selectivity discussed, the most important property is the grain size, because it determines contrast and resolution and it is closely connected with sensitivity. The mean grain size varies from 0.7 to about 7 μm (nuclear emulsions have still finer grain, by as much as a factor of 100, and a correspondingly higher density of grains). In the manufacturing process, there is a period of ripening at about 60°C, whose duration is responsible for the mean grain size. The finer the grain, the slower the emulsion and the higher the contrast, under the same conditions of processing. In most spectrographic work, high contrast is not desired because many weak lines would then be lost between overexposed stronger lines; in addition, the illumination usually does not permit slow emulsions to be used, so that most work is being done with the coarser grain emulsions.

In the region of natural sensitivity, about 1 in 1000 quanta will activate an atom in the AgBr. All activated atoms constitute the latent image which can exist for years but is not visible, and barely detectable.[203] It is the processing which causes the silver to be deposited over entire grains, even if there is only one activated molecule in it. This causes the blackening (Fig. 20),

---

[203] See, e.g., C. F. Powell, P. H. Fowler, and D. H. Perkins, "The Study of Elementary Particles by the Photographic Method." Pergamon, Oxford, 1959.

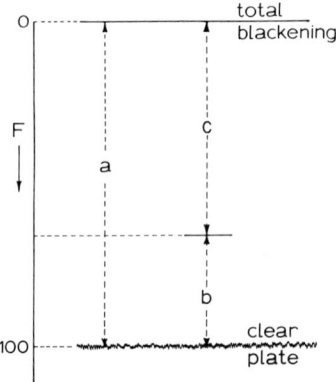

Fig. 20. Illustrating quantities in photographic microdensitometry: F = flux passing through the plate in arbitrary units; blackening $B = b/a$; transmission $T = c/a = 1 - B$; opacity $O = 1/T$; density $D = \log O = -\log T = -\log(1 - B)$.

which is quantitatively defined as the flux *removed* from a narrow pencil of white light passing through the plate with respect to the flux passing through a clear portion of the plate. At moderate blackening, the density of developed grains in a fast emulsion may be of the order of $10^5/\text{mm}^2$. Hence, developing can be considered as a chemical amplification enhancing the latent image millions of times. The reduction of the silver is not restricted to the original grains but it also affects neighboring smaller AgBr islands, and some adjacent grains will fuse, so that the average grain size after processing is greater than it was before. The graininess after development determines the resolving power of the emulsion, defined as the number of lines per millimeter that can be separately distinguished in the image. This number is between 160 and 40 for ordinary spectral plates, corresponding to distances of 6.25 and 25 μm, respectively. It depends not only on the emulsion type, but also on the nature of the developing agent and the developing time. Fine-grain developers usually do not have much effect on the fastest emulsions; they prevent the grains from growing too much in the slow ones, but this also decreases blackening. A photographic resolving width of 15–20 μm, as we have assumed in various examples (Sections 3.4.3.3.3 and 3.4.3.4.1), is quite in agreement with these data.

A spectral-line image is an assembly of silver grains showing highest-average density along the central maximum. When viewed through a traveling microscope, the precision of setting is determined by final graininess and the length of the image part within the field of view. In the most favorable case, it is about 1 μm, and this matches the precision of conventional reading systems. When the lines are straight (Fig. 1), adjustment between two parallel

3.4. OPTICAL REGION

hairs is most adequate. Most unfavorable is the situation for interferograms obtained with crossed-dispersion apparatus because the features are short stretches of fringes (Fig. 3), but this is compensated by the large dispersion and the repetition in several orders. In other cases, it may be the nature of the lines themselves which does not permit adjustment to an accuracy better than 10 μm (Fig. 5).

Spatial reliability is about 1 μm. By this the following is meant: the exposed emulsion undergoes a complex treatment: first the development, then fixing and rinsing and, finally, drying. When it is soaked with a solution, it swells strongly and becomes soft. So it is not without relief that one finds that the location of images in a negative on glass is correct to about 1 μm. The most rigorous test of spatial reliability was made in particle detection by nuclear emulsions. Here, the developed grains, marking the path of a particle, turn out to be in position to within a fraction of 1 μm, in three dimensions![203] However, this can only be achieved if the plates have been handled properly. Especially, the drying process should be performed with caution; it should not be speeded up by chemicals or heat, only a gentle circulation of clean dry air is permitted. Furthermore one should stay about 15 mm away from the plate edges, as the shrinkage of the gelatin is obviously met here by asymmetric forces; sometimes a narrow strip of glass along the rim is bared of the emulsion and irregularities in image position can be found over more that 10 mm inside. Films are less reliable than plates with respect to fidelity to the image. The material is inferior to glass as regards the resistivity against the chemical solutions; in precision wavelength spectroscopy, plates are used almost exclusively.

Unique advantages of the photographic emulsion over other detectors are the ability of collecting light for many hours, and so producing a reasonable signal even in very faint lines (effect of integration), and of giving a permanent record of the spectrum. This latter property has a profound influence on the experimental methods: the procedure of evaluation and measurement can be completely separated from detection, and spectrograms can be stored indefinitely for comparison purposes and later additional observations. The effect of integration has to be considered in connection with saturation. It is clear that the blackening reaches an absolute maximum ($=1$) when the developed plate has become opaque in the area concerned, but already before the image of a strong spectral line reaches that stage, the blackening will grow at a smaller pace, while the weak lines grow faster, so that the differences are leveled off as the exposure is continued. However, the stronger line images will grow in width since the wings are in a similar situation as the weak lines; hence, the width becomes an additional criterion for estimating the intensity. As a result of the plate inertia, intensities below a certain threshold do not easily reach a stage of fast growth. This threshold can be crossed by preexposing the plate to a uniform level of illumination which causes a very slight fog. In this way, the faint lines find the emulsion prepared to a more sensitive state, corresponding to the steep part of the characteristic curve. This conjecture is apparently an oversimplification, though, because postexposure is reported to have a similar effect.

The nonlinear response which is reflected in these favorable features of the photographic emulsion is, at the same time, its greatest drawback in comparison with other detectors when intensity measurements are concerned. These require perfect knowledge of the characteristic curve, in which photographic density D is plotted against log E, where E is the exposure, i.e., the product of illumination and time; D is essentially proportional to the amount of metallic silver per unit area and is measured as $-\log T = -\log(1 - B)$, where B is the blackening as defined before (Fig. 20). The characteristic curve depends on wavelength, type of emulsion, its age, chemical composition of developing agent, concentration and temperature of the solution, and duration of development. It can vary from plate to plate in one package, and, on one plate, from center to rim. Therefore, when intensity comparisons have to be made, each plate has to be calibrated on places as near as possible to the features studied, so that the calibration mark will undergo exactly the same treatment.[204] For intensity measurements, films are often more reliable than plates; their coatings are more uniform in sensitivity as a result of the manufacturing procedure. A number of other defects, such as failure of reciprocity, halation, solarization, turbidity, and Eberhard effect (to be found described in the literature quoted), for which allowance has to be made, contribute to the conclusion that photographic intensity ratios can seldom be trusted to within 5%.

The lowest intensity for lines to produce a measurable blackening is determined by the general background which is exhibited even by unexposed plates and which is enhanced by stray light. In this respect, also, the photographic emulsion is at a disadvantage compared with photoelectric devices which not only have a 100-fold efficiency for photons without intensity threshold, but permit, in addition, a considerable reduction in background.

The primary action of light on the photographic emulsion can be regarded as an internal photoelectric effect whereby electrons are displaced. Owing to the complex nature of the AgBr-gelatin agglomerate, the energy threshold for this process is not fixed as sharply as for the ordinary photoelectric effect. The upper limit of 480 nm given for the natural AgBr (i.e., not sensitized by dyes), therefore, is not very well defined; actually, the sensitivity drops drastically between 480 and 500 nm.

---

[204] L. S. Ornstein, W. H. J. Moll and H. C. Burger, "Objektive Spektralphotometrie." Vieweg, Braunschweig, 1932.

3.4.4.3. The Photoelectric Cell.[†]

The affinity between the photographic and the photoelectric effect, pointed out in the foregoing section, gives rise to the expectation that photoelectric detectors will not go much beyond the photographical wavelength range. This is confirmed by experience, although there is a great variety of tubes available which are each sensitive to a specific wavelength region. Photoelectric devices have been developed for spectroscopic use since about 1930 and have been continuously improved in sensitivity and stability, while new cathode materials pushed the long wave limit further out step by step, up to about 1200 nm. The immediate response invited using ac sources, or chopping the light from stationary sources, because the ac output could then be amplified conveniently. Selective amplification was introduced to suppress unwanted frequencies, thus reducing the noise. The noise is due partly to thermal agitation in the cathode, and it manifests itself as dark current when no radiation is incident. It can be practically eliminated by cooling to liquid nitrogen temperature whereby the threshold sensitivity is strongly improved. However, a certain amount of noise is added by the amplifying unit, due to Brownian motion in coils and resistors which cannot be cooled. A quantum efficiency of $1-20\%$ is normally achieved.

The greatest single improvement was the introduction, after World War II, of the photomultiplier tube, which makes use of secondary emission of electrons for internal amplification. Each dynode releases four or five electrons for each one impinging, so electron multiplier is a more correct designation. While the current is amplified by a factor of 10^3 to more than 10^6, depending on the number of stages, the amazing fact is that this does not destroy the linearity of the response, provided the potential differences (of the order of 100 V) between the successive stages are stabilized rigorously. Moreover, it does not deteriorate the signal-to-noise ratio since the little noise originating in the process itself is always amplified one or more fewer steps than the noise in the primary current. Photomultipliers have entirely superseded the simple photocells for spectroscopic research. The internal amplification cannot improve the intensity threshold so that selective external amplification may be required. This is not difficult since multipliers are fast enough to apply modulation of the incident beam at frequencies well above 100 Hz. However, the smaller the bandwidth, the longer the time constant. This is an inverse proportionality, whereas the noise is proportional to the square root of the bandwidth only. If the recording instrument is slow, e.g., a galvanometer or recording potentiometer, this will impose a narrow band, and filter out all fluctuations below a certain period of, say, 1 msec. As a result, the signal-to-noise ratio is much improved, without selective

---

[†] See also Chapter 11.1 of Vol. 2B (Electronic Methods) in this treatise.

amplification being needed. In case the noise level is not critical, external amplification is unnecessary, as the output of the multiplier is sufficient to activate a modern recording unit. Then, there is no point in modulating the incident light. A representative application has been treated in Section 3.4.3.5.3; with the Hypeac, the scanning normally is a slow process requiring a stable light source. Except for observing the internal structures of selected lines, the multiplier and recording apparatus can also be used conveniently for obtaining an overall view of a not too complicated spectrum by means of the Ebert monochromator preceding the etalon. But the principal advantage is in the measurement of intensity ratios, which is based on the linear response. Of course, this is not an absolute law, as one may expect a saturation effect when the flux becomes too high. But under the necessary precautions, it is accurately valid for a wide range of intensities (at least $1:10^4$). One of those precautions is to ensure that the cathode intercepts the beam and the same part of the cathode surface is always used, because different parts generally have different sensitivities. Furthermore, there may occur a gradual drift in the sensitivity when the observations take a longer period of time, so that test readings must be made now and then in order to obtain eventual corrections. In monochromatic intensity comparisons, a precision of 10^{-3} can be reached easily, provided the source is stable enough. In heterochromatic comparisons, i.e., between wavelengths more than, say, 2.5 nm apart, the difference in color response has to be taken into account. In rough determinations, the information from the manufacturer may be used, but for precision work, calibration is necessary since each tube has its proper characteristics which, moreover, are subject to changes in the course of time. Calibration can be made by means of light sources with a known spectral distribution. The tungsten ribbon lamp satisfies this requirement, but it is applicable only in the visible and infrared regions.

At various times, we pointed out the advantage of measuring flux rather than illumination, once the spectral frequencies are isolated, and we have also mentioned a special application in which the sensitivity of a multiplier was modulated (compare Section 3.4.4.1).

While in most scanning instruments the position of the photomultiplier is fixed, it is sometimes impractical to move the optical parts. In a series of investigations on spectral absorption in solutions of rare earth compounds by means of a 3-m Paschen–Runge grating mounting, it was found practical to move the exit slit and detector along the focal plane. As only simple photocells were available at the time, selective amplification was applied to produce an output that could be measured by means of an ac galvanometer.[205, 206] Using a $6\frac{1}{2}$-m Wadsworth grating mounting, a photomultiplier,

[205] J. Hoogschagen, A. P. Snoek, and C. J. Gorter, *Physica* **10**, 693–698 (1943).
[206] J. Hoogschagen, *Physica* **11**, 513–517 (1946).

and a microammeter, in experiments aiming at spectrochemical application, it was shown that external amplification could be dispensed with for detecting most emission lines in arc spectra.[207] For intensity measurements, the source instability is a handicap. In order to overcome this, a method was introduced by which two units of one multiplier and one exit slit each could be adjusted to two lines simultaneously, behind a 5-m Paschen–Runge mounting. In a search for oscillator strengths in Fe I, light from an iron arc was chopped to make ac amplification possible, and automatic recording was applied to measure a great number of line pairs.[208] The problem of correctly illuminating the cathode surface has been discussed; a diffuser to spread out the light evenly was found to be very useful. Of course, the weak point is that no two multipliers behave alike. In order to circumvent this in a series of experiments to determine oscillator strengths of Fe I, Mg I, and Cs I lines in furnace spectra, only one multiplier tube was used and the light admitted alternately from the two exit slits. To this end, use was made of the stigmatism of a $6\frac{1}{2}$-m Wadsworth mounting by situating these slits at different heights corresponding to the two positions of an optical switch in front of the entrance slit, with a frequency of 50 Hz. The chopping frequency was 800 Hz, and the two signals were divided over two channels for selective amplification, with bandwidths of 25 Hz, and recorded. The photomultiplier remained in a fixed position while the lines were scanned by the two slits. This required careful optical alignment, including reflection over 90°, in order to ensure that the cathode was illuminated in exactly the same way for both beams, for different positions of the exit slit. A diffuser proved to be of great value again.[30] In this system, source fluctuations have been rendered harmless as a result of simultaneous measurement, while one has done away with the problem of differences in sensitivity and its change in time between two multipliers.

3.4.4.4. *Photoconductive Detectors.* These make use of the change of resistivity of semiconductors as a result of irradiation. The effect has been known for pure elements such as Se for more than 100 years, but the first practical Se-resistor cell appeared in 1930. Among compounds, the silver halides show the effect of photoconduction. In the last 25 years, materials have been found which react to infrared light, such as PbS. They have become very important because it became possible to explore spectra between 1000 and 4000 nm with a much higher sensitivity than the thermal detectors could offer. In the region covered by photoelectric cells, there is no practical use for photoconductive devices, since they are less sensitive, nonlinear, and relatively slow. The time constant is of the order of milliseconds so that light chopping or modulation at some 100 Hz for amplification is

---

[207] G. H. Dieke and H. M. Crosswhite, *J. Opt. Soc. Amer.* **35**, 471–480 (1945).
[208] R. Hefferlin, *J. Opt. Soc. Amer.* **49**, 680–685 (1959).

admissible.[116, 209] Noise in the detector can be minimized by cooling to liquid nitrogen temperature.

Results in atomic spectroscopy have been remarkable. In the noble gases, new wavelengths were measured with great precision against higher-order lines, for use as secondary standards in the far infrared.[111] In spite of many earlier investigations in this region, a number of new Hg I lines were found between 1300 and 2000 nm.[210] A survey of these and other results is given by Humphreys.[112] The spectra of many lanthanides and actinides have undergone important extensions by using PbS cells in conjunction with conventional spectrometers at the Harwell establishment,[211] and in combination with SISAM and Fourier spectrometers at the Laboratoire Aimé Cotton.[212] The work at Harwell started with the detection of the infrared resonance lines in Lu I[213] at the wavelengths 1.34, 1.83, and 2.42 μm, in agreement with a prediction from optical analysis.[214] It culminated in high-resolution studies, using a PbS detector in conjunction with a recording Fabry–Perot spectrometer.[215] The observations at Aimé Cotton are still continuing and being analyzed in various laboratories. In this way, an extremely rich wavelength region of I and II spectra has been made accessible by the PbS detector. Recently, this has been extended to III spectra; reports on Ce III,[216] Gd III,[217] and Th III[218] excited in a pulsed hollow cathode discharge are available, covering the region 1100–2600 nm. The latter paper solved a long-standing problem by observing interconnection lines between the lowest configurations, predicted around 2000 nm.[56] These were used to establish the "system difference" at 63.2 cm^{-1}, which is the smallest known in any spectrum.

Strictly speaking, here we are beyond the optical region as defined in Section 3.4.1, but the subject of study is the same, the methods are just extensions of the optical techniques, and without the optical data the infrared observations would not have been interpretable. In all this work, the emphasis is on wavelengths because these are new and accurately measurable, even

[209] U. Litzén, *Ark. Fys.* **28**, 239–248 (1965).

[210] C. J. Humphreys, *J. Opt. Soc. Amer.* **43**, 1027–1029 (1953).

[211] L. F. H. Bovey *et al.*, Rep. AERE C/R2827, R2977, R2976, R3118 (1959); R3225, R3226, R3398 (1960).

[212] J. Blaise, J. Chevillard, J. Vergès, J. F. Wyart, and Th. A. M. van Kleef, *Spectrochim. Acta* **26B**, 1–34 (1971).

[213] L. F. H. Bovey, E. B. M. Steers, and H. S. Wise, *Proc. Phys. Soc. (London)* **A69**, 783–784 (1956).

[214] P. F. A. Klinkenberg, *Physica* **21**, 53–62 (1955).

[215] R. Beer and L. F. H. Bovey, *Proc. Phys. Soc. (London)* **A76**, 569–574 (1960).

[216] S. Johansson and U. Litzén, *Physica Scripta* **6**, 139–140 (1972).

[217] S. Johansson and U. Litzén, *Physica Scripta* **8**, 43–44 (1973).

[218] U. Litzén, *Physica Scripta* **10**, 103–104 (1974).

with unsteady sources. Actually, the PbS cell is also very useful for heterochromatic intensity comparisons of reasonable accuracy because of its relative neutrality, at least in the region 1000–2500 nm. It is in this respect, however, that thermal detectors are matchless.

3.4.4.5. Thermal Detectors. These are strictly aselective as long as it is ascertained that the incident radiation is completely absorbed. That is because the signal depends on energy alone. Hence, they can be used also for determining absolute intensities. In addition, thermal detectors are linear over a wide scale of intensities; namely, as long as the temperature increase is proportional to the energy absorbed. This applies to bolometers, in which resistance changes are proportional to the temperature, and to thermocouples or thermopiles, where an electric potential difference is generated which depends linearly on a temperature difference. Total absorption can be realized in any wavelength region, although it may demand special surface treatment. Hence, the thermal detectors are universally applicable for calibration of other detectors and for measuring the efficiency of monochromators. They are not free from drift so that regular zero readings have to be made. Since it is a slow change, its effect can be eliminated by selective amplification of a modulated signal of a frequency not higher than about 15 Hz. This is because thermal detectors are slow, especially when mounted in a vacuum to suppress gaseous-conduction heat loss, as needed for higher sensitivity. Yet, the absolute sensitivity is always small in comparison with other detectors, except that there are none beyond 4000 nm. Hence, the thermal detectors are hardly used at all in atomic spectroscopy. An early example, however, was the pioneering work by Paschen in the region 1000–4000 nm in various spectra.[219] Furthermore, the discovery of the first member of the Humphreys series in hydrogen, $7i\,^2I \to 6h\,^2H^0$, at $\lambda = 12.37\,\mu m$, is a memorable event.[220]

Some 30 years ago, a pneumatic detector, the Golay cell, was developed to a high degree of perfection. This thermal detector, operating on the principle of increase of gas pressure by heat, is more sensitive and faster than bolometer and thermopile devices. The time constant is of the order of 15 msec only so that it is of increasing importance, especially in Fourier transform spectrometry of molecular infrared spectra.

3.4.5. Evaluation of Spectra

While in direct recording the evaluation procedure is inherent in the observation itself, it is entirely separated from experiment when the photographic method is employed. Common to both methods is the dependence

[219] F. Paschen, *Ann. Phys.* **27**, 537–570 (1908); **29**, 625–663 (1909).
[220] C. J. Humphreys, *J. Res. Nat. Bur. Std.* **50**, 1–6 (1953).

on standards, except in cases like hyperfine structure research, where only wavelength *differences* and intensity *ratios* matter. Although the following discussion will be chiefly confined to the evaluation of spectrograms, much of it will be applicable also to the data obtained through scanning procedures.

3.4.5.1. Wavelengths. 3.4.5.1.1. STANDARDS OF WAVELENGTH. As mentioned in Section 3.4.1.1, wavelengths are determined with respect to the primary standard. However, only by means of the interferometric method used for the determination of the primary standard itself can absolute measurements be made with a precision matching the quality of the natural spectral frequencies. In other cases, direct comparison with the primary standard would lead to large errors. In prismatic spectra, the dispersion curve cannot be derived from geometry; in grating spectra, the geometric formulas (Section 3.4.3.4) cannot take account of deviations of the focal curve due to grating errors, or of the possible failure of the plates to follow the focal curve exactly. In both cases, one needs additional, secondary standards. Prior to 1960, lines in the iron arc spectrum were used for this purpose. Their wavelengths had been measured interferometrically, to seven significant figures. This system was still considerably improved by utilizing the Ritz combination principle, a procedure of fitting the levels until the lines are reproduced best, and then recalculating the wavelengths to eight figures from the level scheme. Ultimately, this system of Ritz standards embraced 1016 Fe I lines in the region from 208 to 1200 nm.[221] However, the Fe arc lines are intrinsically not sufficiently well-defined and reproducible to match such a precision. Other authors concentrated their efforts on interferometric wavelength measurements in the dc hollow cathode spectrum of Fe,[45, 151, 222] but finally it was decided to abandon Fe in favor of the thorium lines from a microwave-excited electrodeless discharge lamp under specified conditions.[223] Thorium has the natural advantage of being monoisotopic and heavy, while the density of Th I and II lines is very high in the optical region. In addition, the intensities are very much less different than in the Fe spectrum, which helps to increase the number of lines with a photographic density favorable for accurate measurement (compare Section 3.4.4.2). Both Th I and Th II are well-analyzed spectra[224–226] so that the test of internal consistency

---

[221] B. Edlén, *Trans. Int. Astron. Un.* **9**, 201–227 (1957).

[222] R. W. Stanley and G. H. Dieke, *J. Opt. Soc. Amer.* **45**, 280–286 (1955).

[223] W. F. Meggers and R. W. Stanley, *J. Res. Nat. Bur. Std.* **61**, 95–103 (1958); **69A**, 109–118 (1965).

[224] R. Zalubas, *J. Res. Nat. Bur. Std.* **63A**, 275–278 (1959); *J. Opt. Soc. Amer.* **58**, 1195–1199 (1968).

[225] T. L. de Bruin, P. F. A. Klinkenberg, and Ph. Schuurmans, *Z. Phys.* **121**, 667–678 (1943); **122**, 23–35 (1944).

[226] J. R. McNally, G. R. Harrison, and H. B. Park, *J. Opt. Soc. Amer.* **32**, 334–347 (1942); **35**, 390–398 (1945).

can be applied. The measurements have been pushed to the limit of eight significant figures though, and finally most of the wavelengths were recalculated from the computer-fitted level systems. The result is a list of 1821 lines, of which 1556 satisfy the conditions for adoption as secondary standards. They cover the region 256.5–1238.1 nm.[227]

It should be remarked that, for analysis of spectra, the precision required in the long-wave region is lower than in the short-wave region, because it is the relative energy we are concerned with in quantum theory. A constant energy difference is $h\,\Delta\nu$ or $hc\,\Delta\sigma$, and because $\sigma = 1/\lambda$, we have $\Delta\sigma = (-)\,\Delta\lambda/\lambda^2$.

3.4.5.1.2. CONDITIONS FOR CORRECT WAVELENGTH DETERMINATIONS. The problem of determining wavelengths in any spectrum is now reduced to interpolating between standard lines in a spectrogram which contains the unknown and reference lines. In precision work, several precautions have to be obeyed for preparing such a spectrogram, since it is necessary that the comparison spectrum be observed in exactly the same circumstances as the unknown spectrum. This means that barometric pressure should be the same because the wavelengths in air depend on the refractive index, and that the temperature does not change since this changes optical and mechanical properties of all spectrograph parts. In addition, the spectrograph should be illuminated in the same way with both light sources, as the spectral line is an image and an image is the result of the interference of many rays which have traversed different sections of the optical elements; any change of the relative amplitudes of various parts of the ultimate wave front displaces the image. The latter condition can be fulfilled strictly only when the standard and unknown lines are emitted in the *same* light source, which, by the way, also would eliminate the other errors mentioned. However, this clearly is impossible in many cases and impractical in others. Actually, one imitates the situation by carefully placing light sources in the same positions or by introducing the light from the comparison source by means of a mirror or 90° prism in a predetermined position. Naturally, the spectrograph should not be touched in between, plate racking is not permitted, and diaphragms in front of the slit should not be changed, even when they are not connected with the spectrograph body, because this would change the internal illumination. Screens immediately in front of the focal plane do not interfere with the formation of the image; they may be used to discriminate the standard spectrum from the unknown one, provided there is a common aperture where the measurements can be made (Fig. 1). However, there is no construction which does not introduce the risk of changes in the adjustment as a result of shocks or vibrations. Even when the two spectra are excited in one and the

[227] A. Giacchetti, R. W. Stanley, and R. Zalubas, *J. Opt. Soc. Amer.* **69**, 474–489 (1970).

same light source, one has to consider that the emission can be located in different parts of the source. For instance, in a hollow cathode lamp provided with a carrier gas, the emission of the gas lines takes place mostly in front of the cavity, whereas the metal lines originate in the plasma inside the cavity.

So, in practice, although one can try to minimize the errors, one has to cope with them. They are discovered only when different spectrograms are compared, but this does not yield the correct values. In many cases, however, the absolute wavelengths are not as important as the wavelength differences, so that a constant shift is not harmful, provided it be small. A constant wave number shift will then be equivalent to a constant wavelength shift and this will not influence the wave number intervals. The instrument normally used for wavelength comparisons is the grating, and this tends to exhibit constant wavelength shifts, if any, over a complete spectrogram. However, since several sets of spectrograms have to be made for a full investigation, there are always overlapping regions on different plates. Now, the shifts are seldom the same, so that the regions of superposition must be used to bring the results into line; which set is adopted as the "correct" one is immaterial for analysis.[228] If the error is too large, it is found that a constant interval is a little different at the two ends of the spectrum. When second and first order measurements are available on one set of plates, it is not unreasonable to assume that the shifts in first order are twice as great as they are in second order; it is equivalent to saying that the metric shift should be about the same. In that case, the true wavelengths are found by correcting the first order wavelengths by twice the difference found. This was applied in one case and indeed found to improve the internal consistency with respect to the line classifications.[229]

It should be emphasized in this connection that different orders should not be used indiscriminately. Measurements have to be made preferably in the same order for both the unknown and reference lines. Kayser discussed the validity of the coincidence method. First he found that the differential influence of changes of temperature and pressure on the wavelengths in air did not lead to shifts larger than 10^{-4} nm, in all practical circumstances.[230] But later he found much bigger discrepancies to reside in optical defects.[231] Indeed, it can hardly be expected that the various optical elements are neutral; more specifically, that the grating efficiency does not vary over the surface in a way which depends on wavelength, so that the image formation is different in two orders.

A different solution to the problem of accurate wavelength measurements is the use of internal standards which have been determined in a separate

[228] P. F. A. Klinkenberg, *Physica* **32**, 1113–1147 (1966).
[229] R. Hoekstra and R. Slooten, *Spectrochim Acta* **26B**, 341–348 (1971).
[230] H. Kayser, "Handbuch der Spectroscopie," Vol. I pp. 716–720. Hirzel, Stuttgart, 1900.
[231] H. Kayser, *Astrophys. J.* **19**, 157–161 (1904); **20**, 327–330 (1904).

experiment, preferably via the interferometric method. This comes down to repeating the work of absolute determinations for a different system of lines, which is a tedious procedure. A completely automatic system, called WINMAC, has been devised for the purpose,[232] but did not find much application, apparently. A routine technique for measuring Fabry–Perot patterns, in which a minimum of data taking is involved without sacrificing precision, was applied to the spectrum of uranium.[233] The use of internal standards is always to be recommended when the spectra are so dense that hardly any secondary standard line is unperturbed. The intermediate standards can then be obtained from underexposed spectrograms, after which they can serve as internal standards on normally exposed "clean" spectrograms of the element in question. Here again, one has to be on one's guard against possible shifts: many gratings possess errors which cause the lines to broaden asymmetrically, as a result of which the "center of gravity" of the photographic profile shifts with increasing exposure. This effect can be detected as a systematic displacement of strong lines with respect to weak lines.[228]

3.4.5.1.3. METHODS OF EVALUATION. The sources of error discussed above do not usually result in wavelength errors of more than 10^{-3} nm, if spectroscopic rules are not violated. Such errors are immaterial for line identification purposes. The uncertainty added by the measurements themselves depends on dispersion and resolving power and on the number of standards available. Between close lying standards, linear interpolation is often permissible, especially in grating spectra. In prismatic spectra, the scale is far from linear, and interpolation is mostly performed by means of a Hartmann formula

$$\lambda = \lambda_0 + [C/(x_0 - x)^\alpha], \tag{3.4.21}$$

where λ_0, C, x_0, and α are constants, and x is the scale reading (i.e., line position) belonging to λ. For reasons of simplicity, α, which is not very different from unity, is usually given the value 1, with the result that λ_0 is no longer a constant over a long region and the calculated wavelengths other than the three starting values are not correct, even when the interpolation does not comprise more than a dozen nanometers. Usually, however, there are, in this stretch, ten or more standard lines available to determine a correction curve which should be smooth enough to ensure an ultimate accuracy equivalent to the adjustment error of 1 μm at best. When no better precision is needed, this serves for identification since for practically every element, there are more accurate lists available. As far as the stronger arc and first spark lines are concerned, they are compiled in the M.I.T. Wavelength

---

[232] G. R. Harrison, *J. Opt. Soc. Amer.* **36**, 644–654 (1946).
[233] D. W. Steinhaus, *J. Opt. Soc. Amer.* **50**, 672–675 (1960).

Tables.[234] Other tables also list lines from higher-ionization stages.[235, 236]

In analysis of complex spectra, one needs a very complete recording of the spectrum and a precision better than 10^{-3} nm, which requires the use of large grating spectrographs. When several standard lines are present in a grating spectrogram, a second-degree polynomial is quite adequate for a region of, say, 50 nm. The coefficients should be adjusted by means of a least squares fitting procedure. In

$$\lambda = \lambda_0 + (x - x_0) \, \partial\lambda/\partial x + (x - x_0)^2 \, \partial^2\lambda/\partial x^2, \qquad (3.4.22)$$

the three coefficients are λ_0, $\partial\lambda/\partial x$, and $\partial^2\lambda/\partial x^2$, while x_0 is the reading associated with the standard that comes closest to the fiducial standard λ_0. Both in this and in the prism formula, we have assumed that the greater λ belongs to the greater x, otherwise one should write $x_0 + x$ in Eq. (3.4.21) and $x_0 - x$ in Eq. (3.4.22). Higher-order polynomials can account for irregularities, but they do not necessarily contribute to precision. It has been remarked already that irregularities occur at the edges of plates; experience teaches that they can best be met by narrowing the region of interpolation. This is, of course, only possible if more than one standard line is present on the extreme 15 mm of the plate. If there is but one standard, it is better to use the grating equation, eventually with adjusted parameters.[101] Obviously, standards falling within the suspect region should not be used in the interpolation procedure for the rest of the plate. Overall application of the grating equation, with only one standard line per plate, has been successful occasionally,[101] but it requires a perfection of the apparatus, especially with regard to the positioning of plates along the focal curve, and the stability of this adjustment, which is beyond normal technical possibilities.

3.4.5.1.4. COMPARATORS. Both grating and prismatic spectra require the determination of image positions, with an accuracy of the order of 1 μm, by means of a spectrum comparator. Traditionally, two types of apparatus are commercially available; viz., the screw and scale comparators. For a description and discussion of advantages and drawbacks, the various technical realizations, details of viewing, projecting and measuring microscopes, the method of use, and testing, the reader may consult Sawyer[18] and the producer's manuals. In the last decades, the trend has been towards automation, to which the screw comparator lends itself more readily than the scale comparator. A completely automatic machine was built at M.I.T. for the wavelength program,[234] which required millions of measurements to be

---

[234] G. R. Harrison, "M. I. T. Wavelength Tables." Wiley, New York, 1939.

[235] H. Kayser/R. Ritschl, "Tabelle der Hauptlinien der Linienspektren aller Elemente." Springer-Verlag, Berlin and New York, 1939.

[236] A. R. Striganov and N. S. Sventitskii, "Tables of Spectral Lines of Neutral and Ionized Atoms." Atomizdat, Moscow, 1966.

made.[237] Periodic and cumulative errors of the screw could be eliminated by means of an ingenious compensatory motion of the plate carriage controlled by a specially made correction cam.[238] Apart from line data, the machine produced a photometric record on film, from which, sometimes, details could be read which had been missed by the automatic photoelectronic pointing method. In studies of specific spectra, it was generally felt that the experienced observer should not be sidetracked; personal judgment in setting on a line is especially important in case of narrow structures, asymmetric profiles, and irregular intensity over the length of the lines. Much of the strain connected with this work in complex spectra could be removed by the introduction of a photoelectric setting device,[239] reducing the setting procedure to bringing two oscilloscope images into coincidence. This was adopted in many laboratories and further improved.[240] By making use of a commercial analog–digital encoder disk attached to the comparator screw, it became possible to record settings on punched cards, at the observer's command, for computer processing.[101, 241]

Although the quality of screws had improved, still the reliability left something to be desired. In addition to departures of the screw, one depends on the lubricant and, consequently, on its temperature and the variable pressure exerted on it by the motion, with varying speed, of the carriage; the backlash makes it necessary to approach settings systematically from one side, and deterioration as a result of wear is to be expected in the long run. A project to investigate such effects interferometrically resulted, after several years, in the construction of an entirely new instrument, called COSPINSCA, an interferential comparator in which the function of the screw is reduced to transporting the carriage.[242] This instrument is in regular operation in the Zeeman laboratory. It measures distances in terms of the wavelength of the 632.8-nm neon–helium laser line by automatic counting of fringes from a Michelson interferometer whose moving mirror (a cat's eye) is attached to the plate carriage. The counting is direction-sensitive so that the operator can shift the carriage back and forth as he wants, over the total length of the trajectory of 250 mm; a certain position always corresponds to a unique number which is punched on paper tape at his command. At the same moment, the apparatus punches an intensity number (between 0 and 99) which is obtained from the transparency of the plate. Tests of the screw revealed periodic departures of 2 μm and long range errors of up to 5 μm

---

[237] G. R. Harrison, *Rep. Progr. Phys.* **8**, 212–230 (1941).
[238] G. R. Harrison, *Rev. Sci. Instrum.* **9**, 15–18 (1938).
[239] F. S. Tomkins and M. Fred, *J. Opt. Soc. Amer.* **41**, 641–643 (1951).
[240] J. M. Bennett and W. F. Koehler, *J. Opt. Soc. Amer.* **49**, 466–470 (1959).
[241] G. H. Dieke, D. Dimock, and H. M. Crosswhite, *J. Opt. Soc. Amer.* **46**, 456–462 (1956).
[242] R. Poppe, R. Hoekstra, and P. F. A. Klinkenberg, *Appl. Sci. Res.* **B11**, 293–309 (1964).

under conditions in which all additional causes of error could be considered negligible with respect to genuine screw errors, confirming other reports.[243] The intensity numbers cannot be converted to true relative intensities when the plates are not calibrated, but they represent a much more objective measure of photographic density than visual estimates. The operator watches the spectrum in projection as it passes by, and can characterize the spectral features by means of a predetermined code, while the setting is performed photoelectrically. This apparatus is capable of a higher precision than can be warranted by a photographic emulsion. Consequently, it is rather slow, although provision is made for rapid bridging of "empty" regions when the spectrum is not very complicated. The tape obtained is computer processed and a list of wavelengths and vacuum wavenumbers with associated intensity marks and other notes is produced in little time.

A completely automatic and fast measuring machine known as ZELACOM has been developed in recent years.[229] This uses the principle of the scale comparator. However, the scale is a 250-mm long grating on glass, having an approximately rectangular transmission function with a grating constant of 10 μm. This is attached to the carriage holding the plate, and reading is done by means of a small stationary grating with the same periodicity, perfectly aligned with the long one. Scale and reading system are commercially available. The light beam traversing the spectrogram produces a photoelectric signal that is digitized every 5 or 2.5 μm, and recorded on computer-compatible magnetic tape. To detect spectral lines, the programs make use of derivatives and second derivatives,[244, 245] while they are safeguarded against finding fake lines. This system operates satisfactorily for low and medium plate densities, but does not always find narrow structures in the saturated region which can be observed visually. Therefore, it is necessary to scan a number of spectrograms of different density in order to catch each spectral line in a favorable situation. In case of spectrograms showing interesting intensity distributions over the lengths of the lines, it is also necessary to scan various strips in one spectrogram in order to assemble the data, making it possible to find the different catagories of lines which a human observer can recognize from the plates by visual inspection. Double-track scanning has been introduced, but this is needed to identify reference lines mainly. Hence, part of the speed advantage must be sacrificed in favor of multiple scanning. However, with a sampling step of 5 μm, the scanning capacity is 2 mm/min at present, and the operator is free, meanwhile, to do other things. The step depends critically on the detail present on the plate. If it were made too large, narrow lines could be missed, especially in absorp-

[243] J. M. Bennett, *J. Opt. Soc. Amer.* **51**, 1133–1138 (1961).
[244] D. W. Steinhaus, R. Engleman, and W. L. Briscoe, *Appl. Opt.* **4**, 799–807 (1965).
[245] R. Hoekstra, *Appl. Opt.* **6**, 807–811 (1967).

tion spectrograms at low dispersion; if it were too small with respect to the line images, a lot of time would be wasted in recording granularity and other accidental fluctuations. The program still has difficulties with asymmetric patterns such as those caused by intrinsic hyperfine structure. A new program is in preparation which analyzes the curves obtained into pseudo-Gaussian profiles. It will be applied to a new extended version of the equipment, in which a 50-mm scale will be mounted, having a periodicity of 8 μm; the two possible steps will then be 4 and 2 μm. The program should roughly take account of the nonlinear response of the photographic emulsion, and the profile function should be made to describe, as close as possible, the density curve of an isolated sharp spectral line.

3.4.5.1.5. INTERFEROGRAMS. For measuring interferograms, techniques not nearly so precise are needed. First, the patterns do not cover an extensive length of plate and, secondly, the fringes are not comparable to spectral lines in sharpness, as a result of the enormous dispersion. Especially, the innermost rings of a Fabry–Perot pattern are often rather vague. The accuracy is considerably improved, however, when several orders are measured and evaluated.[42] This is sufficient as far as investigations of internal structures are concerned. The high precision which can be attained in the determination of absolute wavelengths lies in the correct evaluation of the order number. It can be established from measurements of a few interference patterns of known wavelengths according to the method of exact fractions of Benoît,[246] combined with the value of the spacing from direct micrometer reading. A correction for phase dispersion may be needed when the lines are far apart; it is reliable only for metal coatings.[247] The special methods of Harrison[232] and Steinhaus[233] have not found wide application.

3.4.5.1.6. DIRECT RECORDING. The evaluation of wavelengths in scanning methods does not require comparators at all because recordings can be made on a scale allowing comparison with simple rulers. However, an intrinsic calibration is often more convenient and more reliable, to which end white light fringes have been successfully used.[248] They are produced by applying an etalon as a filter for discrete wavelengths. The fundamental equation (3.4.14) reduces to $m\lambda$ = constant if $n = 1$ and i is constant, and there will be sharp transmission peaks at integral values of the order m. Hence, the intervals between successive wavelengths will be $\Delta\lambda = \lambda/m = \lambda^2/2t$, which is precisely the free spectral range [Eq. (3.4.16)]; in units of wavenumber, it is a constant, viz., $\Delta\sigma = 1/2t$. In photographic spectroscopy, these fringes have been used frequently for calibration in the infrared where standards were scarce. In order to observe them, the spectroscope must have a resolving

[246] R. Benoît, *J. Phys.* (*Paris*) **7**, 57–68 (1898).
[247] W. F. Meggers, *J. Opt. Soc. Amer.* **5**, 308–322 (1921); **6**, 135–139 (1922).
[248] R. A. Fisher, *J. Opt. Soc. Amer.* **49**, 1100–1104 (1959).

limit smaller than $\Delta\lambda$; one obtains a so-called channeled spectrum.[84] For $\lambda = 1000$ nm and $t = 5$ mm, one finds $\Delta\lambda$ to be 0.1 nm. With an astigmatic spectrograph, a slightly different arrangement can be used in which the interference pattern is formed on the slit in the usual way, but only the light going to the center is admitted to the spectrograph.[249]

In the scanning Fabry–Perot spectrometer, an internal scale has been obtained by imprinting on the recordings monochromatic fringes from an auxiliary two-beam interferometer in direct communication with the etalon chamber.[250] This is a most elegant way of circumventing the trouble of irregular gas flow (compare Section 3.4.3.5.3).

3.4.5.2. Intensities. In line classification work, intensities are not of prime importance, but they tend to become more important as theory is shaped more and more to cope with them. The photographic procedure is notoriously complicated, but since wavelengths are obtained from spectrograms almost exclusively, spectroscopists have contented themselves with rough estimates of intensities from visual observation of photographic densities and broadenings. The situation is chaotic in the sense that everyone follows his personal scale, sometimes pseudo-logarithmic and sometimes pseudo-linear, with large variations in the total spread. Within one spectrum (i.e., one element and one ionization stage), the intensity marks from one author have a certain relative value, especially when the lines are not very far apart and lie in the region below 480 nm. In the region of sensitized emulsions, however, the sensitivity depends on wavelength in such a capricious way that visual comparisons without calibration are meaningless, except for very close lines.

Even when the influence of sensitivity has been properly taken into account, one can be led astray easily. For example, self-absorption can considerably diminish photographic blackening, while a line broadened by internal structure, or by external causes like pressure effects, will distribute its intensity over a large area of the photographic plate. In both cases, the density may be equivalent to that of a much weaker but very narrow line. The experienced observer is aware of these effects but, nonetheless, his estimate contains a considerable additional uncertainty. He usually is not too worried about it because, for analysis, he has more useful criteria at hand.

The situation is the reverse in spectrochemistry, where intensity ratios are essential for determining compositions of alloys or other substances, while absolute intensities may be used to establish the quantity of a tracer element

[249] For example, F. S. Tomkins and M. Fred, *J. Phys.* (*Paris*) **19**, 409–414 (1958).
[250] J. M. Gagné, S. Gerstenkorn, and J. M. Helbert, *C. R. Acad. Sci. Paris* **259**, 3479–3482 (1964).

in a sample. The wavelength serves identification only and its accuracy is not very interesting; likewise in astrophysics, where abundancies and several other physical parameters of stellar atmospheres can be evaluated only from spectrograms which have been completely reduced to intensograms. The interpretation of these, however, requires knowledge about oscillator strengths obtained from laboratory experiments in which true intensity ratios have been determined.

The methods for measuring intensity ratios from photographic densities have been well-known for more than 40 years[204]; some aspects have already been mentioned in Section 3.4.4.2. For a discussion of the techniques of calibration and microdensitometer measurements, the reader is referred to the textbooks.[19] A fast, but rough method to obtain objective intensity estimates from spectrograms is in the study of lines excited under standard conditions as a function of the concentration of the element in question in a certain environment: one determines that concentration at which the line becomes just invisible against the background. This method has been utilized on a large scale to obtain intensity values for thousands of lines in seventy elements[251, 252] in an all-out effort to bring some order to the chaotic situation pointed out, not to provide spectroscopy with accurate oscillator strengths. This has been overlooked by a few scientists who pointed out a number of serious errors in the tables concerned.[253]

The photographic method is applicable with all kinds of light sources. Due to the panoramic and the integrating properties of the emulsion, stability of the source is immaterial. Most of the greater accuracy inherent in methods using linear detectors is lost when the light source is ill-controllable. Some of the techniques have already been discussed in Section 3.4.4.3. When stable sources are available, the direct detection methods are always superior. With linear detectors, the measurements become very simple, and small departures from linearity, as well as drift in the cells or the light source, can be accounted for without excessively impeding the accuracy of the results. The greatest care has to be paid to the conditions of the source in order to eliminate the influence of self-absorption, which is the major factor falsifying the true intensity ratios.[52] In case absolute intensities are to be measured for determining oscillator strengths and transition probabilities, one must have information on the numbers of atoms or ions in each energy state under the conditions of the experiment. This still further restricts the usable types

---

[251] W. F. Meggers, C. H. Corliss, and B. F. Scribner, "Tables of Spectral-Line Intensities," Vols. I and II. Nat. Bur. Std. Monograph 32 (1961).

[252] C. H. Corliss, Supplement to Nat. Bur. Std. Monograph 32 (1967).

[253] C. H. Corliss and W. R. Bozman, "Experimental Transition Probabilities for Spectral Lines of Seventy Elements." Nat. Bur. Std. Monograph 53 (1963).

of sources. In principle, only the furnace (Section 3.4.2.3.) comes into consideration,[30] but in practice, stabilized arc discharges can give results which are reasonably reliable.[208] The dc hollow cathode discharge brings out the brightest III lines, but for higher ionizations, one has to use disruptive discharges. Fortunately, the development of beam-foil spectroscopy in the last decade, the subject of Chapter 5.1, has improved the situation for highly ionized systems. By means of this method, it is possible to determine radiative lifetimes of levels directly. Absolute measurements made this way can serve as standards, and relative intensity measurements by means of conventional optical spectroscopy can thus be converted to absolute oscillator strengths and transition probabilities.

A bibliography of atomic transition probabilities is being critically compiled at the National Bureau of Standards, in the framework of the National Standard Reference Data Series.[254]

3.4.6. Analysis of Atomic Spectra

3.4.6.1. Introduction. A great deal of the precision work in spectra is done with the aim of unraveling the structure of atoms and ions, or, less ambitiously, of explaining the observed spectral lines as transitions between energy levels. As there are fewer levels than spectral lines, this alone is a simplification of the description, which is rewarding in itself. But the levels belong to states which have quantized properties expressed in the quantum numbers which are found as a product of the analysis. These give the key to understanding the origin of the levels in the framework of the quantum theory of electronic interactions in the nuclear field. A quantum mechanical treatment can lead to a more accurate description of the states by means of wave functions. A complete and accurate set of wave functions, including all levels, accounts not only for the observed spectral wavelengths, but also for the intensities, the g-factors, and, if the nuclear moments and sizes are known, for the hyperfine interval factors and isotope shifts. A series of levels makes it possible to establish the ionization potential and the binding energies of single electrons, which can be compared with the results of *ab initio* calculations.

In hydrogen atoms and all hydrogenic ions, these problems are solved completely by relativistic quantum mechanics so that little can be gained from observing the optical spectra any more. However, even in a two-electron system, no exact solution is possible, but the numerical approximation can be made indefinitely precise. For instance, in He I the binding energy of the ground state has been calculated at 198, 310.687 cm^{-1} on the Dirac theory,

---

[254] W. L. Wiese *et al.*, Atomic Transition Probabilities. Nat. Std. Ref. Data Ser., NBS-4 (started 1966).

using a 1078-term expansion,[255] while the experimental value was found to be 198, 310.82(15) cm^{-1}. The accuracy was high enough to prove the influence of the quantum electrodynamical correction (the Lamb shift), which was included in the theoretical value.[256] However, in systems containing 20 or more electrons, large discrepancies occur between the results of calculations from first principles and values obtained from observation, especially when unsaturated d or f electrons are involved. In such systems, the spectra also from the standpoint of the observer have aspects which are totally different from those of systems which contain only s and p electrons outside of closed shells: a high density of spectral lines over a wide spectral region, with comparatively little variation in intensity, and, at least in emission, no conspicuous series formation. That means that the cornerstone of old-fashioned analyses of spectra of the alkali and alkaline earth type is absent. For various reasons, the new spectra have been termed complex, while the former ones are denoted as series spectra, but there are, of course, intermediate cases, and sometimes it is found that one and the same species (e.g., the Ca atom) emits both a series and a complex spectrum. There is no complete unanimity as regards this nomenclature because the distinction can be made according to different points of view. Theoreticians tend to regard the coupling conditions as a criterion. No matter what definition is preferred, there are always intermediate cases. An example will make the point much clearer.

The ground configuration of Ta I is

$$(1s^2 2s^2 2p^6 3s^2 3p^6 3d^{10} 4s^2 4p^6 4d^{10} 4f^{14} 5s^2 5p^6) 5d^3 6s^2;$$

i.e., a large number of closed subshells plus three electrons in the 5d state. These three electrons, each having $l = 2$ and $s = \frac{1}{2}$, can combine their quantum vectors in many ways and give rise to just as many different energy states. Excluding all those combinations which are not allowed by the Pauli principle, these are ^{4}F, ^{4}P, ^{2}H, ^{2}G, ^{2}F, ^{2}D, ^{2}P, and another ^{2}D,$^{2, 3, 4}$ or, in total, 19 states (^{4}P has not more than three states because L = 1). The promotion of an electron from 6s to 6p will multiply this number, because one now has one s and one p electron to add to the terms mentioned. Adding an s electron to ^{4}F gives a ^{5}F and a ^{3}F; ^{5}F + p yields (^{5}F)p ^{6}(GFD) and ^{4}(GFD), while ^{3}F + p produces (^{3}F)p ^{4}(GFD) and ^{2}(GFD). So, one obtains 47 levels from the ^{4}F parent term alone. Similarly, the ^{4}P parent term gives 30 levels, etc., so that, in total, the whole excited configuration contains 213 levels. This means that, in Ta I, an electron jump 6p → 6s represents a multitude of transitions from an (odd) system of 213 levels to an (even)

---

[255] C. L. Pekeris, *Phys. Rev.* **115**, 1216–1221 (1959).
[256] G. Herzberg, *Proc. Roy. Soc. (London)* **A248**, 309–332 (1958).

system of 19 levels, whereas, in Cs I, it means just two transitions $^2P_{3/2,1/2} \rightarrow$ $^2S_{1/2}$. The number of possible transitions is limited by the selection rules, of which the ΔJ rule is the only strict one in complex spectra: $\Delta J = \pm 1, 0$; 0–0 forbidden. Applying this, the possible number is found to be 2082! It should be remarked that all transitions have to share the amount of energy radiated away when all excited atoms return to the ground configuration, an energy which is accumulated in only two lines in Cs I. Many lines will be very weak but, generally speaking, the intensity is more nearly uniformly distributed over the lines in a complex spectrum than in a series spectrum. It will be obvious that as far as Ta I is concerned, the configurations in question may be abridged as $5d^36s^2$ and $5d^36s6p$, but this is not the end of the story. It so happens that the binding energies of 5d and 6s electrons—in general $(n-1)d$ versus ns—are very similar, so that the even configuration $5d^46s$ is quite low; in fact, mixed up with $5d^36s^2$. This is true even for the lowest term of $5d^5$. Similarly, there is a second odd configuration, $5d^46p$, intermingled with $5d^36s6p$. As a result, the number of possibilities is at least tripled. As they also occupy the wavelength regions where one should normally expect transitions involving configurations of higher principal quantum numbers, which are much weaker, these higher series members are difficult to find. Analogous situations exist in Nb I ($5p \rightarrow 5s$) and V I ($4p \rightarrow 4s$), but still more complicated cases arise in spectra of elements having unsaturated f electrons.

With thousands of lines associated with just one electron jump, it is obvious that the analysis has to proceed in a way very different from the methods used in simple spectra. Because theory cannot predict anything quantitatively here, one has to rely on statistical considerations, with the Ritz combination principle as the only certain guide (compare Section 3.4.1.1).

3.4.6.2. *Methods of Analysis.* Given an experimental uncertainty in the wavelengths, hence in the (vacuum) wavenumbers, *any* interval will be found a number of times between pairs of lines, a number which can be estimated from Schuster's formula[257]

$$n = \{1 - \Delta\sigma/(\sigma_f - \sigma_i)\}N^2 2\delta/(\sigma_f - \sigma_i), \qquad (3.4.23)$$

where N is the total number of lines in the region of wavenumbers $\sigma_f - \sigma_i$, $\Delta\sigma$ the interval, and δ the error in the wavenumbers. Even if the lines are randomly spread over the region, n shows large fluctuations according to the Poisson distribution. It usually is safer to determine n from a few tests with arbitrary values of $\Delta\sigma$, because the line density will change with wavenumber. In addition, the imminent regularities will be reflected in recurrent

---

[257] A. Schuster, *Proc. Roy. Soc. (London)* **31**, 337–347 (1881).

intervals, which may be disclosed by a systematic search. If an interval is found that occurs 10 times more than the average value of n, it might correspond to a difference between two levels in the atom or the ion. For large numbers of lines, giving $n \geqslant 100$ according to Eq. (3.4.23), a computer is needed, but the result is seldom significant because the "real" line pairs disappear in the background of "fortuitous" pairs. From Eq. (3.4.23), it follows that n decreases linearly with δ, hence the importance of precision in complex spectra. Some spectra do not allow the precision to get better than ± 0.050 cm^{-1} as a result of intrinsic line broadening; e.g., by unresolved hyperfine structure. With 20,000 lines in a region of 20,000 cm^{-1} (case of Tb I), we find n to be 2000. Therefore, it is of primary importance to decrease the noise by reducing N, without losing the signal; i.e., by physical methods warranting the identification of related lines. This can be done on the basis of intensity estimates, line profiles, hyperfine structure, Zeeman effect, etc., but most effectively—although practicable almost exclusively in neutral atoms—by probing the sensitivity to absorption, resp. self-absorption or self-reversal. Since 1000 K corresponds to $kT = 0.086$ eV or ≈ 700 cm^{-1}, only the very lowest levels in the atoms in the vapor are sufficiently populated to absorb light in measurable quantities. That the lines belonging to different stages of ionization should be separated is a triviality, but experimentally it is not quite so easy, because light sources do not usually emit just one spectrum. Especially for the weak lines, the allocation may be ambiguous, while self-absorption in an arc may sometimes cause a I line to mimic a II line in its intensity behavior.

When n can be decreased to $\leqslant 5$ this way, there is a reasonable chance that one or more real intervals show up in a statistical search; confidence about the reality can be obtained only by finding links between such intervals. Two linked intervals encompass line pairs with one common partner. The probability of fortuitous coincidences becomes smaller the more connections are involved, permitting one to set up a level scheme. The probability a coherent level scheme of Sm II, when it explained 1200 lines between 41 low and 200 high levels, was still fortuitous was calculated to be 10^{-40} by Albertson.[258] Usually one does not have to go that far to be sure. Additional information, e.g, from the Zeeman effect and the identification of multiplet terms in conformity with the vector model, produce decisive evidence also. Once a small consistent level scheme is established, extension can be obtained rapidly by computer searching of several coherent intervals throughout the whole spectrum. It should be remarked, however, that, in most spectra, there are levels (usually with extreme values of J) which

[258] W. Albertson, *Astrophys. J.* **84**, 26–72 (1936).

combine rarely and which therefore can be missed easily in a statistical exploration. In such cases, additional information is indispensable. A theoretical estimate is also very useful, but this can be made with sufficient precision only when the analysis is in an advanced stage.

3.4.6.3. Interpretation. The interpretation stage cannot be strictly separated from analysis because certain theoretical features are apparent even at the start, and, inversely, the course of the analysis is often influenced strongly by the advancing interpretation. This is made on the basis of qualitative quantum theory, comparison with related spectra, approximate knowledge of the relative strengths of various interactions, and the determination of J-values by using selection rules and intensity estimates. The Zeeman effect can be decisive for establishing at least one J-value absolutely; it is indispensable for determining the g-factors, which often give information on the other quantum numbers as well, and which always check the correctness of classifications, if not suggesting new ones. To some extent, hyperfine structure data can be used for the same purposes. An additional piece of information is obtained from isotope shift studies. Because the so-called field effect is electric in nature, while the Zeeman effect is magnetic, the isotope shift distinguishes between configurations such as $d^n s^2$ and d^n, which give rise to the same spectroscopic terms.

The detection of configurations and terms leads to the conclusion that certain levels are missing. Some of them may remain obscure, even after extensive searches, owing to the scarcity of combinations or as a result of unfavorable positions of the levels concerned, so that the crucial transitions are in inaccessible wavelength regions.[56] High levels may be insufficiently populated to give observable lines; moreover, they may be subject to Stark broadenings which cause the already weak lines to spread over considerable distances so that they become unobservable against the background (compare Fig. 5). These are additional reasons why series are seldom observed in complex spectra. On the other hand, there is always a residue of lines which cannot be classified, although their assignments to the element and the particular ionization stage are certain; also, "miscellaneous" levels may be found which are not uniquely explainable in terms of the quantum theory without introducing new configurations. If such configurations are assumed, they necessarily are incomplete. Complete analyses do not exist for complex spectra, and the residue of unclassified lines can be considerable. The degree of completeness cannot be established objectively, because expectations depend on the type of spectra concerned, on the techniques available, and on personal judgment with regard to what seems to be within reach. An attempt to classify spectra in a scheme presenting a bird's eye view on the status for every atom and ion emanated from the National Bureau of Standards, following a survey by Shenstone.[259] The scheme is being retouched from time to time, and has been adopted by the Committee on Line Spectra.[260]

The personal judgment includes an appreciation of the available data and, therefore, it is time-dependent. In Section 3.4.2.4, it has already been pointed out, using Hf I and Hf II as examples, that the introduction of a new light source can change the picture completely.[36] Similarly, the opening of a new wavelength region may alter expectations. A case in point is Th III, a two-electron system well-analyzed 25 years ago,[56] with 314 lines classified between two nearly complete level systems. These groups of levels could not be positioned with respect to each other, and the interconnecting lines were estimated to lie in the inaccessible infrared, around 2000 nm, from comparison with Th II[225, 226] and Th IV.[55] Although an extensive configuration calculation was available,[261] this could only confirm the rough guess about the relative position of the two systems. As already discussed in Section 3.4.4.4, this problem was solved recently[218] by observation of 10 interconnecting lines. It turned out that $5f6d^3H_4^0$ is the ground state, lying just a little below $6d^2\ ^3F_2$, contrary to the 1950 anticipation. The interconnection has now been confirmed by means of the Zeeman effect in lines below 750 nm, and this gave rise to a minor revision of the previous analysis. It should be added that 130 observed Th III lines are still unclassified. Of these, 94 are peculiar in comparison with the classified lines: they are very hazy and the wavelengths are strongly influenced by electric fields. Probably, they are connected with high configurations.[56]

With such problems unsolved in relatively simple spectra, it is small wonder that in very complicated analyzed spectra the residue of unclassified lines may comprise 50% of all lines known, though they usually represent less than 10% of the total intensity.

3.4.6.4. Zeeman Effect; Nuclear Effects. For observation of the Zeeman effect, the light source should be in a magnetic field. With an iron-core magnet, the required field intensity of a few tesla is available in relatively small spaces. Suitable devices have been dealt with in Section 3.4.2.4 (electrodeless discharge) and 3.4.2.8 (intermittent arc). Not only should the light source be small-sized, but also it should give rise to sharp lines and be steady in a strong field. The latter requirement is a severe limitation, for the discharge takes place in a plasma, and the charged particles, especially the electrons, are very much perturbed by the strong Lorentz force. Therefore, the light source is not expected to operate in the same manner in the field

[259] A. G. Shenstone, *Rep. Progr. Phys.* **5**, 210–227 (1938).
[260] Committee on Line Spectra of the National Research Council, "Research in Optical Spectroscopy." Publ. 1699 of the Nat. Acad. of Sci., Washington, D.C. (1968).
[261] G. Racah, *Physica* **16**, 651–666 (1950).

as it does without field; extreme cases of different behavior have been discussed in Section 3.4.2.8. The dc hollow cathode tube requires more space so that the field intensity is decreased, but on the other hand, the line sharpness makes it possible to use high-resolution equipment to study the narrower Zeeman patterns (compare Section 3.4.2.6 and Fig. 3). High-intensity, stationary magnetic fields have been made by means of a Bitter solenoid, energized with a current of 10^4 A, at M.I.T. Field strengths of 8–9 T could be maintained in a relatively large volume so that arcs would burn in it parallel to the field lines.[262] Recently, superconducting coils have been introduced for producing fields of up to 6 T, in which hollow cathode tubes have been used for studying the Zeeman effect in U I and U II.[263]

By means of capacitor discharges through coils, magnetic fields up to 13 T were obtained 50 years ago for Zeeman-effect work.[264] Recently, this principle has been successful in combination with a sliding spark for Zeeman-effect observations, in a field of 15 T, of Cu IV lines in the vacuum ultraviolet region.[265]

The information presented by the Zeeman effect is unique. If the Zeeman pattern of a line is completely resolved, it yields the J-values and the g-factors of the two combining levels. Unresolved patterns, pseudo-triplets, and pseudo-quartets permit the evaluation of J and g of one level when these properties are known for the partner level. The percentage of lines measurable in the magnetic field depends on the wavelength and the character of the spectrum. It may be some 10% on the average for I, II, and III spectra in the region above 250 nm. Of these, around 10% may be of the completely resolved type. When the g-values are very close, a sharp triplet is seen. Because g-values can be infinitely close, there is no limit to the magnetic field strength desired; however, in a rich spectrum, adjacent patterns will overlap and present great difficulties in disentangling. Incipient Paschen–Back interaction will also distort certain patterns by changing separations and intensities of components, thus destroying the beautiful symmetry of the unperturbed anomalous Zeeman pattern.

Hyperfine structure is due primarily to the interaction of the nuclear magnetic moment with the outer electrons. It is absent in isotopes having even–even nuclei because these have zero spin. Additional shifts can be explained by an interaction with the nuclear electric quadrupole moment. While these phenomena have been used extensively for determining nuclear momenta, it is also possible, when the nuclear properties are known, to exploit them for spectral analysis. The interaction is proportional to the

[262] H. E. White, *Rep. Progr. Phys.* **6**, 145–154 (1939).
[263] L. Radziemski (Los Alamos), private communication (1975).
[264] P. Kapitza and H. W. B. Skinner, *Proc. Roy. Soc.* (*London*) **109**, 224–239 (1925).
[265] R. Poppe (Amsterdam), private communication (1974).

square of the amplitude of the electronic wavefunction at the nucleus; hence, the splittings measured in a line can give information on the nature of the electronic states involved. In a few instances, analyses have been started by studying hyperfine structure. In these cases, the Zeeman effect was useless because of the dominating natural splittings from which there was no escape, as the elements in question are monoisotopic and have an odd-proton nucleus; viz., $^{141}_{59}$Pr, $^{159}_{65}$Tb, and $^{231}_{91}$Pa. In all three cases, the analysis was concerned with the second spectrum. This is not accidental since the ionized system possesses an unsaturated s electron in the lowest configuration. Statistical studies of lines showing wide hyperfine structure patterns indeed revealed the first regularities in Pr II,[266] Tb II,[267] and Pa II.[268]

Isotope shifts cannot contribute in the same way as hyperfine structure, but should not be neglected. The information is not very specific since all levels in one configuration have comparable shifts. Thus, they give information on the configurations and their degree of mixing. The shift can be made imperceptible by using a separated isotope, which is done for extreme precision of wavelength measurement. For studying the shifts, an arbitrary isotope mixture can be prepared. By means of the source exchange technique (Section 3.4.3.5.3), shifts can be measured which are smaller than the instrumental width by an order of magnitude. The interpretation of the shifts is obscure, as long as the influence of the specific mass shift is unknown. Slow progress has been made in recent years as a result of *ab initio* calculations and of observations of muonic and x-ray isotope shifts.[269]

3.4.6.5. Parametric Calculations in Atomic Spectra. It might seem out of place in a book dealing with methods of experimental physics to include a discussion on quantum mechanical calculations. However, in the analysis of spectra, it is necessary to continuously compare the intermediate results with the theoretical picture. In its crudest form, this involves taking account of predictions from the vector model and of various semiempirical rules about related spectra. A review of such relations can be found in Edlén's Handbuch article.[270] Apart from the direct relations which can be seen most clearly in isoelectronic sequences, there are hidden relationships that only show up when term systems are treated with regard to basic interactions. This requires the application of group theory to configurations, or mixtures of configurations, in order to eliminate the obscuring effects of differences in the number of optical electrons. Adjustable parameters describing the

---

[266] Ph. Schuurmans, Doctoral thesis, Amsterdam (1946).

[267] E. Meinders, *Physica* **42**, 427–438 (1969).

[268] A. Giacchetti, *J. Opt. Soc. Amer.* **57**, 728–733 (1967).

[269] R. Bruch, K. Heilig, D. Kaletta, A. Steudel, and D. Wendlandt, *J. Phys. (Paris)* **30**, C51–58 (1969).

[270] B. Edlén, *Ency. Phys.* **27**, 80–220 (1964).

effects of Coulomb and exchange interaction, spin-orbit and magnetic contributions, etc., can be determined from least-squares fitting procedures of experimental levels. These parameters turn out to change in a very regular fashion in a sequence of spectra, regardless of the number of electrons.[271,272] As a result, in the case of a sequence of spectra of the same stage of ionization, an isoionic sequence as we shall call it, the levels of all related configurations throughout the sequence can be described by a small number of parameters.[273] Recently, a total of 344 experimental levels belonging to the configurations $3d^{N-1}4p$ in the spectra V IV, Mn IV, ..., Ga IV could be represented by means of 32 parameters, the mean error being only ± 113 cm^{-1}.[274] Similar results had been obtained previously in the II spectra[275] and III spectra[276] of these transition elements. The accuracy of the descriptions has gained significantly from the introduction of effective operators to account for the influence of remote configurations connected with excitations of the core. This is a new formulation of an idea incorporated much earlier in the Bacher–Goudsmit relations.[277]

The main contribution to the energy resides in the Coulomb interactions (including exchange). The exact calculation is impossible because the repulsion between the electrons is of the same order of magnitude as the attraction between the electrons and the nucleus. The problem is reduced to a pseudocentral potential problem by assuming that the average effect felt by one electron in an open shell as a result of all other electrons contains a large central part. This is justified by the rapid motion of all electrons around the nucleus, and it is equivalent to what the old spectroscopists meant when they described the optical electron as moving under the influence of a reduced nuclear charge, the parameter being the screening constant. When splitting off that central part and subtracting it from the nuclear potential, one is left with corrections small enough to be treated as perturbations. The perturbation potential will be

$$H - E = V = \sum_{i=1}^{N} \{\xi(r_i)\, \mathbf{s}_i \cdot \mathbf{l}_i - (Ze^2/r_i) - U(r_i)\} + \sum_{i>j=1}^{N} (e^2/r_{ij}),$$
(3.4.24)

---

[271] M. A. Catalán, F. Rohrlich, and A. G. Shenstone, *Proc. Roy. Soc. (London)* **A221**, 421–437 (1954).
[272] G. Racah, *Bull. Res. Council Israel* **3**, 290–298 (1954).
[273] P. E. Noorman and J. Schrijver, *Physica* **36**, 547–556 (1967).
[274] R. Poppe, *Physica* **75**, 341–350 (1974).
[275] G. Racah and Y. Shadmi, *Bull. Res. Council Israel* **8F**, 15–46 (1959); Y. Shadmi, J. Oreg, and J. Stein, *J. Opt. Soc. Amer.* **58**, 909–914 (1968).
[276] Y. Shadmi, *Bull. Res. Council Israel* **10F**, 109–132 (1962); Y. Shadmi, E. Caspi, and J. Oreg, *J. Res. Nat. Bur. Std.* **73A**, 173–189 (1969).
[277] R. F. Bacher and S. A. Goudsmit, *Phys. Rev.* **46**, 948–969 (1934).

where $E = \sum_i \{(1/2m)\mathbf{p}_i^2 + U(r_i)\}$, $U(r_i)$ being the unknown, fictitious, spherically symmetric potential which is specific for a configuration. The interactions splitting up the configurations are connected with the angular momenta (orbital and spin) of the electrons; the group theoretical treatment[7, 8, 278] is the most efficient method to deal with that part of the problem.

The elements of the perturbation matrix being expressed in the parameters, the least-squares fitting procedure also yields the wave functions associated with the characteristic energy values. These wave functions determine g-factors and line intensities. In principle, the best values of the parameters determined by level fitting should also give the best approach to the wave functions, but, owing to the simplifications made, this is not always the case. A slight modification may be in order when this appreciably improves the description of the other properties without too greatly affecting the level fit.

The importance for the analysis of spectra is not only in the interpretation of the levels found, but also in the prediction of levels and their properties. Usually, this is applied to single spectra; however, in isoionic sequences, it is possible to predict entire spectra with an amazing accuracy. Hence, the quantum mechanical description is used as a tool, which becomes more reliable as analysis proceeds because each new level established serves to improve the parameter values. Just by the close association with theory, the empirical analysis can be pushed through. As it may be necessary to diagonalize large matrices, computers are indispensable in this stage of the work also.

In complicated spectra, especially the low ionization spectra of the 4f and 5f shells, configuration calculations are still in a somewhat primitive stage. It is obvious that many simplifications have to be made to render the problems manageable, while in most cases, the empirical levels, though exceedingly numerous, belong to a few parent terms only and so provide insufficient information to determine all parameters. Yet it has been demonstrated that even in rare earths there are marked regularities in the values of most parameters of related configurations,[279, 280] which may be very helpful in future analyses.

The physical significance of parameters, even when they describe structures accurately, is not always unambiguous. Different kinds of interactions may assume the same mathematical form, at least for several states. Small effects may then be absorbed by the "wrong" parameter. On the other hand, it has been found that the isolation of the main effect can lead to the detection

---

[278] D. M. Brink and G. R. Satchler, "Angular Momentum." Oxford Univ. Press (Clarendon), London and New York, 1968.

[279] J. F. Wyart, J. J. A. Koot, and Th. A. M. van Kleef, *Physica* **77**, 159–164 (1974).

[280] P. F. A. Klinkenberg, *Physica* **13**, 1–15 (1947).

of new interactions. The Trees correction $\alpha L(L + 1)$ was established this way,[281] by calculations and analyses of single spectra of the 3d and 5d transition elements, before its nature as one of the effective interactions had been surmised. The most recent example is the detection of an extended correction of similar nature in configurations $3d^{N-1}4p$; this was only possible thanks to the parametric treatment of the isoionic sequence as a whole.[274]

Hence, analysis of spectra is greatly promoted by using the theoretical tools, and, inversely, theory is refined thanks to the profound study of spectra by means of the methods of experimental physics.

[281] R. E. Trees, *Phys. Rev.* **84**, 1089–1091 (1951).

AUTHOR INDEX

Numbers in parentheses are footnote numbers. They are inserted to indicate that the reference to an author's work is cited with a footnote number and his name does not appear on that page.

A

Aarts, J., 261, 323(30), 336(30)
Aberg, T., 178, 179(72), 180
Abraham, D. R., 129, 130, 137(16)
Adriaens, M. R., 233, 239
Ahrenkiel, R. K., 29
Aita, O., 180
Albertson, W., 339
Alburger, D. E., 133
Alder, K., 145
Ali, A. W., 221, 222(36)
Alkhazov, G. D., 151
Allan, N. L., 258
Allison, S. K., 148, 149(4)
Alvarez, L. W., 296
Ames, J. S., 6
Andersen, J. V., 203
Arakawa, E. T., 240
Archer, J. E., 298, 300
Arecchi, F. T., 55, 79(45)
Arfken, G. B., 139
Armbruster, P., 183, 203
Arroe, O. H., 267
Astoin, N., 273
Azäroff, L. V., 148, 149(2)

B

Babcock, H. W., 258
Bacher, R. F., 344
Backenstoss, G., 190
Baird, K. M., 256
Baker, A. D., 250
Baker, C., 250
Baker, S. C., 298
Ballofet, G., 210, 273

Band, I. M., 119, 120
Bandócz, A., 270, 271, 340(56), 341(56)
Barrett, A. H., 257
Barrus, D. M., 162, 163
Bartholomew, G. A., 131
Bartoe, J. D. F., 226, 296
Basbas, G., 200
Bashkin, S., 274
Basov, N. G., 221, 224
Bates, B., 238, 287, 305(93, 94)
Baun, W. L., 168
Bearden, A. J., 162
Bearden, J. A., 149, 168
Beardsworth, E., 119
Bearse, R. C., 139, 140
Bednar, J., 203
Beer, R., 324
Bell, P. R., 16
Benczer-Koller, N., 88, 119
Bennett, J. M., 232, 331, 332
Benoit, R., 333
Bent, R. D., 133
Berman, S. S., 296
Berry, H. G., 104
Bertin, M. C., 119
Bethe, H. A., 105
Beutler, H. G., 229, 291
Bhalla, C. P., 116
Biedenharn, L. C., 139
Bissinger, G., 183
Bjorken, J. D., 64, 94
Blaise, J., 275, 324
Blake, R. L., 151, 162, 163, 170, 197
Blatt, J. M., 116, 117(1)
Blaugrund, A. E., 145
Bloemhof, H. F., 311
Blokhin, M. A., 150

Boehm, F., 146
Bogdankevitch, O. V., 224
Bohr, A., 105, 145
Bohr, N., 64, 66(50)
Boland, B. C., 300
Boldt, G., 224
Bolger, J., 172, 174
Bomke, H., 205
Bonani, G., 183, 185
Bonner, T. W., 133
Boon, M., 32
Born, M., 45, 46, 47, 53, 54, 108, 280, 281(88), 282(88), 284(88), 304(88)
Bosch, G., 261, 323(30), 336(30)
Boursey, E., 210
Bousquet, P., 278, 279, 282(79), 284(79), 304(79), 305, 309(79), 310(79)
Bovey, L. F. H., 324
Bowen, I. S., 14
Bowyer, C. S., 218, 219
Boyer, T., 67, 70
Boyle, A. J. F., 147
Bozman, W. R., 262, 335
Bradford, A. P., 232, 233
Bradley, D. J., 238, 287, 305(93, 94), 310
Bragg, W. L., 15
Braithwaite, W. J., 178
Brand, L., 34
Brandt, W., 183, 200
Braun, W., 220
Breene, R. G., Jr., 50, 81, 87(64)
Bretz, N., 197, 198
Briand, J. P., 173, 174
Briggs, J. S., 183
Brink, D. M., 94, 96, 97, 104, 134, 136, 139, 345
Brinkman, H., 159
Briscoe, W. L., 332
Brochard, J., 282
Brode, W. R., 256, 286(17), 316(17)
Broida, H. P., 220
Brouers, F., 180
Brouwer, J. H., 311
Brown, M. D., 183, 201
Bruch, R., 343
Brueckner, G. E., 296
Brundle, C. R., 250
Brytov, I. A., 234
Buisson, H., 294
Bunsen, R., 7
Burch, D., 170, 171, 197

Burek, A. J., 151, 162, 163
Burger, H. C., 303, 314, 320, 335(204)
Burhop, E. H. S., 189
Burkhalter, P. G., 170
Burnett, C. R., 296
Burns, K., 294
Burridge, J. C., 305
Burrus, C. A., 11
Burton, W. M., 244, 300
Butkov, E., 33, 50, 51, 53, 55(38)
Byram, E. T., 208

C

Cairns, J. A., 183
Cairns, R. B., 242, 248
Campion, P. J., 131
Camus, J., 300
Candler, A. C., 254
Candlin, D. J., 170
Canfield, L. R., 242, 245
Carlson, R. W., 249
Carlson, T. A., 147
Carruthers, G. R., 238
Carruthers, P., 79
Caruso, A. J., 210
Caspi, E., 344
Catalán, M. A., 344
Cauchois, Y., 131, 162
Caudano, R., 250
Chabbal, R., 309, 313, 315, 316
Chandrasekharan, V., 239
Chang, W. Y., 190
Charreton, Ph. Bied, 316
Chase, L. F., 133, 183
Chevallier, P., 173, 174
Chevillard, J., 324
Chiao, T., 183, 201
Clarke, D., 45
Class, C. M., 137
Coblentz, W. W., 9
Cocke, C. L., 165, 166, 203
Codling, K., 211, 212, 215(21), 237, 245
Cohen, M., 244
Compton, A. H., 15, 148, 149(4)
Condon, E. U., 96, 99, 100, 254
Connes, J., 275
Connes, P., 275, 314
Cook, R. L., 11
Cooper, J. W., 179, 261
Cooper, L. N., 190

AUTHOR INDEX

Corben, H. C., 190
Corliss, C. H., 262, 335, 341(36)
Corson, D. R., 31, 32, 96
Costa Lima, M. T., 180
Couderc, A., 282
Cowan, R. D., 268
Cox, J. T., 232, 233
Crasemann, B., 200, 203
Crosswhite, H. M., 266, 271, 315(59), 323, 326(45), 331
Cunningham, M. E., 190
Curnutte, B., 29, 165, 166, 203
Curran, S. C., 150
Currie, W. M., 145
Curtis, L. J., 104
Czerny, M., 296

D

Dalgarno, A., 177
Damany, N., 205, 210, 239, 249, 251(2a)
Damany-Astoin, N., 210
Danilychev, V. A., 221, 224
Darcey, W., 119
Darwin, C. G., 162
Davis, C. K., 183, 184(100), 185
Davis, D., 220
Davis, S. P., 264, 298, 301, 304
Davisson, C. M., 122, 123
Dearnaley, G., 153
DeBenedetti, S., 147
de Boer, J., 145
de Bruin, T. L., 326, 341(225)
Debye, P., 12
Dekker, J. W. M., 311
de Kluiver, H., 258
Delouis, H., 275
Delvaille, J. P., 188, 189
Deme, S., 153
Demidov, A. M., 133
Dennery, P., 33
de Reilhac, L., 249
de Sabbata, V., 139
Deslattes, R. D., 149
Dettman, J. W., 40
Deutsch, M., 16
Devons, S., 146, 189
Devyatkov, A. G., 224
Dieke, G. H., 266, 268, 271, 315(59), 323, 326(45), 331, 335(52)
Dietrich, H., 237

Dimock, D., 197, 198, 331
Dimond, R. K., 180
Dirac, P. A. M., 64
Ditchburn, R. W., 25
Dolejsek, V., 159
Dönszelmann, A., 274
Drake, G. W. F., 177
Drell, S. D., 64, 94
Dreyfus, R. W., 221
Drude, P., 12
Duane, W., 15
Duckworth, H. E., 144
Duerdoth, I., 189
Dufour, Ch., 304, 306
DuMond, J. W. M., 16, 131, 159, 162
Dunning, K. L., 170

E

Eagle, A., 293
Ebert, H., 235, 295
Ebert, P. J., 224
Ederer, D. L., 244
Edlén, B., 247, 255, 326, 343
Edmonds, A. R., 99, 100
Elgin, R. L., 152
Elliott, D. O., 176, 178
Elliott, L. G., 16
Ellis, D. G., 104
Ellsworth, L. D., 129, 130, 137(16), 141, 146
Elton, R. C., 196, 221, 222(36)
Elwert, G., 196
Endt, P. M., 141, 146
Engelhard, E., 256, 265(13)
Engleman, R., 332
Engstrom, S. F. T., 300
Ercoli, B., 261
Erman, P., 147
Ershov, O. A., 234
Evans, R. D., 16
Evenson, K. M., 220

F

Fabian, D. J., 180
Fabry, Ch., 294, 304
Fairchild, E. T., 238
Fairhead, M. J. B., 244
Fano, U., 104, 151, 182
Fastie, W. G., 26, 235, 295, 296
Fawcett, B. C., 175, 221

Fehsenfeld, F. C., 220
Feldman, L. C., 183, 203
Fellgett, P., 11
Ferderber, L. J., 224
Ferguson, A. J., 134, 137
Fermi, E., 64, 190
Fearerbacher, B., 233, 239
Feynman, R. P., 18, 39, 67
Fieffe-Prevost, P., 225
Finkelnburg, W., 268
Fisher, G. B., 246
Fisher, R. A., 266, 333
Fitch, V. L., 190
Flammersfeld, A., 119
Ford, K. W., 190
Fortner, R. J., 201
Fowler, P. H., 317, 319(203)
Fraenkel, B. S., 196, 197
Franks, A., 231, 232(66), 297
Fransson, K., 147
Frauenfelder, H., 134, 138, 139, 147
Fraunhofer, J., 5, 6
Fred, M., 262, 272, 292, 330(101), 331(101), 333
Friedrich, W., 15
Froese Fischer, C., 171
Froome, K. D., 11
Fry, A. S., 266
Fulcher, L. P., 184
Furuta, T., 170

G

Gabriel, A. H., 191, 193, 221
Gagné, J. M., 334
Gale, B., 297
Gallman, A., 133
Garbasso, A., 10
Garcia, J. D., 176, 200, 201
Gardner, J. L., 251
Garret, D. L., 300
Garton, W. R. S., 210
Gaupp, A., 175
Gaydon, A. G., 257
Gedanken, A., 207, 209
Gehroke, E., 314
Gerlach, W., 256
Gerstenkorn, S., 334
Giacchetti, A., 275, 327, 343
Giacomo, P., 305
Gisolf, J. H., 303

Glad, S., 272, 273(62)
Glasoe, G. N., 272
Glauber, R. J., 77
Glick, A. J., 180
Godfrey, G. L., 190
Godwin, R. P., 211
Goertzel, G. H., 119, 132(10)
Goldstein, H., 60, 63
Gollnow, H., 267
Good, R. H., Jr., 33, 35, 36, 38, 40, 41
Goodman, L. S., 274
Goodrich, G. W., 243
Gordon, R. G., 93
Gordy, W., 11
Gorter, C. J., 322
Gottfried, K., 64, 70, 71
Gottlieb, M., 29
Goudsmit, S. A., 344
Graeffe, G., 179
Grainger, J. F., 45
Grandy, W. T., Jr., 61
Greenberg, J. S., 183, 184(100), 185
Greenberger, A., 197, 198
Griem, H., 81
Greiner, H., 229
Greiner, W., 184, 185
Griffin, P. M., 178, 201, 300
Grodzins, L., 140
Groshev, L. V., 133
Guelachvili, G., 275
Guenther, A. H., 296
Gunnerson, E. M., 153

H

Haber, H., 229
Haddad, G. N., 242, 244
Hall, H. E., 147
Hall, J. M., 181
Hameka, H. F., 32
Hamermesh, M., 32, 97, 100
Hamm, R. N., 240
Hanes, G. R., 256
Hansen, J. E., 311
Hanson, W. W., 10
Harr, J., 119, 132(10)
Harris, G. I., 137
Harrison, D. H., 238
Harrison, G. R., 256, 282(19), 296, 298, 300, 301, 316(19), 326, 329, 330, 331, 333, 341(226)

AUTHOR INDEX

Hartman, P. L., 210, 213
Haselton, H. H., 178
Hass, G., 232, 233(72, 73)
Hatter, A. T., 244
Heath, R. L., 125, 126
Hecht, E., 49
Hechtl, S., 129, 130, 137(16)
Heddle, D. W. O., 244
Hefferlin, R., 323, 336(208)
Heilig, K., 311, 343
Heitler, W., 32, 61, 63, 64, 66(50), 81, 82, 122
Helbert, J. M., 334
Henins, A., 149
Henins, I., 162
Henke, B. L., 152
Henley, E. M., 190
Henry, J. C., 257
Hensler, R., 119
Herman, F., 171
Hermansdorfer, H., 220
Heroux, L., 244
Herschel, W., 8
Herzberg, G., 254, 337
Hill, D. L., 190
Hinnov, E., 197, 198
Hinson, D. C., 241
Hirsh, F. R., Jr., 170
Hirschberg, J. G., 310
Hlučka, F., 305
Hobbs, L. M., 313
Hodge, B., 180
Hodge, W., 174
Hodgson, R. T., 221, 224
Hoekstra, R., 328, 331, 332(229)
Hofstadter, R., 124
Holland, L., 299, 305(136)
Hollander, J. M., 118, 119(6), 141(6)
Hoogschagen, J., 322
Hopfield, J. J., 206
Hopkins, F. F., 176, 177, 178(68), 180, 198(86), 199
Horák, Z., 170
Horton, V. G., 240
Houghton, J. T., 92
Houston, V. W., 312
Huang, K., 108
Hudson, R. D., 249
Huffman, R. E., 206, 207, 208, 209
Hughes, T. P., 258
Hulthén, E., 296
Humphreys, C. J., 295, 324, 325

Hunt, F. L., 15
Hunter, W. R., 233, 238
Hutley, M. C., 297
Huus, T., 145
Hyder, A. K., 137

I

Ichinokawa, T., 233
Ingalls, R., 147

J

Jaccarino, V., 147
Jackson, D. A., 302, 305, 313, 314
Jackson, J. D., 33, 36
Jacobus, G. F., 233
Jacquinot, P., 277, 279, 282, 304, 305, 309, 311(82)
Jahoda, F. C., 162
Jamison, K. A., 176, 177, 178(68), 181
Jansen, L., 32
Jarrell, R. F., 296
Jeffreys, B. S., 34, 36, 38, 41
Jeffreys, H., 34, 36, 38, 41
Jenkins, F. A., 280, 281(84), 282(84), 284(84), 296, 304(84), 334(84)
Johann, H. H., 159
Johansson, S., 324
Johansson, T., 159
Johnson, R. G., 245
Johnson, R. L., 230
Johnson, W. C., 239
Johnston, R. G., 242, 246(118)
Jones, B. B., 300
Joos, P., 215
Jordan, C., 191, 193
Jordan, W. H., 16
Judd, B. R., 254, 345(8)
Judge, D. L., 249
Jundt, F. C., 202
Juulman, C. O. L., 287, 305(94)

K

Kadesch, R. R., 310
Kaletta, D., 343
Källen, G., 64, 66(50)
Kapitza, P., 342
Karllson, E., 140
Kashnikov, G. N., 224

Kastler, A., 314
Kato, R., 237
Kauffman, R. L., 174, 176, 177, 178(68), 181
Kaufman, V., 246, 247, 272, 311
Kavanagh, T. M., 201
Kavttman, R. L., 181
Kayser, H., 328, 330
Kelly, R. L., 247
Kennard, E. H., 170
Kent-Smith, R., 185
Kepple, P., 221
Khodkevich, D. D., 221
Kienle, P., 184, 188
Kim, J. S., 133
Kim, Y. N., 189
King, A. S., 260, 261
King, R. B., 261
King, W. C., 11
Kirchhoff, G. R., 6, 7
Kirkpatrick, H. A., 159
Kistner, O. C., 147
Kittel, C., 82
Klein, M. V., 47, 54
Klein, O., 133
Kline, J. V., 297
Klinkenberg, P. F. A., 264, 266, 269, 270, 272, 311, 324, 326, 328, 329(228), 331, 340(56), 341(56, 225), 345
Knipping, P., 15
Knowles, J. W., 131, 132
Knudson, A. R., 170
Koehler, H. A., 224
Koehler, W. F., 331
Kohno, T., 287, 305(94)
Kohra, K., 233
Kolata, J. J., 133
Kolb, A. C., 221
Koot, J. J. A., 345
Kopfermann, H., 259
Korff, S. A., 150
Kostkowski, H. J., 295, 324
Kraft, G., 185
Kraus, A. A., 139
Krause, M. O., 162, 164
Krzywicki, A., 33
Kubo, H., 202
Kubo, R., 93
Kuhn, H. G., 254, 305, 310, 313
Kumar, S., 218, 219
Kunz, C., 237
Kunzl, V., 159, 180

L

Laegsgaard, E., 203
Lallemand, A., 316
Lang, G., 147
Langley, S. P., 9
Lantsov, N. P., 224
Lapson, L. B., 243
Larrabee, J. C., 206, 207
Larsson, A., 157
Lau, E., 303, 314
Laubert, R., 182, 183(88), 184, 186(88), 200
Laue, M., 15
LaVilla, R. E., 179
Lavollee, M., 236
Lazarus, S. M., 183
Lebacqz, J. V., 272
Lecomte, J., 316
Lederer, C. M., 118, 119(6), 141(6)
Lee, L. C., 249
Lee, Y. K., 88, 133
Legg, J. C., 129, 130, 137(16), 139, 140
Leighton, R. B., 39, 96
Leithold, L., 45
Levy, M. E., 208, 209
Lewis, H. W., 183
Lichten, W., 182, 184
Lind, E., 296
Lindhard, J., 144
Lindner, J. W., 264
Lindsey, K., 232
Linkaoho, M., 179
Litherland, A. E., 137
Little, W. A., 183
Littlefield, T. A., 302
Litzén, U., 296, 324(116), 341(218)
Lockyer, R., 267
Longe, P., 180
Longhurst, R. S., 280, 281(86), 282(86), 284(86), 304(86)
Loofbourow, J. R., 256, 282(19), 316(19)
Lord, R. C., 256, 282(19), 316(19)
Lorrain, P., 31, 32, 96
Loudon, R., 61
Lucas-Tooth, H. J., 310
Lukirskii, A. P., 234
Lummer, O., 9
Lumpkin, A. H., 133
Lutsenko, V. N., 133
Lutz, H. O., 203
Lyman, T., 14, 205, 208

M

Macdonald, J. R., 119, 144, 165, 166, 183, 201
McDowell, C. A., 250
Macek, J., 104, 183
Mack, J. E., 176, 267, 313
McKernan, P. C., 246
McNally, J. R., 300, 326, 341(226)
McNutt, D. P., 313
MacRae, R. A., 240
McWeeny, R., 32
McWherter, J., 172, 174
Madden, R. P., 211, 215(21), 242, 245, 246(118)
Maecker, H., 268
Maillard, J. P., 275
Majumder, K., 229
Malherbe, A., 238, 239
Malmfors, K. G., 146
Malo, S. A., 238
Malov, A., 133
Mandel, L., 54
Mandelstam, S. L., 271
Mandl, F., 64
Manning, G., 131
Manson, J. E., 218
Marmier, P., 117, 146
Marr, G. V., 237
Marrus, R., 100, 165
Martinson, I., 175
Marwick, A. D., 183
Mason, M., 39
Mathieux, J. P., 25
Matthews, D. L., 178, 180, 198(86), 199
Matthias, E., 140
Matsu, A., 241
Meeks, M. L., 257
Meggers, W. F., 261, 262, 294, 326, 333, 335, 341(36)
Megli, D. G., 129, 130, 137(16)
Meijer, F. G., 272
Meinders, E., 343
Meissner, K. W., 274, 303, 304, 311, 313
Melvill, T., 5
Menzel, D. H., 94, 97, 98, 99, 254
Merzbacher, E., 31, 64, 70, 72(1), 79, 81, 84, 97, 98, 99, 100, 109, 110, 111, 112, 183
Meservey, E., 197, 198
Meyerhof, W. E., 183, 202
Michel, G., 275
Michels, A., 258

Michelson, A. A., 7, 301
Mickelwait, A. B., 190
Middelkoop, D., 258
Middleton, A., 302, 326(151)
Millikan, R. A., 14
Minnhagen, L., 246, 265
Mitchell, K. D., 266
Mitchell, P., 237
Miyake, K. P., 237
Mokler, P. H., 183, 185, 203
Moll, W. H. J., 320, 335(204)
Moore, C. E., 172, 174, 178, 180, 181, 198(86), 199, 247
Moore, M. S., 137
Morgan, I. L., 170
Moseley, H. G. J., 15
Mössbauer, R. L., 17, 147, 255
Mottelson, B. R., 105, 145
Motz, H. T., 133
Mowat, J. R., 178, 201
Muggleton, A. H. G., 153
Müller, B., 184, 185
Mullin, H. R., 268
Murty, M. V. R. K., 229

N

Nachtrieb, N. H., 272
Nagakura, I., 180
Nagel, D. J., 168, 170, 177, 196
Nagel, M. R., 21
Namioka, T., 229, 230, 236, 291
Neito, M. M., 79
Nelson, T. J., 33, 35, 36, 38, 40, 41
Neuhaus, H., 296
Newton, I., 4
Nicodemus, F. E., 20
Nichols, E. F., 10
Nielsen, H. H., 107, 110(95)
Nishina, Y., 133
Noda, H., 230
Noorman, P. E., 344
Norris, R., 82
Northcliffe, L. C., 144
Northrop, D. C., 153

O

Ogawa, M., 249
Oitjen, J., 181
Okabe, H., 220

Olness, J. W., 133
Olsen, D. K., 174, 180, 181
Onderdelinden, D., 183
Oppenheimer, J. R., 108
Oreg, J., 344
Ormrod, J. H., 144
Ornstein, L. S., 320, 335(204)
Osantowski, J. F., 232, 233(73)
Ott, W. R., 225, 226

P

Padur, J. P., 233
Palmer, E. W., 229, 297
Palms, J. M., 155
Palumbo, L. J., 247
Paresce, F., 218, 219
Park, D., 82
Park, H. B., 326, 341(226)
Parkinson, J. H., 191, 192
Parratt, L. G., 168, 169
Parrish, W., 152
Paschen, F., 272, 325
Passin, T., 162
Pauls, E., 279, 290(81), 299(81)
Peach, G., 50
Pearlman, M. L., 214
Pearse, R. W. B., 257
Pegg, D. J., 178, 201
Peisakhson, I. V., 229
Peitz, H., 184
Pekeris, C. L., 337
Pelletier, R., 315, 316
Penkin, N. P., 235
Pereskok, V. F., 246
Perkins, D. H., 317, 319(203)
Perlman, I, 118, 119(6), 141(6)
Perot, A., 304
Pery, A., 305
Peterson, R. S., 201
Petersson, B., 265
Peticolas, W. L., 82
Pfund, A. H., 26
Phillips, G. C., 139, 140
Pica, R., 306
Pieper, W., 184
Pittenger, J. T., 266
Pitz, E., 226
Pivovonsky, M., 21
Poletti, A. R., 137, 138
Popov, V. S., 184

Popov, Yu. M., 221, 224
Poppe, R., 331, 342, 344, 346(274)
Pouey, M., 236
Powell, C. F., 317, 319(203)
Pringsheim, E., 9
Prins, J. A., 162
Prosser, F. W., 139
Pugh, E. M., 31
Pugh, E. W., 31
Purcell, J. D., 300
Purser, K. H., 202

R

Racah, G., 341, 344
Radziemski, L. J., Jr., 247, 342
Rafelski, J., 184
Rainwater, J., 190
Ramberg, E. G., 170
Ramsden, S. A., 313
Ramsey, J. B., 233, 310
Randall, R., 165, 166, 203
Rank, D. H., 296
Rao, P. V., 155
Rasmussen, E., 306
Raz, B., 207, 209
Reay, N. K., 300
Richard, P., 170, 171, 172, 174, 176, 177, 178(68), 180, 181, 197, 198(86), 199
Redhead, D. L., 224
Ribe, F. L., 162
Richards, P. L., 12
Richtmyer, R. D., 10
Ridgeley, A., 244
Rieckhoff, K. E., 82
Ritschl, R., 302, 330
Ritter, E., 314
Ritter, J. W., 13
Robertson, C. W., 29
Robertson, H. J., 301
Robin, S., 236
Roentgen, W. K., 14, 148
Roesler, F. L., 313
Rohrlich, F., 344
Roizen-Dossier, B., 277
Rollefson, A. A., 139, 140
Romand, J., 205, 210, 251(2a), 273
Roncin, J. -Y., 210
Rooke, G. A., 180
Rose, H. J., 94, 96, 97, 104, 134, 136, 139
Rose, M. E., 94, 119, 132, 139

Rosenfeld, L., 64, 66(50)
Rosner, H. R., 116
Rossi, B., 280, 281(85), 282(85), 304(85)
Rouard, P., 305
Rousseau, M., 25
Rowe, E. M., 214
Rowland, H. A., 7, 14, 227, 287, 291
Roy, G., 139, 140
Rozet, J. P., 173, 174
Rubens, H., 9
Rudolph, D., 230
Russell, D. S., 296

S

Sagawa, T., 180
Sai, T., 291
Sakayangi, Y., 229
Sakurai, J. J., 82
Salle, M., 236
Salmon Legagneur, C., 266, 326(45)
Salpeter, E. E., 105
Samson, J. A. R., 14, 155, 205, 206(3), 211(3), 216, 217, 221(3), 228(3), 229, 231, 233(3), 236(3), 238(3), 242, 243(3), 244, 248, 249(3), 250, 251
Sands, M., 39
Sandström, A. E., 148, 167
Saris, F. W., 182, 183, 184, 186
Satchler, G. R., 139, 345
Saunders, P. A. H., 221
Savinov, E. P., 234
Sawyer, G. A., 162
Sawyer, R. A., 14, 22, 25(60), 256, 280(18), 282(18), 292(18), 316(18), 330
Sayce, L. A., 231
Saylor, T. K., 183
Schaffer, E., 229
Scharff, M., 144
Schawlow, A. L., 11, 111
Schectman, R. M., 104
Schiff, L. I., 31, 72(1), 79, 87
Schiffer, J. P., 137, 139
Schilling, R. F., 144
Schiott, H. E., 144
Schmahl, G., 230
Schmidt, Th., 259
Schmieder, R. W., 165
Schnopper, H. W., 187, 188, 189
Schönheit, E., 229, 236
Schrijver, J., 344

Schüler, H., 259, 267
Schumann, V., 14, 204
Schuster, A., 278, 338
Schulz-Dubois, E. O., 55, 79(45)
Schuurmans, Ph., 326, 341(225), 343
Schweber, S., 69
Schwob, J. L., 196, 197
Scribner, B. F., 268, 335
Seaman, G. G., 119
Seigbahn, M., 168
Sellin, I. A., 178, 201
Senemaud, C., 180
Seya, M., 230, 236, 291
Shadmi, Y., 344
Shardanand, 210
Sharma, A., 233
Sheldon, E., 117, 139, 146
Shenstone, A. G., 263, 266, 294, 341, 344
Shepelev, Yu. F., 234
Shinoda, G., 233
Shipman, J. D., Jr., 221, 222(36)
Shirley, D. A., 162, 250
Shore, B. W., 94, 97, 98, 99, 254
Shortley, G. H., 96, 99, 100
Shortley, G. S., 254
Shreider, E. Ya., 205
Shukhtin, A. M., 235
Siegbahn, K., 140, 162, 252
Siegbahn, M., 148, 157
Siegman, A. E., 55, 79(44)
Siivola, J., 179
Simons, P., 88
Singh, M., 229
Singh, S. P., 180
Sinton, W. M., 296
Sippel, R. F., 133
Sirks, J. L., 294
Skillman, S., 171
Skinner, H. W. B., 342
Slater, J. C., 61, 63, 83, 91, 94, 108, 254, 345(7)
Sliv, L. A., 119, 120
Slooten, R., 328, 332(229)
Smith, P. B., 139, 146
Smith, S. D., 92
Smith, W. V., 11
Snoek, A. P., 322
Sommer, L. A., 314
Soppi, E. J., 146
Sorokin, P. P., 224
Speer, R. J., 230
Spicer, W. E., 246

Spieger, M. R., 38
Spiller, E., 239
Spinrad, B. I., 119, 132(10)
Srivastava, K. S., 180
Srivastava, R. L., 180
Stakgold, I., 33
Stanley, R. W., 326, 327
Steers, E. B. M., 324
Steffen, R. M., 134, 138, 139, 140
Stein, H. J., 183, 185, 203
Stein, J., 344
Steinhaus, D. W., 271, 297, 315(59), 329, 332, 333
Stenstrom, W., 168
Steudel, A., 311, 343
Stigmark, L., 265
Stöckl, M., 183, 185
Stodiek, W., 197, 198
Stokes, G. G., 13
Stoller, Ch., 183, 185
Stratton, J. A., 39
Streater, R. F., 65, 66(51)
Striganov, A. R., 247, 330
Stroke, G. W., 289, 291, 296, 297, 298
Stroke, H. H., 296
Strong, J., 12, 280, 281(87), 282(87), 284(87), 299, 304(87), 316
Strong, P., 119, 132(10)
Struve, O., 316
Stuck, D., 226
Sturgis, N., 298
Sugar, J., 268, 272, 273
Sunyar, A. W., 147
Suter, M., 183, 185
Sventitskii, N. S., 247, 330
Swaine, D. J., 274

T

Tanaka, Y., 206, 207
Tape, J. W., 119
Tarnakin, I. N., 229
Tavernier, M., 173, 174
Tawara, H., 182, 183(88), 184, 186(88)
Tayerle, M., 159
Taylor, J., 152
Taylor, J. M., 153
Tear, J. W., 10
Teller, E., 190
Thoe, R. S., 178, 185, 186
Timothy, J. G., 243

Tinkham, M., 32
Tobocman, W., 139
Toft, A. R., 232, 233(73)
Tolansky, S., 265, 276, 304, 306(42), 309(42), 316(42), 333(42)
Tomboulian, D. H., 155, 210, 213, 244
Tomkins, F. S., 261, 262, 272, 292, 330(101), 331(101), 334
Tousey, R., 232, 233, 300
Townes, C. H., 11, 111
Trambarulo, R. F., 11
Trees, R. E., 346
Triplett, B. B., 183
Turnbull, D. T., 302
Turner, A. F., 296
Turner, D. W., 250
Tymchuk, P., 296

U

Utriainen, J., 178, 179(72).

V

Vanasse, G. A., 12
Van Cittert, P. H., 278, 303, 314
Van der Berg, C. B., 159
van der Weg, W. F., 182, 183(88), 184, 186(88)
van Kleef, Th. A. M., 324, 345
Van Regemorter, H., 196
Vanyek, U. M., 270, 340(56), 341(56)
Varsányi, F., 271
Verbist, J., 250
Vergès, J., 324
Verrill, J. F., 229, 297
Vidal, C. R., 261
Vieweg, R., 256, 265(13)
Vincent, P., 185
Vodar, B., 205, 210, 236, 251(2a), 273
von der Leun, C., 141
von Godler, S., 197, 198
Von Hámos, L., 158
Voros, T., 270, 340(56), 341(56)

W

Wadsworth, F. L. O., 294
Wagner, P., 133
Walker, L. R., 147
Walker, W. C., 241
Walsh, A., 267, 281

AUTHOR INDEX

Walther, H., 311
Wangsness, R. K., 43
Wanner, S. J., 246
Warburton, E. K., 133, 137, 138
Watson, L. M., 180
Watson, R. E., 214
Waylonis, J. E., 232, 233(72)
Waynant, R. W., 221, 222
Weaver, W., 39
Weinreb, S., 257
Weisskopf, V. F., 116, 117(1)
Weissler, G. L., 204
Welford, W. T., 229
Wende, B., 226
Wendlandt, D., 343
Wertheim, G. K., 17, 147
West, D., 150
West, J. B., 237
Westfall, F. O., 261, 262
Westfold, K., 215
Weyl, H., 35
Wheaton, J. E. G., 210
Wheeler, J. A., 190
White, H. E., 234, 280, 281(84), 282(84), 284(84), 304(84), 334(84), 342
Wick, G. C., 97
Wiegand, C. E., 190
Wiese, W. L., 225, 336
Wiggins, T. A., 296
Wightman, A. S., 65, 66(51)
Wigner, E. P., 97
Wilets, L., 189
Wiley, W. C., 243
Wilkinson, D. H., 117, 118
Wilkinson, P. G., 208

Williams, D., 29
Williams, M. D., 221
Williams, M. W., 240
Williams, R. V., 258
Williams, W. E., 301, 302(147), 326(151)
Winters, L. M., 201
Winther, A., 145
Wise, H. S., 324
Wittke, H., 272
Wolf, E., 45, 46, 47, 53, 54, 281(88), 282(88), 284(88), 304(88)
Wölfli, W., 183, 185
Wollaston, W. H., 5
Wolter, H. H., 178
Wood, R. E., 155
Wood, R. W., 10, 280, 281(83), 282(83), 299, 304(83)
Woods, C. W., 176, 177, 178(68), 181
Wu, C. S., 88, 189
Wyart, J. F., 324, 345
Wybourne, B. G., 254
Wynne, J. J., 224

Y

Yamada, Y., 298
Yamashita, H., 237

Z

Zalubas, R., 326, 327
Zastawny, A., 151
Zeeman, P., 294
Zeidel, A. N., 205
Zernike, F., 284

SUBJECT INDEX FOR PART A

A

Absorption
  coefficients, 22, 110
  spectra, 22, 248
  theory, 87
Alternate decay modes, 118
Angular
  correlations, 134
  momentum of e-m fields, 40
  momentum of molecules, 106
Arcs, 267, 269
Aspherical gratings, 229
Atomic spectra, 6, 100ff, 148ff, 204ff, 253ff
  analysis of, 336
  nuclear effects in, 341
Astigmatism, 229
Auger transitions, 178

B

Bands, 21
Beta–gamma correlations, 178
Blazed gratings, 230, 288
Bremsstrahlung, 186

C

Classical theory of e-m fields, 32ff
Coherence, 50, 75
Collision broadening, 23
Collisions, ion–atom, 200
Cold-cathode discharge, 215
Comparators, 330
Concave grating mounts, 291
Continuum, 21, 206
Cornu prism, 285
Coulomb excitation, 145
Counters, 150
Crystal properties, 161

D

Decay rates, 100, 104, 115, 118
Detectors, 26
  gamma-ray, 121
  for optical region, 314
  photoconductive, 323
  photoelectric cells, 321
  solid state, 153
  thermal, 325
  ultraviolet, 241, 244
  x ray, 149
Diffraction gratings, 156, 160, 227, 287
Dipole, 115, *see also* Multipoles
Dispersive devices, *see also* Gratings, Prisms, and Interferometers
Doppler broadening, 23
Doppler shift, 144
Doppler-tuned spectrometer, 164

E

Eagle mounting, 293
Ebert-Fastie spectrometer, 235
Echelle, 299
Echelon, 301
Electrodeless discharge, 261
Electromagnetic fields, 32, 38, 61
Electromagnetic spectrum, 2
Emission spectra, 20, 87
Etalon, 304, 306, 309, 314

F

Fabry–Perot interferometer, 304
Far-ultraviolet region, 204
Flames, 260
Fluorescence, 145
Fourier representations, 55
Free e-m fields, 60
Frequency, 2, *see also* Wavenumber

G

Gamma-ray region, 16, 115
  detectors, 121
  spectrometers, 131
Gamma–gamma correlation, 138
Gauge transformations, 32
Geissler tubes, 265
Grating, 227, 287
  aspherical, 229
  concave, 226
  crystal, 156, 160
  fabrication, 297
  holographic, 229
  mounts, 291, 293
  phase, 231

H

Hamiltonian formulation, 60
Handedness, 45
Helicity, 45
History, 3
Holographic gratings, 229
Hollow cathode discharge, 218, 272
Hot filament discharge, 219
Hydrogen continuum, 206
Hyperfine structure, 342
Hypersatellite lines, 172

I

Induced emission, 87
Infrared region, 2, 8
  near, *see also* Optical region
Intensity, 20, 73
Interaction of light and matter, 79
Interferograms, 333
Interferometers, 304

K

King's furnace, 260

L

Lasers, 221
Lifetime measurements, 141
Light, 32
  intensity, 20, 73
  interaction with matter, 79
Line sources for uv, 215
Linear momentum of radiation, 39
Lines, 20, 24, 254
Littrow mounting, 286
Lorentz line shape, 24
Luminosity, 311

M

Mechanical properties of fields, 38
Methods of spectroscopy, 19
Microwave, 10
  discharge, 220
  excitation, 261
Molecular states, 106
Molecular x rays, 182
Monochromators, 234
Mössbauer effect, 25, 146
Mountings, *see also* Gratings
Multipass systems, 296
Multipoles, 93, 115
Muonic x rays, 189

N

Natural broadening, 23
Nuclear, *see also* Gamma rays
  effects in atomic spectra, 341
  reactions, 137
  transitions, 100

O

Optical region, 253
  detectors, 314
  grating spectrographs, 287
  prism spectrographs, 279
Octupoles, *see also* Multipoles

P

Parametric calculations, 347
Paschen–Runge mounting, 293
Perturbation calculations, 79
Perturbed correlations, 140
Phase gratings, 231
Photoconductive cells, 321
Photoelectric cells, 321
Photoelectron spectroscopy, 162, 250

Photographic emulsion, 316
Photon, 67
Photon sources, 205
Photon states, 73
Plane waves, 41
Plasmas, 191, 208
Plasmon satellites, 180
Polarization, 73
Polarized targets and beams, 139
Polarizers, 239
Potentials, 32
Prism, 3, 7, 279
  spectrographs, 279
  constant deviation, 282
  Cornu, 285

Q

Quadrupoles, see also Multipoles
Quantum conditions, 65
Quantum theory of e-m fields, 64
Quasimolecular x rays, 182

R

Radiation theory, 31
Radiative electron capture, 186
Radiative transitions, 31
Radiofrequency region, 12
Rare gas continuum, 206
Rayleigh criterion, 22, 276
Reflective coatings, 232
Resolving instruments, see also Gratings, Interferometers, and Prisms
Resolving power, 21, 276
Resonance fluorescence, 145
Resonant capture, 136
Rowland circle, 227, 292

S

Satellite lines, 166
Selection rules, 97
Sliding spark, 273
Solar x rays, 193
Solid state detectors, 153
Sources, 24, 204, 224, 245, 246, 259
Sparks, 210, 220, 269, 273
Spectral line, see also Line

Spectrographs and spectrometers
  Doppler-tuned, 164
  Ebert-Fastie, 235
  gamma ray, 131
  optical region, 274
  prism, 274
  ultraviolet, 234
  x ray, 149
Spontaneous emission, 87
Standard sources, 224
State vector, 73
Submillimeter waves, 10
Synchrotron radiation, 210

T

Theory of e-m radiation, 31
Thermal detectors, 325
Tokamaks, 191
Transition probabilities, 80, 100, 116, 141
Triple correlations, 139
Twentieth-century physics, 17

U

Ultraviolet region, 13, 204
  absorption spectra, 248
  continuum sources, 206
  detectors, 241
  emission spectra, 247
  gratings, 226
  near, see also Optical region
  sources, 204, 215, 259
  spectra of gases, 247
  spectra of solids, 251
Units, 2, 149

V

Vector potential, 73
Visible region, see also Optical region
Vodar monochromator, 237

W

Wadsworth mounting, 294
Wavelength, 2, 246, 326, 327
Wavenumber, 2
Widths of spectral lines, 23
Windows for ultraviolet, 238

X

X-ray region, 14, 148
  cross sections, 200
  detectors, 149
  focusing, 158
  muonic, 189
  plasmas, 191
  quasimolecular, 182
  solar, 193
  spectra, 166
  spectrometers, 156, 160

Z

Zeeman effect, 341

SUBJECT INDEX FOR PART B

A

Absorption
  coefficient, 46
  enhanced, 336
  index, 46, 135
  laser studies of, 325
  two-photon, 332
Amplitude spectroscopy, 73
Apodization, 62, 64
Astronomical applications, 48, 100, 243
Asymmetric interferograms, 70
Atomic spectra in the infrared, 2
Attenuated total reflectance (ATR), 48
Azbel–Kaner resonance, 97

B

Backward wave oscillators (BWO), 107
Beam foil spectroscopy, 213, 218
  astronomical applications, 243
  high energy beams, 232
  sources, 216
Blazed gratings, 16, 58
Bolometers, 7, 8, 52, 113
Bridge measurements of impedance, 142
Burch's law, 28

C

Capacitance cells, 136
Carbon arc, 5
Carbon rod furnace, 6
Cavity resonators, 161
Cascading, 236
Christiansen filters, 25
Clausius–Mosotti relation, 171
Cole–Cole diagrams, 94, 175
Collisional broadening, 38
Condenser microphone detectors, 114
Correlation spectroscopy, 28
Crystal diode detectors, 111
Cyclotron resonance, 95
Czerny–Turner spectrometers, 23

D

Debye absorption, 92
Debye equations, 171, 174
Deconvolution, 37
Defect diffusion, 185
Depolarization spectra, 167
Detectivity, 12
Detectors
  Condenser microphone, 114
  crystal diode, 110, 111
  far infrared, 51
  infrared, 10, 11, 13
  microwave, 109
  photo, 6
  pneumatic, 8, 51
  quantum, 6
  thermal, 6
Dielectric constants, 92, 134
Dielectric relaxation spectra, 168, 175, 194
Diode injection lasers, 290
Dispersion curves, 15, 192
Doppler broadening, 219
Dye lasers, 305

E

Ebert–Fastie spectrometers, 23
Echellettes, 16, 58
Electrical properties of matter, 134
Electrode polarization, 140
Emission spectra in infrared, 3
Enforced dipole transitions, 41
Etalon, 304
Extinction coefficient, 135

F

Fabry–Perot interferometers, 55, 84, 127
Far-Infrared Region, 50
  astronomical applications, 100
  detectors, 51
  interferometry, see Fourier transform spectroscopy

Far-Infrared Region (*Continued*)
  spectra of gases, 88
  spectra of solids, 90
Fellgett advantage, 53
Filters, 25, 26, 29, 59, 84
Fluorescence, 334
Focal isolation, 85
Forbidden decays, 260
Fourier transform spectroscopy, 53, 60
  amplitude, 73
  noise, 75
Free wave rf techniques, 154
Frequency mixing, 319, 321
Fresnel zone plates, 84

G

Girard grille spectrometer, 85
Globar, 4
Golay cell, 8, 51
Gratings, 20
  blazed, 16, 58
  echellette, 16, 58
  holographic, 22
  lamellar, 80
  replica, 16, 21
  spectrometers, 20, 57
Gunn effect, 108

H

Hadamard spectrometer, 87
Harmonic generation, 128
Heterodyne spectroscopy, 87
Hindered rotation and translation, 44

I

Impedance measurements, 142, 147
Infrared Region, 1
  astronomical applications, 48
  atomic spectra, 2
  detectors, 10, 11, 13
  emission spectra, 3
  far-, *see* Far-infrared
  near-, *see* Optical region
  optical components, 13
  sources, 2
  spectra of gases, 28

spectra of liquids and solids, 44
windows, 14
Interferograms, 61, 68, 70
Interferometers
  Fabry–Perot, 55, 84, 127
  lamellar grating, 80
  Michelson, 62
  Mock, 86
  resolving power, 62
Interferometry, 60 69, *see also* Fourier transform spectroscopy

J

Jacquinot advantage, 52
Josephson junctions, 97

K

Klystrons, 105
Kramers–Kronig relations, 47, 179

L

Lamb shifts, 256
Lamellar grating, 80
Langevin function, 171
Lasers, *see also* Tunable lasers
  application to absorption spectra, 325
  diode injection, 290
  millimeter wave, 54
  molecular, 280
  spin–flip Raman, 299

M

Michelson interferometer, 62, 79
Microwave region, 102
  absorption cells, 116, 127
  cavity resonators, 161
  detectors, 109, 113
  harmonic generators, 128
  modulation techniques, 114
  radio astronomy, 132
  sources, 104, 105, 108
  spectrometers, 121
Millimeter, waves 54, 128, *see also* Far-infrared and microwave regions
Molecular correlation, 175

Molecular spectra, 1
Multiple relaxation, 180
Multiplex advantage, 53
Multiply excited states, 224

N

Nernst glower, 5

O

Onsager equation, 172
Optical constants, 43, 134, *see also* Dielectric constants
  metals, 94
  semiconductors, 94
  solids and liquids, 92
Optoacoustic effect, 338
Oscillator strengths, 239

P

Parametric conversion, 315
Parametric oscillators, 322
Phase modulation, 70
Photodetectors, 6, 52
Plasmas, 97
Pneumatic detectors, 8, 51
Point contact diodes, 111
Prism, 15
  spectrometers, 17

Q

Quantum beats, 244, 245, 249, 252
Quantum detectors, 6

R

Radiation chopping, 7
Radiofrequency region, 134
  astronomy, 132
  capacitance cells, 136
  time-domain techniques, 163
Reflection spectra, 47, 323
Refractive index, 46, 135
Relaxation
  activation energy, 187
  dielectric, 168, 175

flexible molecules, 196
solids, 199
spectra, 168, 198
times, 173, 181, 194
Replica gratings, 16, 21
Resolving instruments, *see* Prisms, gratings, and interferometers
Resolving power of interferometer, 62
Resonance methods in rf region, 147
Resonance processes, 189
Resonant cavity, 55, 127
Responsivity, 12
Reststrahlen, 25
Rotation spectra, 29
  hindered, 44
Rotation–vibration spectra, 30
Rydberg constant, 330
Rydberg states, 231

S

Signatures, 66
SISAM, 85
Sources, *see* appropriate spectral region
Spectrographs and spectrometers
  Czerny–Turner, 23
  Ebert–Fastie, 23
  Girard grille, 85
  grating, 20
  Hadamard, 87
  prism, 17
  submillimeter wave, 129
  Walsh double pass, 19
Standing wave techniques, 158
Stark beats, 245
Stark effect modulation, 115
Stark spectra, 340
Submillimeter waves, *see* Far infrared and microwave region
Superconducting bolometer, 8

T

Thermisters, 8
Thermal detectors, 6
Thermocouples, 7
Transitions
  enforced dipole, 41
  forbidden, 260
  probabilities, 234

Translational absorption, 42
  hindered, 44
Tunable laser spectroscopy, 273
  absorption, 325
  diode injection, 29
  dyes, 305
  millimeter region, 54
  molecular, 280
Two-photon absorption, 332

V

Vibration–rotation spectra, 30
Viscosity, 194

W

Water spectrum, 44
Waveguide techniques, 54, 116, 127, 128, 150
Wavelengths, 26

Z

Zeeman, effect
  modulation, 115
  quantum beats, 249
  tuned lasers, 275
Zero-field quantum beats, 252